Nonlinear Model-Based Control

Using First-Principles Models In Process Control

Nonlinear Model-Based Control

Using First-Principles Models In Process Control

R. Russell Rhinehart

ISA™

Notice

The information presented in this publication is for the general education of the reader. Because neither the author nor the publisher has any control over the use of the information by the reader, both the author and the publisher disclaim any and all liability of any kind arising out of such use. The reader is expected to exercise sound and professional judgment in using any of the information presented in a particular application.

Additionally, neither the author nor the publisher has investigated or considered the effect of any patents on the ability of the reader to use any of the information in a particular application. The reader is responsible for reviewing any possible patents that may affect any particular use of the information presented.

Any references to commercial products in the work are cited as examples only. Neither the author nor the publisher endorses any referenced commercial product. Any trademarks or trade names referenced in this publication, even without specific indication thereof, belong to the respective owner of the mark or name and are protected by law. Neither the author nor the publisher makes any representation regarding the availability of any referenced commercial product at any time. The manufacturer's instructions on the use of any commercial product must be followed at all times, even if in conflict with the information in this publication.

The material and information contained in this book are for general information purposes only. Views and opinions expressed by the author(s) are solely their own and do not necessarily represent those of ISA.

Printed in the United States of America

ISBN-13: 978-1-64331-242-2 (print)
ISBN-13: 978-1-64331-243-9 (ePub)
ISBN-13: 978-1-64331-244-6 (Kindle)

ISA
3252 South Miami Blvd, Suite 102
Durham, North Carolina, USA 27703
www.isa.org

Library of Congress Cataloging-in-Publication Data
Rhinehart, Russell.
Nonlinear model-based control: using first-principles models in process control / Russell Rhinehart.
p. cm.
Durham : International Society of Automation, 2023.
Includes bibliographical references and index.
ISBN: 9781643312422 (paperback)
LLCN: 2024932056
1. Automation and control. 2. First-principles models. 3. Generic model control. 4. Process-model-based control. 5. Predictive functional control. 6. Horizon predictive control.
https://lccn.loc.gov/2024932056

Acknowledgments

I learned a lot about the practice of control from many professional colleagues who shared their expertise. I am grateful to Danny Abramovitch, Joe Alford, Mark Darby, Brian Hrankowsky, Jimmy Key, Greg McMillan, Kip Mercure, Mike Resetarits, Tariq Samad, Greg Shinskey, Carlos Smith, Anand Vennavelli, Dennis Williams, and very especially to Harold Wade.

I was also supported by and benefited from my academic colleagues who shared and collaborated with me. Thank you to John Beasley, Bob Bethea, Robert Farrell, Rich Felder, Jim Ferrell, Paul Gemperline, Hua H. Li, Charlie Moore, Micky Parten, Graham Puxty, Jim Riggs, Rob Whiteley, Ted Wiesner, and Gary Yen.

Most of my funding for control research was through industrial sponsors who supported graduate students, equipment, travel, and summer salary. I could not have maintained my practicable investigations with other forms of academic research support. Much appreciation goes to ABB, Amoco, Aspen Tech, Bartlett Foundation, British Petroleum, Celanese, Conoco, Diamond Shamrock, DMC Corporation, Dow, DuPont, Fina, IMC, Phillips, Setpoint, Tennessee Eastman, and Yokogawa.

I am also extremely grateful for my graduate students. Although the relationship might seem that I trained them, the totality of our relationship includes my benefit from their problem-solving creativity, their mathematical prowess, their revealing relevant techniques that I had not been aware of, and their effort to do both the experimental and theoretical work necessary to evaluate ideas. Appreciatively, I thank S. Samir Alam, Gaurav Aurora, Phoebe Katterhenry Brown, Songling Cao, Chetan Chandak, Sandeep Chandran, Haoxian Chen, Jin Yong (John) Choi, Mehul M. Desai, Sanjeev Dhir, Pryiabratta Dutta, Prithwijit Ghoshal, Sarvesh Gore, Anand Govindarajan, Amit Gupta, Ting Huang, Mahesh S. Iyer, Suresh Kumar Jayaraman, Nitin V. Joshi, Preetica Kumar, Qing Li, S. K. Mahuli, Upasana Manimegalai-Sridhar, Steven M. Mayo, Prabhu

Murugan, Naggappan Muthiah, Ganesh Narayanswamy, Siva Natarajan, Abrahao Naim Neto, Jing Ou, Samuel O. Owusu, Venkat Padmanabhan, Hemant G. Pandit, Maulik Parekh, Venkat P. Paruchuri, Balashakar (Soundar) Ramchandran, Pramod R. Raul, Vijayalakshmi Sethuraman, Nitin Sharma, Vikas Shukla, Vikram Singh, Anthony G. Skach, Ming Su, Hoshang Subawalla, John Szela, Ganesh Venkataramanan, Vidhya Venugopal, Gaylon L. Williams, and Judson Wooters.

Of course, I have benefited in many ways from many others, but gifts from those mentioned provided experiences that support this book.

I am especially grateful to Liegh Elrod, ISA Book Publishing Manager, for her many and excellent edits and suggestions. Several anonymous reviewers also provided meaningful and appreciated feedback, which I used in the final revisions to the book.

Contents

About the Author

Dr. R. Russell Rhinehart, professor emeritus in the School of Chemical Engineering at Oklahoma State University, has experience in both industry (13 years) and academia (31 years), and was head of the school for 13 years. Russ is a past president of the American Automatic Control Council, was editor-in-chief of *ISA Transactions* from 1998 to 2012, and Director of the ISA Automatic Control Systems Division (now Control and Robotics). He is a Fellow of both ISA and AIChE and a Process Automation Hall of Fame inductee. He received the 2009 ISA Distinguished Service Award and the 2013 Fray International Sustainability Award.

Inspired by his industrial experience, his mission has been to bridge the gap between industry and academia. Russ was the codirector of two industrial consortia (one at Texas Tech and a second at Oklahoma State) and built pilot-scale laboratories for dual use in undergraduate education and graduate research. He left industry in 1982 with a vision to use engineers' process models in control and pursued many aspects of doing so in his academic research career. This book is his collection of practicable methods. His goal is for it to be a useful guide to others seeking to use nonlinear models in control.

His 1968 BS in chemical engineering and subsequent MS in nuclear engineering are both from the University of Maryland. His 1985 PhD in chemical engineering is from North Carolina State University.

Russ is the author of *Nonlinear Regression Modeling for Engineering Applications* and *Engineering Optimization*, published by John Wiley & Sons, and he is coauthor of the CRC Press textbook *Applied Engineering Statistics*. He also authored six handbook chapters on modeling, uncertainty, process control, and optimization and has presented numerous tutorials and workshops. He maintains a website (*www.r3eda.com*) to provide open access to software (including simulators to support this text) and technique monographs.

Russ taught modeling, optimization, laboratory, and process control courses and has developed short courses for industrial participants offered through ISA or directly to companies related to statistical process control, instrument and control systems, modeling, model-based control, optimization, and uncertainty analysis.

In "retirement," he offers consulting services related to engineering analysis, provides monthly "Develop Your Potential" articles for *CONTROL* magazine, and serves on several ISA, American Automatic Control Council (AACC), and International Federation of Automatic Control (IFAC) committees.

Rationale/Preface

My aim is to help the application engineer (whether in graduate school or practice) understand, choose appropriate, and be able to implement model-based controllers. The focus is on using process models, characteristically nonlinear, which could be either first-principles models (mechanistic) or empirical models (e.g., neural networks). The intended audience includes development groups and entrepreneurs within the vendor and software community who are providing new control products, experienced users who can translate models and optimization from calculus to executable code, advanced process control (APC) leaders within the process industries who are managing implementation projects, and graduate students seeking to investigate nonlinear control techniques.

Because vendors have not provided model-based controller functions in their products, implementing the techniques in this book requires skill in developing first-principles models and implementing numerical solutions to calculus and differential equations. The skills are relatively elementary, no more complicated than being able to write the code to implement proportional-integral-derivative (PID) control (integrals and derivatives) or to fit first-order plus dead time (FOPDT) models with regression best. The skills should be within the comfort zone of those who are not too long out of school with an engineering MS degree or above, and the methods should be recognizable by those with a recent engineering BS. For those who are long out of college, the required skill will probably have disappeared from disuse, but the examples provided are intended to refresh that ability.

Many nonlinear control approaches have been published. Because they were developed from diverse perspectives, they seem very different. However, I believe they have a common structure, making them relatively easy to understand, with seemingly unrelated diversity of methods simply due to choices in each stage. This book starts with the common structure and develops six approaches to progressive functionality by discussing the choices within each function.

People can learn by memorizing classroom theory, but skill is developed by doing. Both are important. This book provides sufficient theory to support concepts, but the focus is on the how-to of implementation. Case studies and illustrations in this book provide examples, many of which are accessible to the reader at *www.r3eda.com* for personal exploration. The code is open so you can see the details.

This book is based on a graduate-level course that I've taught about 15 times, as well as over 30 years of experimental exploration of using process models (especially first-principles models) in control. I also use this material as the basis of my short course on nonlinear control. I believe the book is appropriate for an upper-level college course, but I have also written it to address the issues that a practitioner, learning independently, would need in implementing nonlinear control.

My investigation of these methods was inspired by a vision for the possible use of engineer's models (first-principles models) in process control. The possibility was revealed when I was in the chemical industry in the late 1970s and early 1980s when process computers became feasible. Simultaneously, many others in either industry or academia recognized the potential for model-based control. Since then, many approaches and commercial products have been developed. I left industry to explore the fundamentals and disseminate the implementation aspects of model-based process control within academia; this has been my primary focus since the early 1980s.

Although developed with chemical processes in mind, the approaches are appropriate for processing in related industries (e.g., pulp and paper, glass, minerals, refining, food, pharmaceutical, and combustion). Additionally, many manufacturers of equipment that contain processes (refrigeration, generators, absorption, internal combustion, ion exchange, etc.) could include model-based controllers on the equipment.

This book seeks to organize concepts and methods and present them in a progressive manner. The most comprehensive control approach is constraint-handling, horizon-predictive, multivariable, linear, model-based control, which is characteristic of the major commercial products. It can be somewhat complicated. Nonlinear control versions are even more so. Within academia, these are termed *model predictive control* (MPC) or *horizon predictive control* (HPC); in industry, they are called APC. However, several approaches are simpler and industrially practicable.

This book starts with generic model control (GMC) using steady-state models. Steady-state models are relatively easy to obtain. In GMC with steady-state models (GMC-SS), the model inverse is used to calculate the manipulated variable (MV) from a desired controlled variable (CV) target. The implementation is equivalent to proportional-integral (PI) control using the model inverse as output characterization. So, it is also easy to implement in existing control computers. GMC handles single-loop nonlinearity very well for open-loop stable processes when it is simple to solve the model inverse.

The second control approach in complexity and performance is process-model-based control (PMBC), which uses an elementary, first-principles, dynamic process model. It uses a one-step-ahead objective to determine the MV and can use feedback to adapt the model online. PMBC can be used in multiple-input, multiple-output (MIMO) constrained applications. Optimization is recommended to solve for the MV value(s) that make the model move along a desired trajectory. It is useful for processes with well-behaved dynamics (not dominated by dead time or inverse action and of low dynamic order).

PMBC is presented in two stages, one without online model adjustment and one with model adjustment (an optional feature).

A feature of PMBC is continuous incremental online updating of model coefficient(s) to keep the model true to the process. Process attributes are always changing (heat exchanger fouling, piping reconfigurations, manual valve positions, screen plugging, catalyst reactivity, environmental loads, etc.). First-principles model adaptation has several benefits: (1) The model remains true to the process, which permits better control action. (2) Adapting model coefficients that relate to process attributes reveals trends that are useful for both process analysis and predictive maintenance forecasting to the operator/engineer. (3) The adapted model can be used to forecast ever-changing constraint conditions (e.g., maximum flow rates or conversion) that are limited by the process, which can be used in both economic set point optimization and balancing control action.

The fourth approach is predictive functional control (PFC). This uses dynamic models and seeks a step-and-hold MV action that would cause the model to reach a target value at some time in the future, a coincidence point. The MV value is not held for the duration; it is reevaluated at each sampling. This is useful for handling delay and inverse action, as well as for MIMO and constrained nonlinear applications.

A single-input, single-output (SISO) version of internal model control (IMC) is the fifth approach presented. Although it uses linear models, it is useful for characterizing desirable control aspects, such as demonstrating zero steady-state offsets. And, because several of the nonlinear control approaches reduce to IMC when linear models are used, the IMC analysis supports the credibility of the nonlinear approaches. IMC can use "gain scheduling" to adjust the model gain, delay, and time constant based on a first-principles model, making it effectively a nonlinear controller. However, MIMO and constraint handling in IMC are impracticable.

Finally, the sixth approach—using nonlinear models within HPC—is presented. This control approach seeks to identify a schedule of future MV actions that steer the model along a desired future trajectory. This has benefits in avoiding constraints and handling ill-behaved dynamics.

Some key supporting techniques are required in implementing nonlinear control. The most important is optimization to solve for the constrained inverse. Also important are steady-state identification, model validation, filtering, and data reconciliation. Sections of this book cover those topics.

In my experience, academic instruction related to commercial APC products often overshadows understanding the concepts due to a focus on how the mathematics are computed. By contrast, vendor-produced instruction keeps the details hidden. Neither is very helpful to those wanting to explore model-based control. This book seeks to be helpful. Although APC can be complicated, the whole consists of relatively simple elements. Further, the elements could be executed in any number of ways. Additionally, the classic focus on MPC (or APC) ignores the host of simpler techniques that are fully adequate and simple to implement in-house for specific applications where APC (or MPC) would be excessive and expensive. The simpler but often fully adequate model-based methods are GMC with steady-state models, PMBC, and PFC.

My experience in implementing nonlinear control includes some industrial applications, but mostly, it has been on pilot-scale units or simulations within academia. My tasks have been to understand techniques and disseminate appropriateness, primarily to students but also to those in industrial practice. This book continues that aim: help people who want to understand the mystery of model-based process control and relay the understanding and skill to implement or choose applications.

Section 1

Introductory Material

1
Introduction

1.1 Model-Based Controllers

Model-based control algorithms were developed when the digital revolution made it possible for them to do more (e.g., override, ratio, feedforward, and gain scheduling) than analog proportional-integral-derivative (PID) and associated advanced regulatory control (ARC) could do. Early model-based techniques include the Smith predictor and Dahlin's algorithm. In the 1970s, some vendors created distributed control systems (DCSs) and programmable logic controllers (PLCs), which were digital computers for control, and simultaneously, other vendors were creating affordable digital computers for off-line use. Then, many visionaries realized that the affordable digital power would support new model-based control algorithms that would have advantages over ARC techniques. After a development phase to test and refine algorithms, the 1980s saw many control approaches begin to be revealed. The advances continue today, revealing more comprehensive and powerful approaches.

These techniques, designed to solve one problem, failed to provide a comprehensive control strategy. For instance, generic model control (GMC) using steady-state (SS) models is good for nonlinear processes. It has the convenience of acting like a proportional-integral (PI) controller with output characterization and only requiring SS models, but it is not good for multivariable or constrained processes or those with ill-behaved dynamics (inverse action or dead time).

Internal model control (IMC) can handle ill-behaved dynamics but uses linear models and is not very good for multivariable or constraint handling.

Natural language processing (NLP), alternately termed *fuzzy logic* (FL) or *expert systems* (ES), and associated techniques like Takagi-Sugeno-Kang (TSK) models can also handle nonlinearities and have the convenience of using the process expert's qualitative understanding of the process response. However, these become quite complicated for multivariable processes and constraint handling.

Neural networks (NNs) provide a model-free (no predefined functionality) way to create nonlinear empirical models of the process. These could be useful in controllers, but they mask any mechanistic understanding of the model, which I believe has much value.

Predictive functional control (PFC) uses a process model to determine a step-and-hold value for the manipulated variable (MV), alternately termed the *controller output* (CO), to make the model match the model set point at a future time, termed the *coincidence point*. This has the advantage of compensating for both nonlinearity and ill-behaved dynamics. With first-principles models, it also preserves process knowledge and appropriately accounts for measurable disturbances. It is good for single-loop control and can be extended to multivariable constraint-handling control.

Dynamic matrix control (DMC) and similar approaches such as identify and command (IdCom) were early multivariable, horizon-predictive, model-based controllers. They can avoid constraints, handle multivariable interactive processes, and compensate for disturbances. Many companies have developed products based on these methodologies. The present industry term for this approach is *advanced process control* (APC), while the academic term is either *model predictive control* (MPC) or *horizon predictive control* (HPC). Many vendors use their own terminology. However, the dynamic (time-dependent) models are linear. Some are FOPDT, some are state-space, but most are formulated in a vector notation, termed *finite impulse response* (FIR) models. Being linear, they do not adapt to process changes and require extensive process testing to obtain confident vector element values. They mask process knowledge.

Process-model-based control (PMBC) was developed to handle nonlinearity, preserve the engineer's first-principles knowledge, and adapt to the process as process attributes change (fouling, reactivity, ambient losses, efficiency, etc.). Tracking process parameter values is beneficial for forecasting constraint conditions and predicting when model or process maintenance will be required. It is a one-step-ahead prediction method for nonlinear processes, and although relatively simple, it does not handle ill-behaved dynamics.

PMBC embodies the "one-model" approach. Currently, steady-state models are used for supervisory set point optimization (real-time optimization [RTO]), process design, and analysis. Dynamic models (often nonlinear) are used in training, and classically linear dynamic models are used for control. When different modeling approaches are used, process engineers must understand each and adapt each as the process changes. It is desirable to use one process model for all applications, and it should be a first-principles dynamic model, which preserves process understanding that is essential for design and troubleshooting.

There are nonlinear approaches to APC (or MPC). In one approach, an NN is used to capture the process SS gain nonlinearity; then, the FIR (a vector of the response to a unit input step) model is used to approximate the linear dynamics. Like APC, this requires substantial process testing over a wide operating range. In another approach, the process expert's judgment is used to assign gain values; then, the FIR model is used to approximate the linear dynamics. This requires much less process testing because only the unknown dynamics are required. Using qualitative gains might seem to generate imperfect models, but all models are imperfect, and as long as a model is passable (perhaps it would get a grade of C in school), it is fully adequate for control.

Finally, using first-principles models in APC with PMBC adaptation has been explored industrially and is getting performance affirmation. It can handle nonlinearities, multivariable interactions, and disturbances and has constraint-handling features. It preserves process knowledge. Online adaptation keeps the model true to the process and reveals process analysis information that is useful for forecasting constraints and maintenance needs.

This book covers the best-in-class approaches (in my opinion) for using first-principles models in control—GMC with steady-state (GMC-SS) models, PMBC, PFC, and nonlinear APC (nAPC).

Why use simple PMBC?

- Unlike an empirical model such as a first-order plus dead time (FOPDT) approximation, it preserves and transmits process knowledge.
- The model only has a few adjustable coefficients to make it fit the process data, which means less process testing to develop the model.
- Once tuned, it remains tuned throughout the operating range.
- There is only one aggressiveness factor per controlled variable (CV), not several.
- When incrementally adjusted, the model parameter value tracks the process behavior to monitor process health.

- The model forecasts constraints that can be useful in higher-level optimization.
- The same model can be used in other applications, such as optimization, training, design, process analysis, constraint forecasting, and maintenance prediction.
- The variables are not deviations from an initial SS, and the units are not percentages. The variables are not transformed or expressed in abstract mathematics. They are familiar variables that engineers use with units of the process variable.
- Regardless of controller type (ARC, APC, nAPC, fuzzy, or NN), if the controller has feedforward, gain-scheduled, and decoupling features, it has the same reduced variance benefits—higher product quality, lower costs, faster throughput, reduced waste, and so on. I think nonlinear model-based control (nMBC) can be justified like ARC is, but nAPC will have a wider operating range and life.

ARC can handle many of the process difficulties within one structure. For instance, classical gain scheduling a PI controller can make it adapt to process nonlinearity. There may be three operating ranges; the control technician tunes the controller gain and integral time constant for each. PI control has two coefficient values for each of the three operating ranges and two breakpoints in the operating range that must be determined. In all, there would be eight tuning values, usually found by testing at each of the conditions. By contrast, a process-model-based controller has only one tuning coefficient that is valid for the entire operating range. For instance, with ARC, including feedforward to give advance accommodation for a PI feedback loop requires six empirically determined coefficient values for each operating range, whereas the feedforward accommodation is naturally included in a process-model-based controller. If one understands ARC, then even with the extra number of coefficients to be tuned, it might be simpler to implement and maintain than model-based control (MBC). However, I believe if the process has two or more CVs, then MBC is simpler.

1.2 Nonlinear and Nonstationary

The term *nonlinear* means not linear. If you were to plot one variable w.r.t. (with respect to) another, linear would mean the relationship is a straight line. One variable responds to the first power of the other. However, there are many nonlinear patterns. They could curve upward, curve downward, be S-shaped, and so on. Nonlinear does not say what it is, just what it is not.

In differential equations, *linear* means that if two functions are a solution, then the sum (linearly added) of the two is also a solution. This is often referred to as the *principle of superposition*. The functions do not need to be linear.

In an open-loop stable process, such as $\tau\frac{dy}{dt}+y=Ku$, where τ and K (time constant and gain) do not change in time, the response is an exponential change in time, $y(t)=Ku+(y_0-Ku)e^{-t/\tau}$ (t is time, $\frac{dy}{dt}$ is the rate of change of the response in time, y_0 is the initial value of the response, u is the controller output, and e is the base of the natural logarithm). The graph of the response in time is not linear. However, the SS response $y_{SS}=Ku$ is a linear response to u with a gain (slope) of K. And if you look at the differential equation, the response variable y is always expressed as the first power, and so is the independent variable u. So, the response y is termed a linear response to u.

Formally and mathematically, linear can be described as the case if the second derivative of the response w.r.t. any two influence variables is zero.

$$\frac{d^2y}{dwdx}=0 \tag{1-1}$$

An example of nonlinearity in process models is the ideal squared dependence of the pressure drop due to the flow rate.

Stationary means that the coefficient values in the model do not change in time or w.r.t. operating conditions. *Nonstationary* means that the model coefficient values are a function of either time or some influence or response variable. If the temperature changes during a reaction and the reaction rate is temperature-dependent, then composition equations will be nonstationary. If heat transfer is dependent on flow turbulence, which depends on flow rate, then the energy equation will be nonstationary.

In general, nonlinear and nonstationary are different, but if the nonstationary aspect is due to an influence or response variable, the response is also nonlinear.

1.3 Keep It Simple and Safe

A guiding principle in industry is to keep it simple and safe (KISS). Use the simplest approach that is effective. Balance perfection with sufficiency. Nonlinear control can become gloriously complicated. Do not seek to show off your ultimate skill.

In many of my studies on which control approach is best for classic CV and MV measures, the conclusion is, "As long as the strategy accommodates the process difficulty, it is as good as any other strategy." I've explored PID, IMC, ARC, FLC (fuzzy logic control), PMBC, APC, and nAPC on diverse pilot-scale processes and many simulators. I recommend choosing the simplest approach that solves the particular process problem.

The classic CV measures are rise time and settling time for a set-point change and the integral of the squared error for the disturbance rejection mode. Classic MV measures are travel and energy. However, be aware that there are other metrics associated with desirables for control, including the engineer's time to formulate a model, the data collection effort to match a model to the process, and the multiple uses of the model. A desirable metric also includes simplicity in tuning; characteristically, the model-based controllers have only one tuning parameter per CV, while a PID has three.

The benefits of first-principles models in control are as follows:

- Process knowledge is preserved.
- The engineer is not distracted by learning other mathematical techniques.
- The one-model approach can be used for many applications (design, optimization, RTO, analysis, predictive maintenance, and training).
- Tuning is relatively simple (one parameter per CV).
- Optimization balances performance when constraints or extra degrees of freedom (DOF) happen.
- Features like ratio and feedforward are handled naturally.

1.4 What Are First-Principles Models?

First-principles models are the process engineer's models for process design, analysis, optimization, data reconciliation, and troubleshooting. They are based on elementary material and energy balances and ideal constitutive relations, and they are characteristic of college classroom teaching models for heat exchange, fluid flow, reaction engineering, separations, and other unit operations. Representing process mechanisms, they are typically nonlinear.

The models must be validated. Usually, the process engineer uses first principles to generate the calculus/algebra version of the process model, then translates that into executable code to create a process simulator (alternately termed the *surrogate model* or the *digital twin*.) However, the model might be wrong or have some missing features. To validate the model, compare it to the process data. There is likely adequate historical data, but this might require some designed trials. The process comparison could reveal which features must be included or whether the model is good enough. Once validated, it can be used in control.

1.5 Concept and Block Diagram

The diverse model-based controllers were developed independently by folks with diverse backgrounds, mathematical approaches, and jargon to solve different problems. As a result, superficially, there does not seem to be a unifying approach to the controller structures. But there is, and this book presents several techniques from that unifying structure.

There are three functions within MBC: Predict, Correct, and Action (referred to in this document as Act). The structure can be represented by the operations in Figure 1-1.

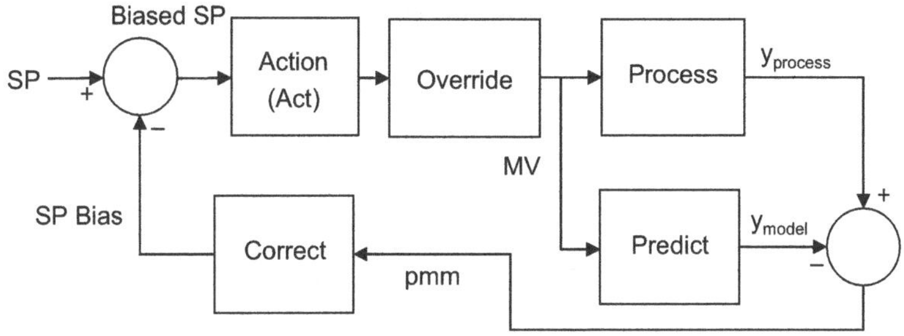

Figure 1-1. MBC representation.

1.5.1 Predict

The first model-based control (MBC) function uses the model to predict or mimic what the process output is expected to do. This function is the lower right box in Figure 1-1. It updates the past model value with past disturbances and control actions to predict the updated (current or new) modeled output. With 10 control actions within the smallest time constant or delay in the model, the control interval, Δt, is usually small enough to permit using Euler's explicit finite difference method to solve the model numerically. More advanced numerical methods are needed if the control scan time is not small relative to the model dynamics.

The acronym for the model in the Predict function is P2N, meaning past-to-now, because it implements one Δt update to reveal what the model predicts the process is doing.

1.5.2 Correct

The second function in MBC is correcting a process-model mismatch. Process-model mismatch, *pmm,* is the difference between modeled prediction and process measurement, as represented by the circle difference operation on the lower right side of Figure 1-1. The modeled value, y_m, will not match the process value, y_p, exactly for a variety of reasons, including model error or unmeasured disturbances. A simple correction approach is to bias the set point, y_{SPbias}, for the model with the process-model mismatch as follows:

$$pmm = y_p - y_m \tag{1-2}$$

$$y_{SPbias} = y_{SP} - pmm \tag{1-3}$$

The logic is illustrated in Figure 1-2. If you aim at the target bull's-eye but hit 3 cm low, aim 3 cm above the bull's-eye next time. If that shot is 0.5 cm above the bull's-eye, the pmm is the difference between where you aimed (+3 cm) and what happened (+0.5 cm). If the pmm is 2.5 cm, aim 2.5 cm above (not 0.5 cm below) the bull's-eye next time.

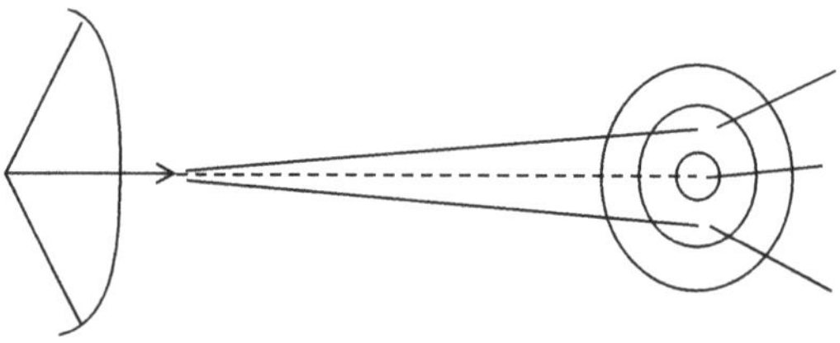

Figure 1-2. The concept of biasing the set point for a model using pmm.

The box labeled "Correct" in the lower left of Figure 1-1 could provide other functions. One of the alternate functions could be related to model coefficient adjustment, which has some managerial information benefits. But in this simple correction approach, it simply is a pass-through of the pmm value to bias the set point.

Notes:

1. The biased set point is the set point for the model, not the set point for the process.
2. pmm is not the actuating error. It is not the deviation from the set point (the target). pmm is the deviation between the actual hit and the model-predicted hit.

1.5.3 Act – an Elementary Choice – Simple PMBC

The third MBC function calculates the control action to determine the controller MV value. This is represented as the box labeled "Act" in Figure 1-1. This function starts with a user-defined, desired performance objective for the controller. A simple desire in the PMBC strategy is to calculate a u-value (the MV in Figure 1-1) that would push the model toward the biased set point at a rate proportional to the deviation from the biased set point. (This is an effective strategy for low-order models with inconsequential delay.)

Mathematically, this simple desire is:

$$\left.\frac{dy_m}{dt}\right|_{desired} = \frac{y_{SPbias} - y_m}{\tau_{want}} = K_c(y_{SPbias} - y_m) \tag{1-4}$$

The single tuning factor, τ_{want}, is the time constant for the desired first-order trajectory for the model to return to the biased set point. Alternately, K_c, the reciprocal of τ_{want}, can be used with a large value, creating a more aggressive controller action. This is more convenient for some, but it masks understanding the desired trajectory.

Notes:

1. The biased set point, y_{SPbias}, is for the model, and y_m is the model-predicted value.
2. $y_{SPbias} - y_m = y_{SP} - pmm - y_m = y_{SP} - y_p + y_m - y_m = y_{SP} - y_p = e$. The right-hand side of Equation 1-4 is $K_c e$, which is the normal model for proportional action. However, rather than stating it as such and masking the concept of the desired model trajectory, I will preserve the Equation 1-4 structure.
3. $\frac{y_{SPbias} - y_m}{\tau_{want}}$ is the user-desired rate of change of the modeled output, y_m. If there were no process-model mismatch, then this would also be the consequential rate of change of the process, y_p. But only rarely would a model exactly match the process behavior.

Now, to determine the action, insert the desired rate of change to the model and solve the model inverse to determine the controller output, u.

Usually, the model is used to calculate the rate of change of the modeled response, dy_m/dt, given process influences and state. However, in control action, the solution must determine the value of the MV, u, that would cause the desired rate of change. This is the inverse of the normal modeling solution procedure, often termed *solving the inverse.*

If you are lucky, the inverse will be mathematically tractable. You will be able to explicitly solve for u. However, it might not be possible to rearrange the model equation to do so, and a numerical method might be needed. Root-finding might be a first thought, and it works. However, optimization also works and has the advantage of being able to cope with constraints. This book will focus on using optimization to solve nonlinear implicit equations.

This set of Predict, Correct, and Act procedures represents the PMBC controller strategy.

Returning to Figure 1-1, each of the three functions (Predict, Correct, Act) could be specified with various options or executed with various techniques. The multitude of choices has led to a variety of controller algorithms and associated acronyms.

If the model is linear and first-order, the three-stage method described here is equivalent to an internal reset feedback version of a standard PI controller with the integral time equal to the process time constant. Note the following two points: (1) It is comforting that idealizations lead to a familiar solution, confirming the correctness of the control steps. (2) If the model is linear and first-order, use the conventional PI control technique, which will be comfortingly familiar to many of your stakeholders.

The acronym for the model in the Act function is N2F, meaning now-to-future. It is used to determine the control action to make the model behave as desired in the future.

Notes:

1. For single-input, single-output (SISO) applications, the controller output, u, could be the signal to the final control element (e.g., a valve, variable speed drive, or electrical heater). This is often termed *direct digital control* (DDC). However, in more complicated applications, it is desirable to have the model-based controller determine set points (e.g., flow rates and temperatures) for secondary loops to manipulate the final control element. The secondary loops are often regulatory PI or PID controlled, but if they have nonlinear and ill-behaved dynamics, the secondary loops could also have SISO model-based controllers.
2. The Predict and Act functions use the same mathematical model. They could use the same function to perform the calculation. However, the P2N model in the Predict function must be initialized with the prior modeled states and influenced by the actual implemented control action. By contrast, the N2F model in the Act function must be initialized with the P2N model states and influenced by the control action being contemplated by the Act function.
3. If the models are not identical in the Predict and Act functions, the states will have a steady-state offset and/or dynamic incompatibility.

1.5.4 Override

The Override function in Figure 1-1 is not necessarily part of the controller, but it could be. For instance, if the controller is tuned to be aggressive, it may calculate the controller output to exceed feasible values. It may ask for 123% or –4% in the signal to the valve. Or if the feasible flow rate range is 0 to 36 gpm, the aggressive controller might calculate 37 gpm. In such cases, the controller-desired output must be limited to what is feasible.

The operator may not want the controller to make large instantaneous changes but incrementally move the MV to a desired value. Large instantaneous changes may create thermal stresses, water hammer, and other undesirable phenomena. To prevent these, a rate-of-change (RoC) constraint can be imposed on the controller action.

Alternatively, the override might be external to the controller. It may be that an operator is controlling the valve locally or that a safety/override controller was selected to send the signal to the process. In this case, the MV value override that the controller desired is external to the controller.

Note: In any override case, the MV value used by the P2N model in the Predict function is the actual MV value, not that calculated by the Act function that may not be aware of constraints or overriding events external to the controller.

1.5.5 Infeasibilities

The controller may have square root, division, or other operations that could lead to an execution error. Usually, the override action needed to prevent the error is obvious, but it must be included in the executable code.

Example 1-1

Create a simple model-based controller for a mixing process. Hot and cold water enters a tank with a constant volume, V, which is reasonably well mixed. The mixed fluid temperature, T, is measured, with minimal sensor lag relative to the mixing time constant, and the hot water flow rate, F_h, is manipulated to control the mixed fluid temperature, T. The inlet cold water flow rate, F_c, and temperature, T_c, are measured again with little sensor delay. The hot water valve also has an insignificant lag and a linear installed characteristic.

Assuming constant fluid properties, the SS temperature, T, model is

$$T_{model} = \frac{F_h T_h + F_c T_c}{F_h + F_c} \tag{E1-1-1}$$

where subscripts h and c represent the hot and cold fluids.

And the dynamic (transient, time-dependent) model is

$$\frac{V}{F_h + F_c}\frac{dT_{model}}{dt} + T_{model} = \frac{F_h T_h + F_c T_c}{F_h + F_c} \tag{E1-1-2}$$

Using a simple finite difference numerical method, the model would be solved in the P2N Predict function as

$$T_{model,\,now} = T_{model,\,prior} + \left[\frac{\Delta t}{V}\right][F_h T_h + F_c T_c - (F_h + F_c)T_{model,\,prior}] \tag{E1-1-3}$$

The process-model mismatch and biased set point, SP, are

$$pmm = T_{process,\,now} - T_{model,\,now} \tag{E1-1-4}$$

$$T_{model,\,SP} = T_{SP} - pmm \tag{E1-1-5}$$

If the control objective is to have the model move toward the biased set point at a rate proportional to its deviation (the first-order CV trajectory with time constant $\tau_{desired}$)

$$\left.\frac{dT_{model}}{dt}\right|_{desired} = \frac{T_{model,SP} - T_{model,now}}{\tau_{desired}} \tag{E1-1-6}$$

Then substituting the desired rate for the modeled rate

$$\frac{V}{F_h + F_c}\frac{T_{model,SP} - T_{model,now}}{\tau_{desired}} + T_{model,now} = \frac{F_h T_h + F_c T_c}{F_h + F_c} \tag{E1-1-7}$$

And solving the inverse (for the F_h value) using the N2F model

$$F_h = \frac{\frac{V}{\tau_{desired}}(T_{model,SP} - T_{model,now}) - F_c(T_c - T_{model,now})}{(T_h - T_{model,now})} \tag{E1-1-8}$$

With a linear installed valve characteristic, the valve stem position x, fraction of full travel, linearly affects the flow rate

$$F_h = xF_{h,max} \tag{E1-1-9}$$

And because the controller output, u, is a percentage of full scale, the action is

$$u = 100\frac{\frac{V}{\tau_{desired}}(T_{model,P} - T_{model,now}) - F_c(T_c - T_{model,now})}{F_{h,max}(T_h - T_{model,now})} \tag{E1-1-10}$$

Here it is assumed that V is constant. An interaction with the level creates a multidimensional control problem. This simplifying assumption is made to illustrate calculations.

Nominally, use Equation E1-1-3 to get the P2N model value and use Equations E1-1-4 and E1-1-5 to determine the model set point. Use Equation E1-1-10 to determine the controller output to the hot water valve.

There is a divide by V in Equation E1-1-3. If $V = 0$, then the modeled T is the SS T from Equation E1-1-1. Use an IF-THEN conditional to select the right equation.

If $\tau_{desired} = 0$, then there is a divide by zero in Equation E1-1-10. This user error would mean a desire to immediately, instantly have the model value become the modeled set point. Use an IF-THEN conditional to switch from Equation E1-1-10 to the inverse using Equation E1-1-1, which becomes

$$u = 100\frac{F_c(T_{model,SP} - T_c)}{F_{h,max}(T_h - T_{model,SP})} \tag{E1-1-11}$$

If $T_h = T_{model,now}$ there is a divide by zero in Equation E1-1-10. If this is the case, no control action is possible. Increasing or decreasing F_h cannot change the modeled temperature. Perhaps the cold water flow is zero, or the hot and cold temperatures are the same. Might as well stop wasting hot water. Use a conditional to set $u = 0$.

The controller output must be limited to within 0% and 100%. If u exceeds the limit, use a conditional to reset u to that limit.

1.6 Multiple Model Use – A Caution

The process model is used in two places in the controller: in the Predict (the P2N calculation) and Act (the N2F calculation) functions.

The Predict function updates the past modeled value with the actual implemented MV to predict what the CV is doing now. This is a one-time-step update using one-step past values. I use the name past-to-now, or P2N. Initialize the one-time-step update with the prior P2N modeled value.

The Act function uses a model to forecast what might happen in the future, associated with a trial solution for the MV value. In the example, the model inverse could be explicitly solved. But an iterative search may be needed in some models: Guess at an MV and see what the modeled CV does. Re-guess at an MV value until the future modeled CV matches the desired rate of change. I call this the now-to-future (N2F) model. Initialize the N2F prediction with the current P2N modeled value. After each MV guess, the N2F model must be re-initialized with the current P2N value.

The P2N and N2F models must be identical. If the model is complicated, the functions can be the same subroutine or function. But they must be initialized as appropriate.

However, there are three process models in control simulation and testing. A process model is also used as a surrogate for the process response. This is the upper right "Process" box in Figure 1-1. This simulation model should be different from the controller models. We never seem to be able to describe Nature exactly, which seems to be more complicated and less ideal than our concepts. Process responses are affected by internal noise (turbulence and incomplete missing) and unmeasurable disturbances (environmental effects). It would be cheating to demonstrate a controller using the same model in the simulator as in the controller. Ensure the surrogate process model is a few steps less ideal than the controller model.

The process model in the controller is not the true process model. Although it is intended to be a model of the process and often is referred to as a *process model,* it is not.

1.7 Model Sufficiency/Perfection

The process model in the controller must adequately represent the process, but it does not need to be an exact (or true or perfect) representation. We use FOPDT models effectively in control. Those first-order plus deadtime idealizations mimic a high-order response with a reasonable (not perfect) fit. Further, the empirical determination of the gain, dead time, and time constant for the "process model" are all somewhat in error, subject to the experimental vagaries. FIR models seem more perfect than the simpler FOPDT models, and they are a better fit for matching experimental data, but they too are linear, and the coefficient values also represent experimental noise and uncontrolled disturbances. Both of these models are adequate for control.

Alternatively, one could use rigorous models in the controller to attempt to explain every process nuance perfectly. However, first-principles models have an appropriate balance of sufficiency with perfection and are fully adequate for control. Contrasting modeling perfection, consider that humans can catch a fly ball or steer a car through a curve without rigorous models, with just intuitive continual correction. Like humans in the previous example, PI controllers, which are based on linear, stationary, simplistic (FOPDT) inexact models, are also adequate for control. Here is a qualitative explanation as to why this is: If there are 30 or so control intervals during a process transient, then even if the model is only 70% correct, leaving a 30% error in its first action, at the next control action the remaining error is 70% corrected, leaving a $0.3 * 0.3 = 0.09$ error. By the time a few control actions have been taken, the error is smaller than the discrimination interval for the digital representation of numerical values, visually appearing to be zero.

1.8 Nonstandard Code

The equations and code for any process-model-based controller, including overrides to trap execution errors, represent the model; this makes the code unique to the process and how it was modeled. Although the controller structure and stages within are the same from one application to another, the code is unique to the application. This would be a disadvantage.

1.9 Characteristics of Process Control Applications

Characteristic difficulties of processes, process control, and the implementation environment within the process industries shape the methods that are practicable. These characteristics differ from other control applications (robotics, flight, etc.). Here is a list of characteristics of control applications in the chemical process industries. See Chapter 5 for more details.

- Nonlinear
    - Gain, or sensitivity, changes with the MV (or CO) or the CV (or PO).
    - Nonstationary: Time constants and parameter values change in time or change with respect to operating conditions.
- Ill-behaved dynamics
    - Large dead time relative to time constants
    - High-order
    - Integrating
    - Open-loop unstable
    - Inverse-acting
- Slow dynamics
    - This is usually a benefit, contrasting the speed of response of electronic, mechanical, and aeronautical applications, permitting less expensive devices.
- Multivariable (Interactions)
    - DOF = 0 – "Square," meaning the same number of MVs and CVs
    - DOF < 0 – more CVs than available MVs
    - DOF > 0 – more MVs than CVs
    - DOF changes with constraints
    - DOF changes with operator action (manual mode [MAN], override, maintenance)
- Constraints
    - Soft or hard due to environmental, health, or safety
    - Soft or hard due to product specification or equipment life
    - Soft or hard due to equipment limitations (e.g., a valve cannot open more than 100%)

- Now or into the future (as current MV or disturbance changes impact the future values of a CV or auxiliary variable)
- Either the rate of change or value or both

- Overrides
  - Safety or constraints
  - Process maintenance taking over an MV in a MIMO system
  - Operator placing some MVs in MAN in a MIMO system
- Disturbances
  - External uncontrolled "environmental" factors. These usually have some persistence but wander about a nominal value. They could, however, progressively change due to fouling, aging, deactivation, or other conditions.
  - Measured or unmeasured influences.
  - Equipment switching, such as pressure-swing absorption and ion-exchange regeneration.
- Noise
  - Seemingly random independent perturbations on the measurement. These are often attributed to the measurement system, not a real process phenomenon. They are often Gaussian distributed.
- Instrument, control system, and process costs
  - To measure everything desired
  - To keep processes linear
  - To meet economic profitability in process design
- Faults
  - Measurement errors due to sensor failure or control implementation due to final control element problems (stiction, steam utility pressure, instrument air pressure, etc.).
  - Measurement errors due to calibration drifts.
  - Process attributes such as bypass, blockage, cavitation, and collapsed internals.

  - Control system capacity such as high traffic-missed transmission or delays, central processing unit (CPU) overload, and alarm priority overload.
  - Spikes in a signal due to electronic interference.
  - Calibration issues such as discrimination errors or a process variable (PV) beyond the calibrated range.
  - Inaccurate input data. Data validation or reconciliation may be required to solve faults. Garbage in, garbage out (GIGO) is an old mantra. If the input data is erroneous, the control action will also be. Perfection is not required, but the data must be free of major faults.
- Models
  - The cost to generate and maintain them is not conventional.
- Resolution
  - Digital discretization
  - Truncation
  - Stiction
  - Dead band, play in the linkage
  - On-off action
- Calibration
  - Measurement devices are presumed linear and calibrated as such.
  - Transducers (current to pneumatic to digital), actuators, sensors, and so on must be calibrated. They are nominal, not precise.
- Capability of operators/managers
  - Operators/technicians/maintenance personnel could have associate degrees, but many have less education. Accordingly, human-machine interface (HMI) understandability and overview, the complexity of the control algorithm, and training are all issues that must be accommodated.
  - Similarly, the engineer will have a BS degree and perhaps some equipment and basic control training courses, so the controller must be understandable and easily adapted by the process control engineer.

- Operation
  - Throughput, inventory level, product grade, equipment reconfiguration, and process attributes (fouling, catalyst activity) are continually changing, effectively changing process gain, time constant, and dead time.
  - Control is frequently switched between MAN and AUTO (automatic) modes, needing initializations for bumpless transfer.
- Slow dynamics – Settling times from minutes to hours
  - The control system cost must be low, but fault-free reliability to handle aberrations must be high.
  - Control action must be gentle for the stability of the operation. Some control strategies switch to new constraints to "save a penny," causing an upset with a much larger impact.
- Batch or continuous
  - Continuous processes, ideally, run continuously under the same conditions. The CV is continuously measured and displayed. Occasionally a set point changes, but mostly control keeps disturbances from causing the CV to deviate from the set point.
  - A batch process, ideally, follows a recipe that guides it through stages. During the batch, the material is moved toward its desired final value. At the end of the batch, the CV can be measured. That batch is complete, and the CV cannot be adjusted. However, controls such as temperature, mixing rate, pH, or oxygenation may be continuously implemented during the batch. Typically, these represent nonlinear control opportunities because the process attributes change during the batch.
  - However, the continuous process might have a large measurement delay, such as with a gamma-ray or chromatograph composition detector. Then, as with a batch, it is too late to change that material once a deviation is reported. Control is needed to adjust the process operation, but continual feedback may be responding to noise and tampering with the process. Similarly, the end-of-batch results could provide feedback information to adjust the recipe. Both could be considered end-of-batch control applications.

1.10 Control Solutions

Often, nonlinearity is the primary problem for SISO chemical process control. One solution is to design a process for control success—design the process for linear responses or large inventories to reduce interaction and temper upsets. However, design is an act of balancing multiple objectives; other desirable issues, such as capital cost, flexibility, resource use, energy integration, and sustainability, would have to be sacrificed to accommodate process control. Practicable techniques for nonlinear control can ease design constraints and enable operating more competitive processes.

With nonlinear processes, gain scheduling is a conventional, control-oriented solution to accommodate nonlinearity. In gain scheduling, the PID controller coefficients are changed to reflect the current operating region. The engineer can use process knowledge to create tuning values (or empirically tune the controller at the different operating regions), which are either placed in a lookup table or expressed in equations. The controller first looks up (or calculates) PI (or maybe PID) coefficient values; then it uses the conventional PI(D) algorithm for control.

However, if the engineer's process knowledge is expressed as either a dynamic or SS nonlinear model, it is about as easy to use a PMBC to implement gain scheduling. Several PMBC approaches have succeeded in industrial applications, and there are several commercial products for relatively simple, nonlinear, process-model-based controllers. Engineers have implemented versions in-house, and control integrators and service providers have been implementing model-based controllers for decades.

SISO or MISO (multiple-input, single-output) PMBC control has several advantages over either PI(D) or gain-scheduled PI(D). PMBC has a single tuning parameter, nonlinear compensation throughout the entire operating range, process knowledge preservation, and continuous process monitoring, which offers improved process health, predictive maintenance, and constraint recognition. For multiple-input, multiple-output (MIMO) processes, PMBC can decouple nonlinear interaction, balance deviations from set points when MV constraints are hit, and determine economic optimum MV values when there are extra DOF.

In contrast to the "large model," multi-step-ahead APC techniques characterized by HPC or MPC, the one-step-ahead controllers presented here can solve many problems and can be implemented by a process engineer.

1.11 Control Frequency

The control frequency is the number of control actions per time interval and is often measured in hertz (Hz), the number of control events per second, also called *scan rate.* Control frequency is also referred to as the *control interval* or alternate names such as *scan time* or *period,* which would be the reciprocal of Hz.

The control interval should be small relative to the process response time but not excessively small to create a computational burden. A rule of thumb has been that there should be 30 or more control actions within the open-loop settling time. Faster execution is better for precise control, but the precision improvement is a diminishing returns effect and barely noticeable. If the open-loop settling time is dead time plus three time constants (an FOPDT model concept), then the control interval should be about $\Delta t_{control} \approx (\theta + 3\tau)/30$. An alternate rule is $\Delta t_{control} \approx \theta/5$. Round this to the nearest convenient interval in your software.

Also, consider the numerical solution of the models using the heuristic rule that the simulation time interval must be small relative to any time constants or delays in the model. The smaller the simulation time interval, the truer the numerical solution to the ideal. However, very small modeling time intervals do not substantially improve the numerical model fidelity to the continuum calculus concept. Sufficiency trumps perfection when balancing functionality with computational speed. The simulation time interval should be less than one-tenth of either the modeled delay or the smallest time constant, $\Delta t_{simulation} < min(\theta_1, \theta_2, \theta_3, \ldots \tau_1, \tau_2, \tau_3)$.

It is convenient to have the control interval match the simulation interval, but if the two are disparate, make the control interval, $\Delta t_{control}$, an integer number, *N*, times the model solution time interval, $\Delta t_{simulation}$, and run the simulator for *N* iterations within each control interval.

1.12 Desired Model-Based Controller Attributes

Desirable attributes of a controller can be partitioned into various categories of being operator-friendly and having technical benefits.

1.12.1 Convenience for the Operator/Engineer

- It is simple, easy, and fast to set up.
- It is simple to tune, debug, understand, and agree with its action. If the operator is uncomfortable, it goes to MAN. This is particularly important when the controller is avoiding future constraints, compensating for interactions, and having variables with delays or that act much more slowly than others. The controller may need to reveal to the humans why it is making its choices.
- Balancing constraint deviations is intuitive.
- An operator can place elements within a MIMO controller in MAN or remove measured variables with a fault, and the rest works.

- Understanding the algorithms and internal functions is simple, and they can be easily maintained.
- There is one model version, not two, and certainly not different forms.
- The one-model approach is the same as that used in the supervisory optimizer that determines the set points for the MPC.

1.12.2 Technical Performance

- Can be tuned to provide a wide range of actions from aggressive to temperate.
- Handles hard constraints on the MV value and rate of change.
- Handles soft constraints on CV and AuxV (auxiliary variable).
- Includes economic factors to determine the best economic action when DOF > 0.
- Does not wind up when on a constraint.
- Reflects the nonlinear character of the processes so that behavior is similar in all operating regions.
- Supports bumpless transfer from MAN to AUTO.
- Supports a zero SS offset.
- Are robust to noise and various disturbance patterns.
- Avoids execution errors. Perhaps it is inherently robust (linear with no divides, square roots, or logarithms that could create an execution error), or there are appropriate overrides or error-trapping logic to prevent those or others such as subscript out of range (if the delay becomes longer than an array has dimension).
- Does not get misdirected by root-finding that is diverted in the wrong direction or a local optima or surface aberration in optimization.
- Is guaranteed to return a good solution within a specified execution time. It must complete all calculations within the sampling interval, which must be at least 1/30 of the process settling time.
- Adapts the model to make it useful for other supervisory process management tasks—forecasting when regeneration is needed (catalyst activity, heat transfer fouling), assessing energy or material consumption per product unit for process analysis, forecasting which operational conditions are possible (maximum conversion, maximum production), or revealing faults or undesired situations (flooding, screen blockage).

- Accounts for multivariable interactions.
- Handles ill-behaved dynamic processes (dead time, inverse, integrating, open-loop unstable).
- Can use empirical models.
- Has a sufficient number of future MV trial solutions to shape the modeled CV to the desired reference trajectory.
- Has a minimal number of measurements to increase reliability and minimize the probability of a data fault.
- Is supervisory and transmits set points to lower-level SISO PI(D) controllers so the lower-level controllers keep the process running smoothly if the model-based controller fails or is taken offline.
- Is devised to handle external safety overrides. This places a constraint on the MV values that the optimizer can choose. The overridden value is also needed in the P2N model update.
- Identify mathematical infeasibility, alert the operator, and revert to regulatory control.

1.13 Modeling Approaches

This book focuses on using first-principles models in control. However, many empirical modeling approaches exist, such as NNs, fuzzy FL, and TSK models. These are useful when the process is too complicated to develop phenomenological models. They are developed from empirical testing, similar to the experimental approaches used to devise FOPDT and FIR models for APC. Such models can be successfully implemented within the general MBC structure shown in Figure 1-1 and several MBC approaches described in subsequent chapters.

Examples include the following:

- Dutta and Rhinehart explored the use of NNs in a simple MBC of a lab-scale distillation process in their paper "Experimental Comparison of a Novel, Simple, Neural Network Controller and Linear Model-Based Controllers" presented at the 1995 American Control Conference in Seattle, Washington.[1]

---

1. P. Dutta and R. R. Rhinehart, "Experimental Comparison of a Novel, Simple, Neural Network Controller and Linear Model Based Controllers," in *Proceedings of the 1995 American Control Conference*, paper TA10-3 (Seattle, WA, 1995).

- M. Parekh, M. Desai, H. Li, and R. R. Rhinehart demonstrated FL (a human heuristic model) as a control approach to lab-scale, in-line pH control, which was later demonstrated on a commercial-scale process in their article "In-Line Control of Nonlinear pH Neutralization Based on Fuzzy Logic" published in June 1994.[2]
- Ou and Rhinehart used multiple NNs to forecast future CV values in horizon predictive control for a lab-scale distillation.[3,4]

1.14 Takeaways

- There are many sorts of model-based controllers. The ones addressed in this book were selected because they (1) embody key approaches and (2) are most practicable for industrial application.
- When you judge a control strategy, do not consider just the CV and MV performance metrics. Consider all the issues summarized in Section 1.12.
- Although it takes engineering analysis to create the equations and code for a model-based controller, the many advantages include those summarized in Section 1.3.

Exercise

1.1 Take a relatively simple process unit that you can model and develop a simple model-based controller, as shown in Example 1-1.

---

2. M. Parekh, M. Desai, H. Li, and R. R. Rhinehart, "In-Line Control of Nonlinear pH Neutralization Based on Fuzzy Logic," *IEEE Transactions on Components, Packaging, and Manufacturing Technology*, part A, vol. 17, no. 2 (June 1994): 192–201.
3. J. Ou and R. R. Rhinehart, "Grouped Neural Network Model Predictive Control," *Control Engineering Practice* 11, no. 7 (2003): 723–732.
4. J. Ou, and R. R. Rhinehart, "Grouped Neural Network Modeling for Model Predictive Control," *ISA Transactions* 41, no. 2 (April 2002): 195–202.

2 Models

2.1 Introduction

Process-model-based control (PMBC) starts with the engineer's process model. Such models are often developed from first principles (elementary, phenomenological, or mechanistic models, not full rigor), but they may also contain empirical features or be purely empirical. Modeling may start with a steady-state or dynamic material and/or energy balances. In contrast to traditional linear control models developed for first-order plus dead time (FOPDT), autoregressive moving average (ARMA), or finite impulse response (FIR) patterns, the models here represent the process nonlinearity and nonstationary attributes.

First principles means to consider only the major mechanisms. This contrasts with rigorous models, which seek to include even minor mechanisms and perfectly model the constitutive relations. First-principles models are characteristic of those used in college courses and introductory textbooks to explain topics such as heat exchangers, reactions, distillation, filters, and such unit operations.

2.2 First-Principles Dynamic Models

To start, consider what accumulates (the inventory) that relates to the response of interest. For instance, a process does not inventory liquid level. Liquid contents within a tank are added or removed, and the inventory is the volume of material. Level is a measure of that inventory. Similarly, a process does not inventory temperature. The inventory is in the form of thermal energy (heat) that is added or removed. Temperature is a measure of the quantity of heat that has accumulated. To model a temperature response, consider the mechanisms for heat addition or removal. To model pressure response, consider the mechanisms for adding or removing gas molecules, the mass inventory.

The quantity being inventoried is often labeled the *conserved quantity*. It does not disappear. Heat can be transferred, and material can be transported or converted into reaction products, but neither disappears in chemical processes. Commonly, mass, momentum, and energy are the conserved quantities.

Specify the inventory that you need to consider:

- Pressure is the response to an inventory of molecules and thermal energy.
- Temperature is the response to an inventory of heat.
- Charge is the response to an inventory of electrons.
- Wealth is the response to an inventory of dollars.
- Level is the response to an inventory of molecules or mass.
- Weight is the response to an inventory of molecules or mass.
- Speed is the response to an inventory of momentum.
- Composition is the response to the inventory of several categories of mass.
- pH is the response to the inventory of several categories of mass.

Constitutive relations relate the inventory quantity to other system variables, the response variables. The ideal gas law, rearranged as $P = nRT/V$, could relate the number of moles of gas (the conserved quantity) within a volume at a particular temperature to the resulting pressure (the response) that might be measured. A tank geometry could relate level to liquid volume, and liquid density could relate volume to mass. For a right circular cylinder $V = \pi r^2 h$ and $V = M/\rho$, so the constitutive relation that relates level to the inventory of mass is $h = M/(\pi r^2 \rho)$.

Next, identify the point (or volume) where the quantity being inventoried is accumulated. This could be a tank or vessel, a volume of liquid in a heat exchanger tube, or a moving fluid.

The following examples illustrate procedures for developing models for control.

2.2.1 Basic Approach – Model Example 1: Thermal Mixing in a Tank

This simple example of mixing hot and cold fluid in a tank illustrates the basic approach to developing a model (see Figure 2-1).

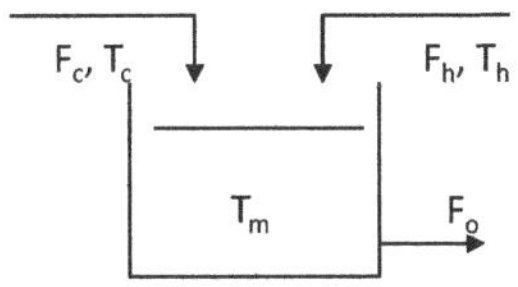

Figure 2-1. Mixing hot and cold fluid in a tank. Here, *F* represents flow rate, *T* represents temperature, subscript *c* represents cold, subscript *h* represents hot, subscript *m* represents mixed fluid, and subscript *o* represents outlet fluid.

1. Define a control volume: the tank contents.
2. Choose a conserved quantity: thermal energy because we are interested in the temperature response.
3. Assume conveniences (aspects that simplify modeling but have inconsequential errors) to balance sufficiency with perfection: The water entering the tank is instantly and perfectly mixed, so the contents are always spatially uniform. There are no evaporative losses. There are no thermal losses. The liquid level is constant. The fluid properties are constant.
4. Apply IOGA: The acronym refers to the inventory going *in* through the control volume boundary, less the inventory going *out* through the control volume boundary, plus the inventory *generated* within the control volume, all of which is equal to the inventory *accumulated* within the control volume over a small increment of time, Δt.

$$\dot{E}_{inflowing}\Delta t - \dot{E}_{outflowing}\Delta t + \dot{E}_{generated}\Delta t = E_{inventory}\Big|_{t+\Delta t} - E_{inventory}\Big|_{t} \qquad (2\text{-}1)$$

In Equation 2-1, $\dot{E}_i$ represents the rate of thermal energy transferred or produced, and $\dot{E}_i\Delta t$ is the quantity transferred or produced in the time interval.

Insert the elementary constitutive relations between measurable variables and the IOGA quantities; that is, $\dot{E}_{out} = F_o \rho c_p T_o$, which assumes the reference temperature is 0 and the coefficients are constants (ρ is fluid density and c_p is fluid specific heat).

There are two sources of heat inflow (the cold and hot water streams), one heat outflow (the exit stream), and no internal reaction to generate heat. The thermal energy IOGA becomes

$$F_c \rho c_p T_c \Delta t + F_h \rho c_p T_h \Delta t - F_o \rho c_p T_o \Delta t + 0 = V \rho c_p T_o \big|_{t+\Delta t} - V \rho c_p T_o \big|_t \tag{2-2}$$

Note: Because ρ, c_p, and V (volume) are considered constants, then $F_{out} = F_c + F_h$.

Divide by ρc_p to simplify and divide by Δt

$$F_c T_c + F_h T_h - (F_c + F_h) T_o = V \frac{T_o \big|_{t+\Delta t} - T_o \big|_t}{\Delta t} \tag{2-3}$$

In the limit of very small Δt, the finite difference is the derivative

$$F_c T_c + F_h T_h - F_o T_o = V \frac{dT_o}{dt} \tag{2-4}$$

In standard ordinary differential equation (ODE) form

$$\frac{V}{(F_c + F_h)} \frac{dT_o}{dt} + T_o = \frac{F_c T_c + F_h T_h}{(F_c + F_h)} \tag{2-5}$$

2.2.2 Including a Delay

Dead time is a delay from the event occurrence to its observation. Common examples are that thunder follows lightning by about 5 s per mile distance. In baseball, the batter hears the "crack" immediately after hitting the ball, but the spectator, high in right field, hears it 2 s after seeing the batter swing.

The constitutive model for transport delay in a pipeline is $\theta = \frac{L}{v} = \frac{L\pi D^2/4}{F}$ (with an ideal plug flow, the transport time is length divided by velocity or volume divided by flow rate). In the mixing tank example in Section 2.2.1, there would be a transport delay if the mixed fluid temperature is measured downstream of the tank. There might also be delays due to sample transport to either the lab or the online analytical instrument, the lab analysis time, or the online analytical instrument cycle time. If there is a dominant lag, but there are several smaller lags, the time constants for the smaller lags can be considered as pseudo-delays. For example, if the thermowell and the control valve have 3 s and 1 s time constants, which are small compared to a 1-min time constant for the tank, these can be modeled as delays. Add all the delays (θ) to get the total. $\theta_{total} = \theta_1 + \theta_2 + \theta_3 + \cdots$ This θ_{total} is the value needed in the FOPDT model.

Typically, delays from sampling intervals or communication devices are inconsequential.

2.2.3 Notes About Model Coefficients

The following are noteworthy points regarding the time constant:

- In Equation 2-5, $\frac{V}{(F_c + F_h)}$ has the units of time. It is the value of τ_p (the process time constant) needed for the FOPDT model. $\tau_p = \frac{V}{(F_c + F_h)}$.
- The time constant is the volume for inventory accumulation divided by the flow rate through the volume. In general, τ_p is the capacity for the control volume to accumulate the relevant inventory divided by the flow rate through it.
- If you know the value of one operating condition, then the equation shows how it scales with alternate operating conditions. It provides the gain scheduling rules: $\tau_2 = \tau_1 \frac{V_2}{V_1} \frac{F_{T1}}{F_{T2}}$. Here, the subscripts indicate two operating conditions, and the F_T represents the total flow rate.
- Even if you do not trust your ability to get the models right, you can probably use the models to scale time constants to other operating conditions.

The following are noteworthy points regarding the process order:

- The one volume of inventory accumulation led to a first-order differential equation. The solution to Equation 2-5 is a first-order response.
- Two volumes of inventory, such as tanks in series, lead to two ODEs—a second-order response.

The following are noteworthy points regarding the gains:

- At a steady state, $\frac{dT_o}{dt} - 0$, so Equation 2-5 shows that $T_{o,SS} = \frac{F_c T_c + F_h T_h}{(F_c + F_h)}$ is the steady-state temperature (if F_c, T_c, F_h, *or* T_h all remain constant).
- The first-principles model indicates the influence variables. On the right-hand side of Equation 2-5, these are F_c, T_c, F_h, and T_h.
- If T_c is considered a disturbance, then the derivative of the steady-state temperature w.r.t. T_c is the gain of the disturbance (K_d). Here, calculus is easy, $K_d = \frac{F_c}{(F_c + F_h)}$.

- If F_h is considered the controller influence, then the derivative of the steady-state temperature w.r.t. F_h is the gain. Here, calculus is not so easy, $K_{F_h} = \frac{(T_h - T_c)F_c}{(F_c + F_h)^2}$. If you are rusty on analytical calculus, use a finite difference numerical approach. Calculate $T_{o,SS}$ for two similar F_h values, then calculate the gain by the ratio of the difference in $T_{o,SS}$ values divided by the difference in the F_h values. The gain from the first-principles model is the gain needed in the FOPDT model.

- Equation 2-5 reveals how gain changes with operating conditions, which is part of a gain scheduling rule.

- If F_c, for instance, is considered a wild flow, and F_h is to be controlled as a ratio, then the nominal ratio value is $r = \frac{T_{SS} - T_c}{T_h - T_{SS}}$. The actual value will likely need to be adjusted by feedback to compensate for sensor errors or influences not included in the model (e.g., ambient heat losses).

- If, again, for instance, F_c and F_h are both being manipulated to manage two controlled variables (CVs), such as total flow rate and temperature, then the gains from the steady-state equation can be used in a relative gain array procedure for loop pairing.

- Your controller outputs are likely to be a percentage signal sent to a control valve, not individual F_c or F_h values. If you know the valve gain, K_v, then the gain experienced by the controller is the $K_{F_h}K_v$ product.

- If you do not trust your ability to get the models right, you can still use the models to scale gains to other operating conditions.

If you have the steady-state model relating the process output to the inputs from the controller and disturbances, $y_{ss} = f(u, d)$, which you might from equipment design, process optimization, or process analysis, then:

- If there is one major volume of inventory, the model is first-order, which means $\tau \frac{dy}{dt} + y = y_{ss} = f(u,d)$.

- With a modicum of process experience, you probably know a reasonable value for the process time constant.

- The steady-state model reveals the gains. Again, if you are rusty with calculus, use a finite difference approach.

The following are noteworthy points regarding the delay:

- The model indicates how delay scales with operating conditions. For instance, if the delay is only because of flow transport and you know the value of the delay at one set of conditions (subscript 1), the delay at a second set of conditions (subscript 2) is easily estimated by the ratio of lengths and flow rates: $\theta_2 = \theta_1 \frac{L_2}{L_1}\frac{F_1}{F_2}$.
- Even if a bit more complicated, the models can be used for gain scheduling rules.
- If you do not trust your ability to get the models right, you can still use the models to scale delays to other operating conditions.

2.2.4 Summary of Creating First-Principles Models

Take each of the following steps. Do not jump to writing the differential equation.

1. Specify the thing to be inventoried; the thing must be something that is conserved (mass, momentum, or energy). Level is not an independent variable. It cannot be moved. Level does not leave a tank. Level is not inventory. Level is the result of a stored quantity of mass or boxes or grain and the physical dimensions of the volume containing it. If you are interested in level, do the IOGA on mass in the tank. Mass is the thing that goes in and out and accumulates. Level is simply a response to the accumulated mass.
2. Specify the control volume where the inventoried quantity accumulates. The control volume must be spatially uniform. Exclude protrusions such as mixer shafts and blades from the control volume. This is often considered a macro-volume, such as a tank or tray. But sometimes, it must be a differential element, such as a small length interval in a packed bed reactor or heat conductor. This instruction will deal with spatially uniformly distributed properties. However, the modeling method used for spatially distributed variables is the same.
3. Specify the time interval.
4. Write out the acronym "IOGA" as a guide.
5. List each mechanism for I, O, G, and A. So far, there have been no equations or symbols, just a clear understanding and statement of the situation. I and O represent the inflow and outflow of the inventory into and out of the control volume. G and A represent generation and accumulation within the control volume. Focus on major I, O, and G mechanisms. For instance, exclude evaporation or ambient heat transfer if these are secondary influences.

6. State idealizations, including mechanisms that are not being considered. This could include the dissipation of mechanical energy to heat, splashing losses, evaporation, and the complete settling of entrained mist to liquid. It could also include ignoring the relatively fast dynamics of valves or sensors, relative to the slow process.

7. State each I, O, G, and A quantity in terms of symbols. (1) Use quantity, not rates. (2) Be sure that the units are correct. I and O represent the quantity that entered or left the control volume by passing through the control volume boundary during time interval Δt. G is the amount generated within the control volume during time interval Δt. A is the amount that accumulated in the control volume during time interval Δt.

8. Divide by Δt and, as in calculus, take the limit as $\Delta t \rightarrow 0$ to convert the finite difference to the derivative notation.

9. Specify the constitutive relations that connect the IOGA inventory terms to the response of interest. For instance, the ideal gas law relates pressure to the inventory of moles; the hydrostatic relation relates pressure to the inventory of liquid; the concentration of a species is calculated from the inventory of masses of A and B; a chemical reaction equilibrium calculation relates the concentration of acid and base to pH, and so on. Begin with ideal relations.

10. State the idealizations, such as those in the chosen constitutive relations, material properties considered to be constant (density, specific heat, etc.), reference values chosen to be zero (e.g., the reference temperature), linear relations between variables, mechanisms that are ignored, or fast dynamics that are ignored. You may want to revisit such assumptions, and others may want to understand such modeling limitations.

11. Substitute constitutive relations into the ODE.

12. Rearrange the ODE to get the differential equation that reveals how the state variable responds to influences.

13. Simplify the equation by regrouping coefficients into common or dimensionless terms and canceling coefficients that appear in all terms.

14. Verify the units. Check to see if the steady-state limit makes sense.

15. State the initial condition.

The following examples illustrate generating explicit, first-principles process modeling for control. They are taken from a variety of common processes to help reveal widespread applicability. Although relatively simple, they are also chosen to reveal nonlinearity, nonstationary, disturbance, and interaction attributes that PMBC can handle.

Explicit models can be written as relatively simple differential equations—the model is mathematically explicit. This is in contrast to "grey-box" models that are also mechanistically derived from first principles, but they may have IF-THEN conditionals (due to transitioning laminar to turbulent flow, nucleate of film boiling, etc.) or may require internal root-finding or optimization algorithms (e.g., associated with chemical or phase equilibrium). Rather than being expressed as an explicit equation, these might be expressed as a procedure (a function or subroutine).

2.2.5 Model Example 2: Flow Rate Response to Controller Output

If a valve is installed in a process line, undergraduate fluid dynamics or process control texts use the pressure drop through pipes, valves, and fittings to derive the steady-state, incompressible fluid flow rate response to the valve stem position:

$$F = Cv \bullet f(x)\sqrt{\frac{\Delta P}{G\xi + \alpha \bullet Cv^2 \bullet f^2(x)}} \tag{2-6}$$

where

F = volumetric flow rate

ΔP = static pressure driving force over the entire pipe section (not just the valve)

$a = \frac{\rho}{2A^2}\sum k$ = the combination of all piping system friction loss coefficients

G = fluid specific gravity

Cv = valve coefficient of discharge (with the same units as F)

$\xi = 1$ = dimensional unifier with the same units as ΔP

x = valve stem position $0 \leq x \leq 1$

$f(x)$ = inherent valve characteristic, which may be linear, $f(x) = x$; or equal percentage, $f(x) = aR^{1-x}$ (R is the valve rangeability coefficient, and a is the capacity coefficient); or other

This equation accounts for the friction losses in the pipe system and across the throttling control valve. The valve stem position, x, is the influence, and the flow rate, F, is the response. Even if the valve has a linear inherent characteristic, F exhibits a nonlinear response to x.

Characteristically, in the relatively short process lines within a typical manufacturing facility, the dynamics of fluid acceleration in response to a valve stem position are much faster than the valve position response to the actuator (which responds to air pressure with a time constant of about 1/4 to 2 s). Accordingly, the dynamics that dominate the fluid flow rate response to the controller output are often from the valve stem position, and it is reasonable to describe the valve stem position as a first-order response to controller output (either from an air material balance on the actuator or from experience watching valves).

$$\tau \frac{dx}{dt} + x = 0.01u \tag{2-7}$$

where τ is the time constant for the valve actuator and u is the controller output in percent. The 0.01 value converts the manipulated variable (MV) (typically 0% to 100%) into the valve stem position, x, as a fraction.

The fluid dynamics may need to be considered for long pipelines that must accelerate or decelerate a large mass of liquid.

The linear dynamics of Equation 2-7 are followed by a nonlinear characterization of Equation 2-6; this assembly is often termed a *Wiener model.* Coefficient values in Equation 2-6 can be evaluated by conducting an engineering analysis of the process, but with uncertainty on the line friction losses, it might be best to adjust a to make the model match the process data. The time constant for Equation 2-7 can be revealed by the observed flow rate transient response to a simple controller output step test.

2.2.6 Model Example 3: Composition Mixing in a Continuous Flow-Through Tank

Consider two streams of concentrations, c_1 and c_2, entering an ideally well-mixed tank kept at a constant liquid level. F_1 and c_1 represent wild flow properties (disturbance inputs), and F_2 is the controlled flow. The net inflow concentration is

$$c_{in} = \frac{c_1 F_1 + c_2 F_2}{F_1 + F_2} \tag{2-8}$$

Note that c_{in}, the net inflow concentration, has a nonlinear response to F_1. Usually, the mixing dynamics in the tank define the outflow concentration as

$$\frac{V}{F}\frac{dc}{dt}+c=c_{in} \tag{2-9}$$

where c is the mixed, in-tank concentration, V is the tank volume, and F represents the total flow rate ($=F_1 + F_2$); the ratio V/F has the units of a time constant, $\tau = V/F$.

This represents a static nonlinearity of Equation 2-8 followed by the linear dynamics of Equation 2-9. Equations 2-8 and 2-9 can be combined to represent a nonlinear *Hammerstein model*:

$$\tau\frac{dc}{dt}+c=\frac{c_1F_1+c_2F_2}{F_1+F_2} \tag{2-10}$$

While developed from combining Equations 2-8 and 2-9, Equation 2-10 could also be derived from a transient material component overall balance on the tank.

2.2.7 Model Example 4: Heat Exchange

Consider that steam is used to heat a process fluid and that the dynamics of the process T response to a steam flow rate can be modeled as first-order. An elementary, steady-state energy balance reveals that the outlet T response will be:

$$T_{out}=T_{in}+\frac{\lambda\dot{Q}_s}{F\rho Cp} \tag{2-11}$$

where T_{in} and T_{out} represent the inlet and outlet temperatures of a process fluid at flow rate F with density and specific heats of ρ and Cp. The steam flow rate is $\dot{Q}_s$ with the latent heat of vaporization λ. This simple model of the primary mechanisms ignores the secondary mechanism, such as incoming steam superheat, condensate subcooling, or ambient heat losses.

Assuming that valve dynamics are negligible compared to the heat transfer dynamics ($x = u/100$ in a relatively short time) and that the installed valve characteristic is $f(x) = f(u/100)$, the steam flow rate response to controller output is

$$\dot{Q}_s = Cvf(u/100)\sqrt{\Delta P/G} \tag{2-12}$$

Combining with first-order process dynamics produces the nonlinear dynamic model

$$\tau\frac{dT}{dt}+T=T_{in}+\frac{\lambda}{F\rho Cp}Cvf(u/100)\sqrt{\Delta P/G} \tag{2-13}$$

2.2.8 Model Example 5: MIMO In-Line Hot and Cold Fluid Mixing

Consider that hot and cold fluids are mixed in-line (like a home shower) and that fluid flow rates F_h and F_c are manipulated by valves with a linear-ish installed characteristic. Their steady-state flow rate response to controller signals, u_1 and u_2, are

$$F_h = k_h u_1 \tag{2-14}$$

$$F_c = k_c u_2 \tag{2-15}$$

The values for valve gains, k_h and k_c, could be empirically determined.

From energy and material balances, the steady-state mixed temperature and total flow rate are

$$T_{in} = \frac{T_h F_h + T_c F_c}{F_h + F_c} \tag{2-16}$$

$$F_{in} = F_h + F_c \tag{2-17}$$

Assuming first-order dynamic response of temperature (due to in-line mixing and heat exchange with the pipes between the mixing point and the delivery point or due to thermowell lag at the measurement point), the net models for the temperature and total flow rate dynamic responses to u_1 and u_2 are

$$\tau_T \frac{dT}{dt} + T = \frac{T_h k_h u_1 + T_c k_c u_2}{k_h u_1 + k_c u_2} \tag{2-18}$$

$$F = k_h u_1 + k_c u_2 \tag{2-19}$$

Including a plug-flow-modeled transport delay $\theta = \frac{V}{F}$, where V is the pipe volume, Equation 2-18 could be written as

$$\tau_T \frac{dT(t)}{dt} + T(t) = \left. \frac{T_h k_h u_1 + T_c k_c u_2}{k_h u_1 + k_c u_2} \right|_{t-\theta} \tag{2-20}$$

Notes:

1. Equation 2-18 could be solved numerically with a simple finite difference method, but Equation 2-20 would need a procedure that stores past values of temperatures and control actions to include the forcing function value (the right-hand side of Equation 2-20) based on conditions θ time delay for the current outflow temperature.
2. This model assumes that the valve response is instantaneous (no actuator lags), that valves have a linear installed characteristic, and that flow rates are independent. (They are not independent. If one flow rate increases, it increases back pressure on the exit line, reducing the other flow rate.) These are relatively simple additions to the model, but these would change the solution procedure from a simple one equation method to needing a procedure for the additional computational steps.

2.2.9 Model Example 6: Car Speed

Start with Newton's law of motion for the linear motion of the car (ignore rotational mechanisms) and express acceleration as the rate of velocity change. This explicitly provides evidence of the time dependence in the state variable (in mathematical language), the CV, which is velocity (from a process view).

$$F = \frac{ma}{g_c} = \frac{m}{g_c}\frac{dv}{dt} \tag{2-21}$$

Here, m is the car mass, a is the acceleration, F is the accelerating force, and g_c is the dimensional unifier (32.17 lbf-s²/lbm-ft, or 1 N-s²/kg-m).

Force will be modeled as having three elements, as shown in Equation 2-22. One accelerating force (or a decelerating force) is from the engine (the first in the series of terms, ku). The second term, av^2, represents resistance to motion, a decelerating drag force, from various aspects (wind resistance, rolling friction, etc.). Gravity helps accelerate when going downhill and decelerate when going uphill. In Equation 2-22, θ is the angle of climb from the horizontal, and g is the gravitational acceleration (at sea level 32.17 ft/s², 9.807 m/s²). For this simple model, consider that the torque from the engine is linearly converted to a force on the car and that torque is proportional to the accelerator pedal position, u. Also, for this simple model, assume that drag forces are proportional to the square of velocity.

$$F = ku - av^2 - \sin(\theta)mg/g_c \tag{2-22}$$

Notes:

1. Drag coefficient a changes over time, partly because the air density changes (pressure, humidity, temperature, moisture).
2. The velocity in the drag element is not the car velocity on the road but the car velocity relative to the air, which changes with wind speed and direction. Also, if one wants to preserve the impact of the sign of the velocity, the drag term should be $-av|v|$.

If you wanted to model rolling friction as linear with velocity and include air velocity, the av^2 term would become $a_1|v| + a_2(v - v_{air})^2$. If you wanted a more rigorous model for the engine impact, considering that there is a force when idling and the force asymptotically diminishes with pedal position, the ku term might be expressed as $k_0 + k_1(1 - e^{-u/k_3})$.

Combining Equations 2-21 and 2-22, then rearranging them into the standard form in which the state variable appears to the first power generates the model

$$\frac{m}{avg_c}\frac{dv}{dt} + v = \frac{k}{av}u - \frac{\sin(\theta)mg}{avg_c} \tag{2-23}$$

$$\tau\frac{dv}{dt}+v=f_1(u,v,a,m,g,\theta,g_c) \tag{2-24}$$

It should be evident that this model has several nonlinear and nonstationary aspects. Even though the accelerating force is modeled as a linear function of pedal position u, the model indicates a nonlinear response of velocity to pedal position in which the gain, k/av, is not a constant. The model is nonstationary (changes in time) for several reasons. Coefficient a and angle θ both represent external disturbances that change in time, and velocity changes in time. Further, because the time constant, m/avg_c, changes with coefficient a and velocity, v, the gain changes with coefficient a, and the gravitational effect changes with a, v, and θ, the model is nonstationary.

2.2.10 Model Example 7: Level in a Conical Tank

Figure 2-2 illustrates a right circular conical tank. The flow rate into the tank is F_{in} and out is F_{out}. The exiting flow rate depends on the pressure difference as one driving force and the hydrostatic pressure as another. The object is to model how the level depends on several influences.

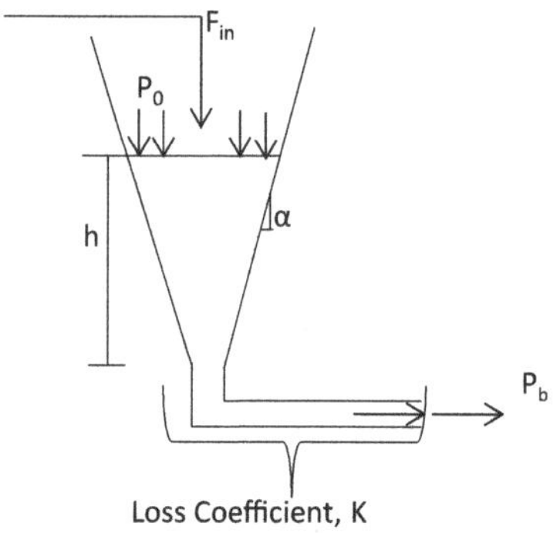

Figure 2-2. Level in a right circular conical tank. Here, P_o represents the overpressure on the tank, h represents the liquid level (height), a represents the slope of the cone, and P_b represents the back pressure from downstream.

Because we are interested in level, the inventory will be the material (mass). Mass, m is the conserved quantity, not volume, V. The inflow quantity is the volumetric flow rate times the density, ρ times the time interval, $F_{in}\rho\Delta t$ [(gal/min) (lbm/gal) (min)]. The outflow of material is similar: $F_0\rho\Delta t$ [(gal/min) (lbm/gal) (min)]. The internal generation is zero. The accumulation is the mass difference over time: $lbm_{now+\Delta t}-lbm_{now}=m_{t+\Delta t}-m_t$. The IOGA balance is then

$$F_{in}\rho\Delta t-F_{out}\rho\Delta t=m_{t+\Delta t}-m_t \tag{2-25}$$

Divide by Δt, and take the limit of very small Δt to determine the mass conservation equation.

$$F_{in}\rho - F_{out}\rho = \frac{dm}{dt} \tag{2-26}$$

Use constitutive relations to relate variables to state variable h. For ideal turbulent flow, the flow rate is proportional to the square root of the pressure drop across the restriction, ΔP.

$$F_{out} = K\sqrt{\Delta P} = K\sqrt{\left(P_0 + \frac{\rho g h}{g_c} - P_b\right)} \tag{2-27}$$

Ideal geometry for a right conical shape relates volume to height (but this ignores ribs in the tank, exit end effects such as the discharge volume and height, etc.).

$$m = V\rho = \frac{\pi\alpha^2}{3}h^3\rho \tag{2-28}$$

The time derivative of Equation 2-28 relates $\frac{dm}{dt}$ to $\frac{dh}{dt}$.

$$\frac{dm}{dt} = \pi\alpha^2 h^2 \rho \frac{dh}{dt} \tag{2-29}$$

Substituting constitutive relations into the mass balance

$$F_{in}\rho - K\rho\sqrt{\left(P_0 + \frac{\rho g h}{g_c} - P_b\right)} = \pi\alpha^2 h^2 \rho \frac{dh}{dt} \tag{2-30}$$

Assuming constant density and rearranging

$$\pi\alpha^2 h^2 \frac{dh}{dt} + K\sqrt{\left(P_0 + \frac{\rho g h}{g_c} - P_b\right)} = F_{in} \tag{2-31}$$

And one more rearrangement step to convert the model to a standard form, revealing nonlinear time constant and gain

$$\frac{\pi\alpha^2 h^3}{K\sqrt{\left(P_0 + \frac{\rho g h}{g_c} - P_b\right)}}\frac{dh}{dt} + h = \frac{F_{in} h}{K\sqrt{\left(P_0 + \frac{\rho g h}{g_c} - P_b\right)}} \tag{2-32}$$

2.2.11 Model Example 8: A Second-Order Flow Rate Model

Reconsider the fluid flow rate model in Section 2.2.5, Model 2. The valve response to the controller output is still Equation 2-7 but rearranged here to isolate the derivative

$$\frac{dx}{dt} = (0.01u - x)/\tau \tag{2-33}$$

If the pipeline is very long, the inertia of the fluid mass in the pipeline will cause it to slowly move to the final F value after a change in valve stem position. Using Newton's second law of motion, $F = ma/g_c$, for the liquid mass, revealing acceleration as the rate of change of average velocity. Using A for the cross-sectional area and representing the forces as $A\Delta P$ and $Av = \dot{Q}$

$$A\frac{dv}{dt} = \frac{d\dot{Q}}{dt} = \frac{g_c A \sum \Delta P_i}{m} = \frac{g_c A}{m}(\Delta P_T - \Delta P_V - \Delta P_L) \tag{2-34}$$

The fluid mass being accelerated is $m = AL\rho$. The three pressure drop values represent the total line pressure drop, the drop through the valve, and the drop due to line "friction" losses. With classic turbulent flow models, $\Delta P_V = \frac{G\xi}{(Cv \bullet f(x))^2}\dot{Q}^2$, and $\Delta P_L = \sum k\frac{1}{2}\rho v^2 = \frac{\rho}{2A^2}\sum k\dot{Q}^2 = \alpha\dot{Q}^2$. For convenience, the combined coefficient $\frac{\rho}{2A^2}\sum k = \alpha$. Inserting into Equation 2-34

$$\frac{d\dot{Q}}{dt} = \frac{g_c}{L\rho}\left[\Delta P_T - \left(\frac{G\xi}{(Cv \bullet f(x))^2} + \alpha\right)\dot{Q}^2\right] \tag{2-35}$$

The reader might want to see that Equation 2-35 reduces to Equation 2-6 under steady-state conditions.

This is a second-order process where Equation 2-33 defines the value of x, which is part of the forcing function in Equation 2-35. There are two states, the intermediate x and the CV, $\dot{Q}$.

2.2.12 Model Example 9: A Simple Higher-Order Model

This example will model fluid composition within a set of two tanks in series, each with a dead zone and otherwise well-mixed contents. Inflow to the first tank will change composition, and the outflow of that first tank will be the inflow to the second tank. The tanks have constant volume. In the well-mixed region, the presumption is that the contents are instantly and perfectly mixed. This is not truly possible, but it is a reasonable assumption if the mixing speed is fast relative to the flow-through time constant. The mixing dead zone is the space behind baffles and such geometric spaces. The dead zone is not inactive. It has internal mixing and exchanges material with the active zone. Both zones will be modeled as ideal mixers with fluid exchange.

The control objective is to add concentrate with a value of c_{in} to the process inflow to the first tank to bring the outflow composition to the set point. The process wild inflow rate is F_w with concentration c_w. The controlled flow rate is u with concentration c_u. The total inflow rate is

$$F = F_w + u \tag{2-36}$$

And the mixed inflow composition is

$$c_{in} = \frac{F_w c_w + u c_u}{F_w + u} \tag{2-37}$$

The mixing model results in

$$\frac{dc_1}{dt} = \frac{F}{V_{A,1}}(c_{in} - c_1) + \frac{f_1}{V_{A,1}}(c_{D,1} - c_1) \tag{2-38}$$

$$\frac{dc_{D,1}}{dt} = \frac{f_1}{V_{D,1}}(c_1 - c_{D,1}) \tag{2-39}$$

$$\frac{dc_2}{dt} = \frac{F}{V_{A,2}}(c_1 - c_2) + \frac{f_2}{V_{A,2}}(c_{D,2} - c_2) \tag{2-40}$$

$$\frac{dc_{D,2}}{dt} = \frac{f_2}{V_{D,2}}(c_2 - c_{D,2}) \tag{2-41}$$

Here c_1 and c_2 represent the compositions in the active zones of the two tanks, and $c_{D,1}$ and $c_{D,2}$ are the compositions in the respective dead zones. $V_{A,1}$ and $V_{A,2}$ are the volumes of the active space in the tanks, and $V_{D,1}$ and $V_{D,2}$ are the volumes of the respective dead zones. The total liquid volume in tank 1 is $V_{A,1} + V_{D,1}$. F is the flow rate through the system, and f_1 and f_2 are the exchange flow rates between the active and dead zones.

This model has four states that respond to process influences. The response states are $c_{D,1}$, c_1, $c_{D,2}$, and c_2, but only c_2 would be the CV. The differential equations are linear, but an input to the first equation, c_{in}, is a nonlinear response to the control action, u. The differential equations are also interactive: c_1 affects $c_{D,1}$, and $c_{D,1}$ affects c_1. Because the wild flow rate and composition change in time, the differential equations are also nonstationary.

2.2.13 Model Example 10: A Heat Exchanger

This example uses a liquid-liquid, countercurrent, single-pass heat exchanger. The hot fluid is in the shell side and modeled as an oil. The process fluid being heated is modeled as water. The control objective is to control the exit water temperature by sending a signal to the flow control valve on the hot fluid entrance line. The control valve to be modeled has a linear inherent characteristic, but installed in the line, it has a quick-opening installed characteristic. The valve is air actuated, with a stem position response that is roughly first-order to the controller output, the MV. Each of the entrance and exit lines has a thermowell to measure temperature. The measurement has a first-order lag and a delay due to fluid transport between the sensor and the heat exchanger.

The flow rate of the hot fluid is controlled. The uncontrolled inputs to the heat exchanger are the cold fluid flow rate, the hot and cold fluid temperatures, and ambient heat losses. These four variables fluctuate continually using an ARMA(1,1) model driven by normal (Gaussian) independent deviations with zero mean and a standard deviation, NID(0,σ) (see Section 3.5). The user can also make intentional changes to these variables.

A key and undesirable characteristic of a heat exchanger is fouling of the heat transfer surface. It might be due to scale buildup, a component that precipitates or chars when hot, or another cause. As fouling progresses, the hot fluid flow rate must be increased to maintain the outlet fluid temperature at its set point. Fouling typically progresses over time; however, it may happen fairly rapidly if the inlet fluid contains a batch of undesirable material. The simulator also changes the fouling factor.

In addition to creating a controller that can respond to the effects of fouling by controlling the fluid flow rate, the model enables process managers to identify when fouling is occurring. This knowledge facilitates scheduling maintenance and forecasting constrained operations. The model will have utility for both.

There are several stages to the model, which will be developed independently.

2.2.13.1 The Control Valve

The controller output, the MV, is u, in the 0% to 100% range. The valve stem position is x, the fraction of full travel, in the 0 to 1 range. Because of the air actuator, x has a nearly first-order response to u with a time constant of τ_v.

$$\tau_v \frac{dx}{dt} + x = 0.01u \tag{2-42}$$

Numerically,

$$x := \lambda_v 0.01u + (1 - \lambda_v)x \tag{2-43}$$

$$\lambda_v = 1 - e^{-\Delta t/\tau_v} \tag{2-44}$$

In Equation 2-43, the ":= " represents a computer code assignment statement (not an algebra equal sign). On the right-hand side, λ_v is a weighting factor, and x represents the valve stem position at the prior time step, Δt, which is used to predict the new value at the end of the time interval. Unlike an algebraic ARMA(1,1) model, subscripts are not needed in computer code. Use Equation 2-43 to determine how x responds to u.

The valve equation describes how the steady-state flow rate responds to the stem position.

$$F = C_v f(x)\sqrt{\frac{\Delta P_v}{G\xi}} \tag{2-45}$$

Here, C_v is the valve capacity coefficient with the same units as F, the volumetric flow rate. ΔP_v is the pressure drop across the valve, G is the fluid specific gravity, and $\xi = 1$ is a dimensional unifier and has the same units as ΔP_v. $f(x)$ is the inherent valve characteristic, which reveals how a steady-state flow rate depends on stem position, $0 \leq f(x) \leq 1$. For a linear valve, $f(x) = x$.

In a long pipeline, the liquid mass needs time to accelerate from a prior velocity value to a new steady-state value. However, typical process flow lines are short enough so that the flow can be considered at the steady state associated with x. This is often termed as a pseudo-steady state.

However, as the flow rate changes, the friction loss pressure drop, ΔP_f, in the line changes. Assuming that the system pressure drop, ΔP_s, does not change

$$\Delta P_s = \Delta P_v + \Delta P_f \tag{2-46}$$

Using a classic turbulent flow model for losses in piping and fittings, with k representing the sum of all pipe and fitting influences,

$$\Delta P_f = k\frac{1}{2}\rho v^2 = k\frac{1}{2}\rho\frac{1}{A_{xc}^2}F^2 \tag{2-47}$$

Solving Equation 2-45 for ΔP_v, inserting Equations 2-45 and 2-47 into 2-46, combining constants into coefficients, $b = G\xi/\Delta P_s$ and $c = k\frac{1}{2}\rho\frac{C_v^{\ 2}}{A_{xc}^2}/\Delta P_s$, and using the linear inherent characteristic, and solving for F, converts the linear inherent characteristic to the installed characteristic.

$$F = C_v x\sqrt{\frac{1}{b + cx^2}} \tag{2-48}$$

Use Equation 2-43 to determine how x responds to u, then Equation 2-48 to determine how the controlled flow rate, F, responds to x.

2.2.13.2 The Heat Exchanger

The steady state design equation for a single-pass, countercurrent heat exchanger is

$$\dot{Q}_t = UA\Delta T_{lm} = UA\frac{(T_{h,i} - T_{c,o}) - (T_{h,o} - T_{c,i})}{ln\left(\frac{T_{h,i} - T_{c,o}}{T_{h,o} - T_{c,i}}\right)} \tag{2-49}$$

Here, $\dot{Q}_t$ is the rate that heat is transferred, U is the overall heat transfer coefficient, A is the heat transfer area, and T represents fluid temperature. The subscripts h and c represent the hot and cold fluids, and the subscripts i and o represent inflow or outflow locations. ΔT_{lm} is called *delta-T-log-mean* and represents the effective average temperature difference along the tube length for heat transfer.

Given values for UA and for $T_{h,i}$ and $T_{c,i}$, there are three unknowns in Equation 2-49. These are solved by using equations for the heat gained by the cold fluid and lost by the hot fluid, ideally modeled as

$$\dot{Q}_c = F_c\rho_c Cp_c\Delta T_c = F_c\rho_c Cp_c(T_{c,o} - T_{c,i}) \tag{2-50}$$

$$\dot{Q}_h = F_h\rho_h Cp_h(T_{h,i} - T_{h,o}) - \dot{Q}_{amb} \tag{2-51}$$

where $\dot{Q}_{amb}$ represents the losses from the heat exchanger to the environment, here credited only to the hot shell-side fluid.

Because $\dot{Q}_t = \dot{Q}_h = \dot{Q}_c$, Equations 2-49 through 2-51 can be solved to determine $\dot{Q}_t$, $T_{h,o}$, and $T_{c,o}$, given values for the inlet flow rates, fluid properties, $\dot{Q}_{amb}$, U, A, and the inlet temperatures. However, the equations do not seem to permit an explicit solution, so a root-finding approach is needed. Guess the value of $\dot{Q}_t$ as $\dot{Q}_g$, solve Equations 2-50 and 2-51 for the associated outlet temperatures, and then use Equation 2-49 to see if the associated $\dot{Q}_t$ is equal to $\dot{Q}_g$. For this application, interval halving (bisection search) from high and low extreme possible values for $\dot{Q}_g$ is stable and fast. The low guess is that the cold fluid only rises by a small amount (1 degree). The high guess would be the minimum of either the hot fluid exiting very close to the cold inlet T or the cold fluid exiting very close to the hot inlet T.

The overall heat transfer coefficient depends on the fluid heat transfer coefficients, the conduction resistance through the tube walls, r_{cond}, and fouling resistance, r_{foul}.

$$\frac{1}{U} = \frac{1}{h_h} + r_{cond} + \frac{1}{h_c} + r_{foul} \tag{2-52}$$

And experimentally, dimensionless correlations such as the Dittus-Boelter relation for turbulent flow show that $h \propto F^{0.8}$, so

$$U = \frac{1}{\frac{1}{h_{h,ref}}\left(\frac{F_{h,ref}}{F_h}\right)^{0.8} + r_{cond} + \frac{1}{h_{c,ref}}\left(\frac{F_{c,ref}}{F_c}\right)^{0.8} + r_{foul}} \tag{2-53}$$

This procedure determines the steady-state exit temperatures of the heat exchanger (HX). Experimentally, the dynamics of the heat exchanger are found to be nearly first-order and small relative to the transport delay of the fluid. The lags in the temperature sensors can be modeled as first-order.

Assuming plug flow (nearly valid for turbulent flow) and adiabatic flow (no heat transfer to or from the piping system), the transport delay from the HX to the sensors is simply $\theta = V/F$, where F is the flow rate and V is the volume from the middle of the HX to the T sensor, and the transport delay is digitized to the number of simulation time steps.

$$N_{delay} = INT\left(\frac{V}{F\Delta t}\right) \tag{2-54}$$

The temperature of the fluid at the ith simulation interval would be the temperature exiting the HX N_{delay} samples prior. This is for both cold and hot fluids. Further, the inlet temperatures are sampled prior to the fluid entering the HX, so the fluid entering the HX at any particular sample is that at the sensor N_{delay} samples after.

In the simulator posted on *www.r3eda.com*, the clock method stores and retrieves delayed values using an array of 100 elements. Instead of viewing it as a linear column array, visualize a circular array, like the numbers on a clock, with the array element 100 adjacent to the 1, like the clock hour 12 is adjacent to the 1. At each sampling, increment the store index, i_{put}, by 1, and if $i_{put} = 101$, then reset it to 1. Store the value in the $x(i_{put})$ array element. Determine the i_{read} index value as $i_{put} - N_{delay}$. If i_{read} is negative, reset it using $i_{read} := i_{read} + 100$.

Finally, all the temperature sensors lag in responding to the process fluid. This is modeled as a first-order lag with the thermowell time constant represented by τ_{tw}

$$\tau_{tw}\frac{dT_{sensor}}{dt} + T_{sensor} = T_{fluid} \tag{2-55}$$

Numerically,

$$T_{sensor} := \lambda_{tw} T_{fluid} + (1-\lambda_{tw}) T_{sensor} \tag{2-56}$$

$$\lambda_{tw} = 1 - e^{-\Delta t/\tau_{tw}} \tag{2-57}$$

All in all, to determine the outlet temperatures, use Equation 2-53 to determine the value of U, then the N_{delay} past inlet temperatures and the interval halving procedure to determine the HX outlet temperatures. Then, use Equation 2-56 on the delayed outlet temperatures to determine the sensor T.

The HX model is not a one line equation. It would be implemented as a subroutine or function.

2.2.14 Models 11 and 12

Appendixes B and C provide details on 2×2 models: T and F targets for in-line mixing hot and cold fluid (model 11) and composition and level in pH neutralization (model 12).

2.2.15 Adjusting Model Coefficients to Better Fit the Process

Most models have coefficients with much uncertainty in their values. These could be the friction loss coefficient in a piping system, the tray efficiency in a distillation column, the ambient heat losses from a unit, the catalyst activity in a reactor, the Btu content in fuel, or the heat transfer fouling coefficient in a heat exchanger. There are many others that are more related to the transient response rates. One of these could be the time constant on a distillation tray because the boiling makes it difficult to know the liquid volume, and the internal reflux makes it difficult to know the exact flow through. Another could be the delay in a piping system for which new configurations make the exact volume unknowable.

You should initialize your models with nominal values for those coefficients, but the process will probably not match those nominal values. Adjust those values to make the model best match the process.

Adjust model coefficients that have uncertain values to match the process data as closely as reasonable. You can probably use historical process data to do this. However, if the historical data does not have enough controlled information or perturbation magnitude, you might need to design experiments to generate the needed data.

Some coefficients will relate to the dynamic response of the process (time constants, rates of change, delays, etc.). Use only transient process periods to adjust these coefficients. The process does not express dynamic response during steady-state periods, so there is no basis for adjusting such coefficients. If you seek to adjust dynamic response coefficients with steady-state data, you will be using noise to adjust them, which will cause the coefficients to random walk into undesirable values. The model will also have coefficients that are related to the gains and steady-state values. These could be adjusted with either transient or steady-state data.

Notes:

1. You should not adjust transient and steady-state model coefficients with the same transient data. The coefficients may have counter-effects. It is best to adjust steady-state coefficients with steady-state data only.
2. You should have an independent measured variable for each model coefficient you adjust.

Often there are just a few of these coefficients with uncertain values, and it may be convenient to adjust them manually. However, you could use nonlinear regression techniques to automate the adjustment.

See Section 12.2 for a method to incrementally update model coefficient values in real time with online data.

2.2.16 Generic Format for Dynamic Model Equations

This work will represent the generic dynamic model as

$$\frac{d\underline{y}}{dt} = \underline{f}(\underline{y}, \underline{u}, \underline{d}, \underline{p}) \tag{2-58}$$

where $\underline{y}$ is the vector of process responses, which are often called *state variables.* At least some of the state variables will be the CVs for the controller. $\underline{u}$ is the vector of the

MV controller influences on the process, $\underline{d}$ represents measurable disturbances, and $\underline{p}$ represents model parameters (also termed *model coefficients*).

2.2.17 Solving Dynamic Model Equations

The solution to the nonlinear dynamic model will probably be analytically intractable for three reasons: the model is nonlinear; the model coefficients change with any of several variables; and the influence (forcing function) will not be a simple step, ramp, or other simple function that permits mathematical analysis. Most likely, the models will be solved by a numerical method approach. Usually, the simplest approach, *Euler's explicit finite difference method,* is fully adequate. This approach uses a forward finite difference approximation to the derivative (undoing the modeling step that takes the limit of infinitely small Δt to get the derivative).

Start with model Equation 2-58. It is valid for all time. Consider the present time, $t = i \bullet \Delta t$, where the current time is the i number of time intervals (Δt) since the initial conditions. The time interval, Δt, counter starts at zero when the y values are the initial condition values. For each y value

$$\left.\frac{dy}{dt}\right|_{t=i \bullet \Delta t} = \left. f(\underline{y},\underline{u},\underline{d},\underline{p})\right|_{t=i \bullet \Delta t} \tag{2-59}$$

The forward difference approximation to the derivative is the future value minus the present value divided by the time interval between them.

$$\left.\frac{dy}{dt}\right|_{t=i \bullet \Delta t} \cong \frac{y(t+\Delta t)-y(t)}{\Delta t} = \frac{y_{i+1}-y_i}{\Delta t} \tag{2-60}$$

Equating the right-hand side of Equation 2-59 to that of Equation 2-60 and rearranging to solve for the next-in-time y value

$$y_{i+1} = y_i + \Delta t \bullet \left. f(\underline{y},\underline{u},\underline{d},\underline{p})\right|_i \tag{2-61}$$

This is termed an *explicit method* because the future y value is calculated explicitly from present values. In an implicit numerical method, some terms on the right-hand side of Equation 2-61 would contain future values.

The numerical method of solving for the model response only provides values at the discrete time increments of $t = i \bullet \Delta t$. And the numerical values presume that the rate of change does not change during the Δt. Accordingly, the numerical solution values are not exactly those that the analytical solution might provide. However, as Δt becomes small, the constant rate basis between intervals is nearly valid, and the numerical solution is numerically close to what might be calculated by the analytical solution.

PMBC, predictive functional control (PFC), and nonlinear model predictive control (N-MPC) will be derived from generic Equation 2-58.

If there is a single response variable, y, this is a multiple-input, single-output (MISO) representation. The multiple inputs are the control action and disturbances.

It is relatively easy to rearrange the model equations into the form of Equation 2-58 by isolating the derivative term on the left-hand side of the equation and following the directions to formulate the model-based controller.

2.3 Simulation Time Interval

Conventionally, a small Δt is defined as being an order of magnitude smaller than the smallest delay in the model or the smallest time constant in the model. The smaller the simulation time interval, the truer the numerical solution to the analytical solution, but an excessively small simulation Δt leads to excessive computational time. Sufficiency trumps perfection. Balance good enough functionality with faster model execution.

The simulation time interval should be less than 1/10 of either the modeled delay (θ) or the smallest time constant, $\Delta t_{simulation} \ll min(\theta_1, \theta_2, \theta_3, \ldots \tau_1, \tau_2, \tau_3)$. If there are disparate values within the delay or time constant, and some are very small compared to others (perhaps more than an order of 10 smaller), those features will rapidly reach their final value long before the slower features do. Consider changing the model of the very fast responses to steady-state versions of the models. These are called *pseudo-steady-state models*. The impact on the modeled output will be negligible, permitting using a larger simulation, Δt, which has the benefit of a faster computation.

Systems of differential equations with disparate time constants are termed *stiff*. Classical numerical procedures to solve them include fourth-order Runge-Kutta and Gear. You could choose such methods. However, they use more computational time than the pseudo-steady state method recommended earlier. Balance perfection with sufficiency. The models (based on human concepts of mechanisms, simplifications, idealizations, assumptions, and elementary constitutive relations) are not completely true to the process. A perfect solution to the idealized model will not match exactly what the process will do. The model choices choose sufficiency, so choose sufficiency (speed and simplicity) in the numerical solution procedure.

2.4 Steady-State Models

To derive steady-state models, use the same IOGA approach but do not include the accumulation term.

2.5 Steady-State First-Principles Models with Presumed Linear Dynamics

It is usually easier to obtain steady-state models than dynamic models. Further, steady-state models were probably used in the initial process design and are currently used in online process analysis and/or real-time supervisory set-point optimization. If so, the steady-state models should be available.

It is often realistic to presume that the process dynamics are low-order and linear and to use the steady-state model with the linear dynamics to create the transient process model.

A Hammerstein model type uses a nonlinear forcing function to lead a linear dynamic response. An example is shown in Equation 2-62, a combination of Equations 2-8 and 2-9. The nonlinear blending leads to a steady-state model.

$$c_{SS} = c_{in} = \frac{c_1 F_1 + c_2 F_2}{F_1 + F_2} \tag{2-62}$$

Assuming an ideal, perfectly mixed, constant volume and a constant flow rate, the mixing dynamics in the tank define the outflow concentration as

$$\tau \frac{dc}{dt} + c = c_{SS} \tag{2-63}$$

Note: Here τ represents the blending time constant, c is the mixed composition, and $\frac{dc}{dt}$ is the rate of change of the composition.

Equations 2-62 and 2-63 represent a static nonlinearity followed by linear dynamics, a Hammerstein model. Equations 2-62 and 2-63 can be combined to represent one nonlinear model:

$$\tau \frac{dc}{dt} + c = \frac{c_1 F_1 + c_2 F_2}{F_1 + F_2} \tag{2-64}$$

In contrast, a Wiener model type uses a linear dynamic model to lead a nonlinear response. An example is provided by placing Equations 2-6 and 2-7 in the generic structure of Equation 2-58. Consider the single-input, single-output (SISO) Wiener model of Equations 2-65 and 2-66, in which the function symbol $g()$ represents a steady-state relationship.

$$\tau\frac{dx}{dt}+x=ku \tag{2-65}$$

$$y=g(x,d) \tag{2-66}$$

Note: Here, y represents a generic process response variable, which is a function of two influences. One influence is the disturbance d, and the other, x, is a lagged response to the control action u.

Take the time derivative of Equation 2-48 using the chain rule

$$\frac{dy}{dt}=\frac{\partial y}{\partial x}\frac{dx}{dt} \tag{2-67}$$

and substitute $\frac{dx}{dt}$ from Equation 2-65 and x from the inverse of Equation 2-66 into Equation 2-67.

$$\frac{dy}{dt}=\frac{\partial y}{\partial x}\left[ku-g^{-1}(y,d)\right]/\tau=f(y,u,d,p) \tag{2-68}$$

Note: Here, the rate of change of the state variable, y, is a response of the state, the process influences, and modeled parameter values, p.

Equation 2-68 is the model representation needed here.

You will need to derive a dynamic model in the form of Equation 2-58 or 2-68 to implement several of the nonlinear controllers described here.

Note: The outcome might not be one line of code even for a SISO model, as implied by Equation 2-58 or 2-68. The inverse of the steady-state model might require a root-finding technique. Or, evaluating the right-hand side might require conditional IF-THEN statements associated with laminar or turbulent flow models or a boiling regime. It might be that getting a numerical value that represents the right-hand side of either equation requires a subroutine.

Alternatively, you may only have steady-state relations. In that case, generic model control (GMC) with a steady-state model may be an option.

2.6 Other Modeling Approaches

Although this book is about using first-principles models in process control, surrogate models (examples are given in this section) could be substituted.

2.6.1 Grey-Box Models

"Grey-box" models are mechanistically derived from first principles. However, they may have IF-THEN conditionals (due to the transition of laminar to turbulent flow, nucleate of film boiling, etc.) and/or may require internal root-finding or optimization algorithms (such as associated with chemical or phase equilibrium, or implicit algebraic equations), and/or may have several internal states. They are not represented by a simple equation but by a computer procedure (subroutine or function) needed to solve for the rate of change of some variable.

There might not be an explicit derivative for your model. However, values for rates of change and sensitivity could be obtained numerically from the procedure. Using a forward finite difference, the rate of change would be estimated by $\frac{dy}{dt} \cong \frac{y(t+\Delta t)-y(t)}{\Delta t}$, and the sensitivity of the modeled y to some variable x would be estimated by $\frac{dy}{dx} \cong \frac{y(x+\Delta x)-y(x)}{\Delta x}$.

Regardless of the method for getting values, grey-box models can be presented in the generic format.

For example, the equation set in Section 2.2.12 could be solved in a function with u as the controller influence and c_2 as the response. If the ODEs are solved with Euler's method, this could be the pseudo-code as follows:

```
FUNCTION C2_OF_U (U, FW, CW)
    F := FW + U
    CIN := (FW * CW + U * CU) / (FW + U)
    C1 := C1OLD + DT * ( F * (CIN - C1OLD) / VA1 + FE1 * (CD1OLD - C1OLD) / VD1)
    CD1 := CD1OLD + DT * FE1 * (C1OLD - CD1OLD) / VD1
    C2 := C2OLD + DT * ( F * (C1OLD - C2OLD) / VA2 + FE2 * (CD2OLD - C2OLD) /
    VD2)
    CD2 := CD2OLD + DT * FE2 * (C2OLD - CD2OLD) / VD2
    C1OLD := C1
    CD1OLD := CD1
    CO2LD := C2
    CD2OLD := CD2
    C2_OF_U = C2
END FUNCTION
```

The ":= " symbol represents a computer assignment statement. The function is structured to have three inputs: the controller influence, U, and the two disturbances, FW and FC. Of course, many variable values must be initialized before this function is executed. Also, the executable function code for complicated processes might be much longer. The complexity of the model is not all that important. What matters is that it can be executed within the required control time interval.

Also, what you choose for the mathematical basis of the function is not important. This example is a Fortran-like translation of the explicit finite difference versions of the equations in Section 2.2.12. If you want to use implicit versions or a predictor-corrector method, that is OK. If you want to use a first-principles model to generate training data for a neural network (NN) model and then place the trained NN model in the function, that is OK.

2.6.2 FOPDT and SOPDT Models

Many modeling approaches are used within chemical process control practice. The most frequently used model is the first-order plus dead time (FOPDT) model that has coefficient values empirically determined by local step-testing the process using the classic reaction curve method from the 1940s. Some use second-order plus dead time (SOPDT) models because they are better at matching realistic dynamics.

2.6.3 ARMA Models

Alternately, some use ARMA-type models. These linear models could be presented in various mathematical ways—time-domain (as a differential equation), Laplace transform, z-transform, back-shift operator, and others. To improve model quality and minimize the process upset from step-and-hold tests, some people are moving toward using least-squares to fit linear models to data generated from multiple perturbations, not just a single step-and-hold.

2.6.4 FIR Models

FIR models are locally linear empirical models used in model predictive control (alternately termed MPC, DMC, or APC).

2.6.5 State-Space Models

State-space models (matrix and ABCD models) are also linear multivariable models used in modern control. If the models are linear, they can still be formulated within the nonlinear model-based control methods. However, if linear models are adequate representations of the process, then linear model-based controllers could be used. These include internal model control (IMC) for MISO, state-space (also called *modern* or *ABCD matrix control*) for unconstrained MIMO, or classic model predictive control (MPC, also called APC or DMC) for constrained horizon predictive control.

2.6.6 *Wiener and Hammerstein Models*

A common modeling approach uses steady-state first-principles models to predict the process gain, combined with elementary dynamic models for delay and time constants. For instance, to relate process dead time to flow rates, use a plug-flow transport delay model, and to estimate the process time constant, use flow rates (such as how velocity affects heat transfer coefficients) or the flow-rate-to-volume ratio. Then, the gain, time constant, and delay can be inserted into a locally valid FOPDT model. As conditions change, the coefficients in the model change and the nonlinear and nonstationary FOPDT model remains locally valid.

2.6.7 *NN Models*

NN models are gaining acceptance. These may be fit to the data generated either by step-testing the process or from the first-principles models. If first-principles modeling is too difficult, and the process nonlinearity cannot be functionally described, then fitting the NN model to the process data is effective. If the first-principles models have computational issues (they take too long to use online or have root-finding requirements that might become unstable), they could be used to generate training data for the NN model.

Dynamic NN (recurrent NN) models are fairly complicated, requiring many neurons (many coefficients) to map the dynamics. However, static models can be used in several ways. In one method, the static NN could represent the process static gain, and simple linear models could estimate the dynamics. (This MPC approach, called the Process Perfecter, was introduced by Pavilion Technologies, which is now owned by Rockwell Automation.) Alternately, dynamic modeling could be obtained using multiple steady-state NNs in a grouped NN concept. Here, one NN forecasts 1 step ahead, another NN forecasts 2 steps ahead, another forecasts 5 steps ahead, another forecasts 10 steps ahead, and so on. Finally, one forecasts the steady-state value.[1] In any case, static or one-point NN models are much simpler than recurrent models.

2.7 Modeling Perspectives

The modeling techniques discussed in the previous section and similar process models are not rigorous; fortunately, the best model accuracy is not needed for control. Consider this: A model is not quite perfect in either dynamics or gain, and the PMBC decides on a control action that is only 75% perfect, leaving a 25% error. In simplistic

---

1. J. Ou and R. R. Rhinehart, "Grouped Neural Network Modeling for Model Predictive Control," *ISA Transactions* 41, no. 2 (April 2002): 195–202.

qualitative reasoning, at the next control action, the PMBC will correct the 25% error by 75%, leaving 25% of the 25% error uncorrected. With a rule of thumb that there should be 30 control actions within a process settling time, the continual "steering" by the controller will have the process at the set point, or close enough, with an undetectable $0.25^{30} \approx 10^{-18}$ fraction remaining to be corrected. Drivers can steer their cars through turns. If the first move of the steering wheel does not align the front tires perfectly to the road curve, continual readjustment will. Balancing precision with sufficiency, a model representing a homework grade of B– or C+ is fully adequate for control.

Common evidence from standard control practice supports this in that a proportional-integral-derivative (PID) controller works adequately over a reasonable range of process conditions. It may be aggressive where either the process gain or dead time is high or sluggish elsewhere, but it can still regulate and track set points. Based on linear FIR models (the dynamic matrix), advanced process control (APC) is a linear controller. However, it, too, works over a reasonable range of nonlinear process characteristics. The controller might be graded a perfect A in one operating region, but it still works in other operating regions where its model is a mediocre representation of the process.

This robustness of control effectiveness to model imprecision has several advantages. It lessens the burden on the control engineer to achieve model perfection. It also permits the controller to remain functional as a process continually changes because of fouling, reactivity, yield, filter/screen accumulation, or piping modifications.

Further, simple process models, such as those expressed in Sections 2.2.1 through 2.2.14, often represent desirable computational features of execution simplicity and low computational complexity and storage.

In addition, such simple models often represent the engineer's understanding of the process and reflect methods already in use for process analysis, device sizing, and analytical troubleshooting. The process engineer can develop such models. Using first-principles models in control also preserves and enhances desirable mechanistic process understanding. By contrast, using linear models (perhaps expressed as z-transforms to satisfy the values of theoreticians) requires the process engineer to learn that type of mathematics, diverts focus, and masks the mechanistic understanding (which is essential to manage the process).

The linear models do not exactly match the process, nor do the relatively simple first-principles nonlinear models. Even the best effort at developing a rigorous model will generate a model that is not perfect for the hidden methods of Nature. There is uncertainty in coefficients representing, for example, heat transfer fouling, catalyst reactivity, tray efficiency, friction losses in pipes and fittings, the concentration of non-key components, and the amount of noncondensable gases. There are also simplifications, such as presumed first-order dynamics, the Bernoulli equation, and the ideal gas law. Plus, there is uncertainty regarding the value of disturbance and influence measurements. Accordingly, the models used for control often benefit from real-time correction.

2.8 Takeaways

- Balance perfection with sufficiency. Use first principles to develop models. If you cannot model the cause-and-effect mechanisms, consider using empirical methods (regression to ARMA or NN models) to match that part of the data. Adjust coefficients with uncertain values to make the model best match the process. If the fit is generally good, you are finished with this part. Congratulations. Getting a first-principles model to match the process means that you understand the process. Confidence in your mechanistic understanding will be an asset in process design, analysis, improvement, and troubleshooting, as well as in control and optimization.
- If a comparison to process data reveals that the model is poor, reconsider the mechanisms or the empirical model functionalities. You need a moderately good process understanding to manage the process. Let modeling success move you to that level of understanding.

Exercises

2.1 Test your ability to develop first-principles models by using the procedure in Section 2.2.4 to derive those models in Sections 2.21 and 2.2.5 through 2.2.13.

2.2 Use a process you are familiar with to derive the transient model equations.

2.3 Use a steady-state model you have access to and the Wiener–Hammerstein technique to create a dynamic model.

2.4 Use the guide from Section 2.2.3 to generate first-order plus dead time (FOPDT) models that are locally true to any of the nonlinear models in Section 2.2.

3 Process Simulation

3.1 Introduction

Dynamic process simulation is an essential tool for validating and adjusting a model and exploring control efficacy. Simulators are also used to test and demonstrate automation algorithms, such as control techniques, controller tuning, data reconciliation, steady-state detection, model adaptation, noise filtering, constraint prediction, fault detection, and neural network (NN) or fuzzy modeling. This section discusses simulation techniques I find useful for methods of exploration and validation.

There are three basic concepts. The first is that the deterministic simulator represents what we model as the truth about Nature. In general, the exact rules and coefficient values Nature uses to generate data are unknowable. But in a simulator, you will have written the equations and know the coefficient values. Knowing the truth about your simulator provides a basis for validating techniques and meeting the goal of reaching that same truth within the noisy data.

If the technique you are exploring works on your simulator, this provides strong evidence that it will work on the real process. However, the evidence from simulation is only strong if the simulator is a valid representation of Nature (see Chapter 4).

The second concept is that the simulation must include mechanisms that generate the vagaries of the experimentation in a manner consistent with Nature. The simulator must add noise to every measurement. It must also have ever-changing environmental influences, input conditions, calibration drifts, and process attributes. Environmental influences could include ambient heat losses due to weather or relative humidity in the air intake. Input conditions could include the Btu content of a fuel, raw material composition, or flow rates. Calibration drifts could be on any process measurement and are related to instrument warm-up, instrument aging, sensor fouling, and other conditions. Process attributes, for example, could include distillation tray efficiency, orifice erosion, screen plugging, or heat transfer area fouling.

The third is that the model you use in a controller cannot exactly match the model used in your process simulator. You cannot know the truth about Nature. Pretending you can and using the same model for your process simulator and control is cheating. Accordingly, the model for control should have similar but slightly different constitutive models and coefficient values. For instance, if the model uses the ideal gas law, the simulator should use a different equation of state model, such as the Van der Waals equation. If the model uses the ideal squared relation between the flow rate and pressure drop (the Bernoulli model), the simulator should use the nonideal power of about 1.8 (which often better matches reality). If the controller model uses a first-order plus dead time (FOPDT) model, the simulator should use a second-order plus dead time (SOPDT) model. If the model thinks a volume is 3 m^3, the simulator should use another value, such as 2.7 m^3.

Processes are subject to many forms of disturbances and other afflictions. Natural sources of raw material (minerals, fruits, etc.) continually change impurities and basic composition; fuel continually changes in Btu content; and air continually changes in humidity, temperature, and pressure. There are many other input disturbances. Also, environmental loads on processes continually change, such as feed temperature and ambient heat losses due to outside temperature, wind, and rain. In addition, internal to the processes, attributes such as catalyst reactivity, heat exchanger fouling, and pressure-loss mechanisms in flow lines continually change. All these inputs, environmental loads, and process attributes change simultaneously and independently.

Consider driving a car. When the road angle rises, you must press harder on the accelerator pedal to maintain speed. If the wind blows from behind, you must back off on the accelerator pedal to maintain speed. If the road roughness changes, the rolling drag changes. Road angle, wind, and rolling friction are external loads. Internal variables that change in time include the engine temperature, which affects fuel combustion efficiency and pressure losses between the piston rings and cylinders.

Whether a chemical, mechanical, or human process, there are diverse uncontrolled input disturbances, external loads, and internal attributes that change. These require a change in the manipulated variable (MV) to keep the controlled variable (CV) at the set point.

Although many control theories place "the disturbance" as a single additive bias to the output of the process, there are many disturbances, and they do not act as a measurement bias to the simulator output. They should be modeled as process input influences.

It is easy to have the computer make step or ramp changes in an input, and some influences switch on and off. But this is not commonly the way Nature devises it. Naturally, the input pattern would be characterized by a random walk about a mean value, a stochastic pattern. This chapter indicates how to include disturbances, more realistically, as autoregressive moving average (ARMA) simulator inputs.

In addition, measurements are subject to noise, which is a random perturbation of the true process output. Noise might be due to turbulence in the flow lines, mechanical vibration of the sensor, incomplete mixing with packets of varying temperature, composition passing by a sensor, or other factors. Noise confounds the controller.

Frequently, the perturbations are short-lived relative to the controller execution time interval, making them seem independent (the prior values do not influence the current perturbation) and normally (Gaussian) distributed. The acronym for normally and independently distributed perturbations with zero mean and a standard deviation is NID($0, \sigma$). This chapter provides a way to model measurement noise.

Sensors are also processes, meaning they have several external influences and internal attributes that affect their calibration. One could also model calibration error as due to ARMA(1,1) behaviors.

However, there are more afflictions than those mentioned. Stiction is a slip-stick phenomenon common to flow control valves. The valve stem must slide through the valve body to open and close the inner valve. This hole in the body could permit process material to leak out, so there is packing material between the stem and the body. If the packing is too loose, process material can slip out. If it is too tight, the packing can grab and hold the valve stem. Usually, when the controller wants to open the valve, the air pressure on the actuator changes, and the force imbalance between the actuator and the spring permits the stem to move to a position that makes the forces balance. However, if the packing grabs the stem, the stem will not move until the force imbalance overcomes the static friction. Then, the valve stem starts moving, and because sliding friction is less than static friction, the stem jumps to a new position, stopping where the static friction again overcomes the actuator-to-spring force imbalance.

From the controller's perspective, if the flow rate is a little too low, it sends a signal to open the valve a small amount. When there is stiction, the valve position does not change and the controller integral action sends a signal for the valve to open more. When the valve stem position does not change, the integral continually ramps up the control signal. Then, the stem jumps to a new position, making the flow rate too high. The integral ramps down until the stem jumps to a position that is too low. This cycling is characterized by jumps in the flow rate (a square wave in time) and ramps in the controller output (a sawtooth pattern).

Lost motion happens when there is play in mechanical linkages. This could be on a rotary or butterfly valve in the process industry or in positioning devices in mechanical or robotic devices. With play or slop in linkage, the final element follows the target position, always behind by a small value. Then, when the target position reverses, the final element does not change until the linkage slop is removed and the final element can begin to move. This is often termed *dead band*, but *lost motion* is more appropriate because the term dead band usually refers to the band around a set point used to trigger on-off control action.

Although there are parallels with a dead time (a delay affecting control action), stiction and lost motion are not due to a time delay in the signal.

Resolution is the smallest reportable change in a signal. Your digital watch might reveal the time in seconds and report that time holds constant at 11:15:26 until it jumps to 11:15:27. Time did not stop for the duration; the display held the last reportable value until it could report a new value. The resolution is 1 s or 1/60 of a minute, or 1.67% of a minute. A digital thermometer may have a resolution of 0.5°C, and as the temperature progressively changes, the display holds the prior value until the change is enough to display the new value. If the thermometer range is 25°C to 45°C, it can only report 41 distinct values and has a resolution of 2.5% of the full range. If a 10-bit processor is in the signal transmission path, it can only hold $2^8 = 1024$ values. If the calibration range is in the middle 64% of its signal range (as 4 to 20 mA is a portion of the 0 to 25 mA range), only 655 values are used. This means that the signal resolution is the CV range divided by 655. The resolution, the ability to detect change, is 0.153% of the full range.

Resolution is detectable as the signal jumps between discrete values. In nonlinear processes or sensors, it can be invisible at one operating range and visible in another. When detectable, poor resolution can cause a controller to cycle in a pattern similar to valve stiction.

3.2 Generating Deterministic Equations for the Simulator

In modeling, the term *deterministic* means "exact" or "unchanging." No matter what time of the day or where you are in the world, when you multiply 3 times 4, you get 12. However, rolling a die might get a 1, 2, 3, 4, 5, or 6. If the outcome is not perfectly reproducible, the process is stochastic. The real world is stochastic; measurements have noise, and experiments cannot be duplicated exactly. For example, if you try to mix a 50/50 solution by metering equal volumes, you cannot make it exactly 50/50 to the molecule. Environmental conditions are continually changing.

Section 2.2.4 outlines the procedure for generating first-principles models. These are deterministic models. Anyone who properly solves them will get the same numerical value answer as anyone else, anytime and anywhere. Follow that outline to generate the deterministic equations for the simulator.

However, because the real world is stochastic, once the deterministic models have been derived, add noise and drifts using the techniques described in the following sections.

3.3 Simulating Noise

Noise on measurements could come from several sources. One is "thermal static" due to the device electronics. Another source is electromagnetic influences from proximity machines on the measurement transmission wires. These are external to the process. However, the process might also generate noise. Flow turbulence causes pressure perturbations on a differential pressure measurement device. Boiling causes pressure perturbations and thermal variations on sensors in the two-phase mixture. Incomplete mixing downstream of an injection point causes composition fluctuations on a sensor. These noise perturbations often appear to be both independently and normally distributed on the time scale of process measurements.

A person could argue that on a microsecond scale, the cause of the perturbation (e.g., a cold fluid packet due to incomplete mixing) persists at the sensor and that the noise is not independent and random. However, at 1/10 of a second time scale, which is relatively fast for process control, the persistence of many events has ended and the perturbations appear random and independent.

The Gaussian (normal, bell-shaped) distribution is commonly accepted as a representative noise model. It represents the confluence of many random influences of equivalent impact. I have been very pleased with the 1958 Box-Muller model[1] of generating Gaussian distributed, independent perturbations with a mean of zero and standard deviation of σ, NID(0, σ).

$$n_i = \sigma\sqrt{-2\ln(r_{1,i})}\ \sin(2\pi r_{2,i}) \tag{3-1}$$

Here, the variables $r_{1,i}$ and $r_{2,i}$ represent uniformly distributed independent random numbers on the 0 to 1 range (excluding 0), and n_i is normally and independently distributed with a mean of 0 and a standard deviation of σ. ln and sin are the natural logarithm and sine functions. The subscript i represents the sampling counter and is shown to indicate explicitly that the perturbation changes with each sampling. But there is no need to include the index, i, in the simulator.

There are several approaches to including noise in a measurement. The most common is to consider that the noise is added to the true measurement. To simulate this, add the noise perturbation, n_i, to the deterministic model output. In Equation 3-1, used to determine the noise perturbation, σ has the same units as the measured variable.

$$x_{measured,i} = x_{deterministic\ model,i} + n_i \tag{3-2}$$

Another approach is to consider that the noise magnitude scales with the variable, which may be relevant in flow rate measurements when the intensity of process turbulence is proportional to the flow rate. Here, σ in Equation 3-1 is dimensionless and represents the ratio of the noise amplitude to the measured variable.

$$x_{measured,i} = x_{deterministic\ model,i}(1 + n_i) \tag{3-3}$$

Another method is to use process gains to characterize the noise. For instance, the measurement noise in pH processes might be due to imperfect mixing, with base-rich and acid-rich fluid packets passing by the sensor. In a strong acid, strong base mixture, the pH sensitivity to composition is very high in the near neutral (pH = 7) region but very low in the far from neutral ranges (pH < 5, or pH > 9). So, one might simulate the noisy signal with a radial basis amplification, Equation 3-4. Here, s is a scale factor that indicates how the noise amplitude diminishes with deviation from neutrality, and σ in Equation 3-1 would represent the noise amplitude in the near neutral range.

1. G. E. P. Box and Mervin E. Muller, "A Note on the Generation of Random Normal Deviates," *The Annals of Mathematical Statistics,* 29 no. 2 (1958): 610–611.

$$x_{measured,i} = x_{deterministic\ model,i} + n_i e^{-((pH-7)/s)^2} \tag{3-4}$$

In some processes, the noise is not symmetric. The + deviations may be larger or smaller than the – deviations, and you may need an alternate function to generate skewed noise. For instance, in turbulent, flow-induced pressure fluctuations, the negative deviations are generally larger than the positive deviations. In a population of humans, the negative height deviations from the average (which includes children) are larger than the positive height deviations from the average.

You must understand the sources of random noise in your process measurement to be able to choose a noise model.

In any case, you must choose a value for σ in Equation 3-1. The 5σ range includes about 99% of all values. So, if you use your experience for the range of normally noisy values when the process is at a steady state, set σ to be one-fifth of that range.

$$\sigma \cong \frac{Range\ at\ SS}{5} = \frac{x_{SS\ noisy\ high} - x_{SS\ noisy\ low}}{5} \tag{3-5}$$

In Equation 3-5, do not use the range of possible process values. Use the range of the perturbations at a steady state. For example, the speedometer on an automobile might have a noisy display with a σ of about 0.1 mph, which means that the display fluctuates between 54.75 and 55.25 mph at 55 mph. The same vehicle might have a speed range from 0 to 80 mph. Incorrectly using the speed range in Equation 3-5 would imply that the measurement noise has a standard deviation of 16 mph, indicating that the display would fluctuate between 15 and 95 mph at 55 mph.

3.4 Simulating Bias

Noise is often termed *random error*. However, there may be a measurement bias, permanent offset, or systematic error. To simulate a systematic error, simply add the error to the measurement.

$$x_{measured,i} = x_{deterministic\ model,i} + bias \tag{3-6}$$

You can simulate both random and systematic errors.

A systematic error might be scale-dependent with the process variable (PV) value justifying a model such as Equation 3-3.

3.5 Simulating Ever-Changing Drifts

Four categories of variables have ever-changing variations. The first is measurements. The common view is that an experiment is controlled, and the measurement contains the noisy fluctuation and/or systematic error. In this view, add the perturbation to the deterministic simulator output to obtain the simulated measurement, as done in Sections 3.3 and 3.4. In addition, the measurement device calibration will drift in time due to any number of influences, such as warm-up of electronics, sensor fouling, ambient humidity, or atmospheric pressure. This section discusses a method for modeling such ever-changing drift. The first variable type to consider is the measurement devices. This includes the process response measurement and every variable that would be measured, such as the flow rate of a wild flow used in a ratio strategy or the inlet temperature of a heat exchanger used to estimate ambient energy losses.

The second category of variables that drift in time are the inputs to an experiment that are not exactly controlled. Inputs result from the final control element (valve, knob, or dial), which is physically set to some position or manipulated to make an input reading (flow rate or temperature) have a desired value. Either way, the true value of the input is not the desired value; it is the measured value adjusted for the unknowable measurement error. Measured inputs are the second category of variables that continually drift.

There are other uncontrolled and unmeasured inputs to a process. These include external influences such as ambient heat load, composition of an input reagent, tire inflation pressure, fuel Btu content, wastewater pH, and air density. These are examples of the third category of variables, those that have ever-changing attributes that drift about some nominal value in time.

The fourth category of ever-changing variables is internal process attributes. These include friction losses in piping systems due to any number of events, such as scale buildup or flaking off, screen plugging, manual valves adjusted, or entrained gases or particles. This category includes the buildup of noncondensable gases in heat exchangers, catalyst reactivity, and many other internal factors.

All four types of variables have variation: inputs, environmental variables, process coefficient values, and measurements.

To create a simulator, first define the equation or procedure to obtain y_{true}, the truth about Nature, from $x_{desired}$, the desired value of the influence.

$$y_{true} = f\left(x_{desired}, p_{ideal}, e_{ideal}\right) \tag{3-7}$$

The $x_{desired}$ represents the controlled and other measured variables. The p_{ideal} represents the ideal values of the process or equipment characteristics (catalyst reactivity, tire air pressure, heat exchanger fouling, fluid flow pressure-loss coefficient, etc.) The e_{ideal} represents the ideal values of the environmental and unmeasured influences (temperature, humidity, raw material composition, etc.). Before using the model, perturb each variable in the three input categories by appropriate disturbances (drifts, *d*, and noise, *n*). Then, use the model to calculate the true response. Finally, perturb the response with measurement drift and noise:

$$x_{actual} = x_{desired} + d_x + n_x \quad (3\text{-}8a)$$

$$p_{actual} = p_{ideal} + d_p + n_p \quad (3\text{-}8b)$$

$$e_{actual} = e_{ideal} + d_e + n_e \quad (3\text{-}8c)$$

$$y_{model} = f(x_{actual}, p_{actual}, e_{actual}) \quad (3\text{-}8d)$$

$$y_{measured} = y_{model} + d_y + n_y \quad (3\text{-}8e)$$

Although the disturbances, *d*, and noise, *n*, are added here, it might be more appropriate to make them a multiplicative factor when you anticipate that the noise level or perturbation magnitude scales with the input value.

To create a simulator using this model, begin by defining the equation or procedure to obtain y_{true}, the truth about Nature, but calculate that value from the actual, not the ideal, influence values. Combining Equations 3-8d and 3-8e:

$$y_{measured} = f(x_{actual}, p_{actual}, e_{actual}) + d_y + n_y \quad (3\text{-}9)$$

Looking at measurement sequence or time, there are two basic concepts of variation. One is independent variation, in which sample-to-sample variation is uncorrelated. This is often termed *noise*. Here, the cause of one perturbation from ideal does not persist, and the next perturbation is the result of independent causes. The other mechanism for variation reflects a cause that has persistence, in which sequential perturbations are correlated. The ever-changing disturbances, labeled as *d* in Equations 3-8 and 3-9, represent autocorrelated, time-dependent trends, or drifts, in an influence or model coefficient. The symbol *n* represents uncorrelated, independent noise.

Independent perturbations have no persistence; the value of one n_i is independent of the proceeding values. However, many influences on experiments have persistence (a short-term persisting influence). For example, if a cloud covers the sun, the shadow persists for a while. If the blocked sunlight would influence a temperature disturbance, then a negative d value at one sampling (in the shadow of the cloud) should generate a negative d value at the subsequent sampling (the shadow persists). In this case, if the clouds were randomly scattered in the sky, with randomly assigned density, the random drivers would influence the persisting effect.

For dynamic, time-dependent simulators, a common method for simulating how disturbances (bias or systematic error in measurements, as well as uncontrolled influences on the process) change in time is to consider that they are driven by random events that have persistence.

A drifting influence on the jth variable with a first-order persistence driven by $NID(0, \sigma)$ noise is modeled as

$$\tau \frac{dd_j}{dt} + d_j = n(t) = \sigma_d \sqrt{-2\ln(r_{1,i})} \sin(2\pi r_{2,i}) \tag{3-10}$$

Again, the subscript i represents the sample number, the time interval. Expanding this differential with a forward finite difference and rearranging leads to the ARMA model for d_j

$$d_{j,i} = (1-\lambda)d_{j,i-1} + \lambda\sigma_d \sqrt{-2\ln(r_{1,i})} \sin(2\pi r_{2,i}) \tag{3-11}$$

where

$$\lambda = 1 - e^{-\Delta t/\tau} \tag{3-12}$$

and Δt is the simulation time step.

If $\Delta t \ll \tau$, which is usually the case in simulation, then the simplified approximation $\lambda = \Delta t/\tau$ can be used. In creating a simulation, the user would choose a time constant for the persistence that is reasonable for the effects considered and a σ value that would give the disturbance a reasonable variability. Because the first-order persistence tempers the response to the random influences, the σ-driver must be greater than the desired response σ_{d_j}.

Equation 3-11 is often called an *ARMA model;* $d_{j,i}$ is the response. The model contains one autoregressive term on the right-hand side containing $d_{j,i-1}$ and one moving average term on the right-hand side, based on the ith influence. Accordingly, it is an ARMA of order 1 and 1, noted as an ARMA(1,1) model.

In creating a simulation with a stochastic influence, the user would choose a time constant for the persistence that is reasonable for the effects considered and a σ_d value that would make the disturbance have a reasonable variability. At each sampling, Equation 3-11 would provide the perturbation value for the variable x, p, or e. The stochastic variable could represent barometric pressure, ambient heat load, raw material composition, or any ever-changing influence or process character. In its first use, the values for d_j in Equation 3-11 should be initialized with zero.

How should one choose values for λ and σ_d? First, consider the time constant, τ, in Equation 3-10. It represents the time constant of the persistence of a particular influence. Roughly, $\tau \approx 1/3$ of the lifetime of a persisting event because the solution to the first-order differential equation indicates that after three time constants, d_j has finished 95% of its change toward the final value. So, if you consider that the shadow of a cloud persists for 6 min, then the time constant value is about 2 min. Or you could use $\tau \approx 1/4$ of the persistence period, representing 98% of the time to change. After you choose a τ value that matches your experience with Nature and a time interval for the numerical simulation, Δt, calculate λ using Equation 3-12.

To determine the value for σ_d, propagate variance in Equation 3-11. The result is Equation 3-13. Use your choice of σ_x (and λ, which depends on your choices for Δt and τ) to determine the value for σ_d.

$$\sigma_d = \sigma_x \sqrt{\frac{2-\lambda}{\lambda}} \tag{3-13}$$

Choose a value for σ_x, the resulting variability on the x variable. To do this, choose a range of fluctuation of the disturbance. You should have a feel for what is reasonable to expect for the situation you are simulating. For instance, if it is barometric pressure, the normal local range of low to high might be 29 to 31 in of mercury; if it is outside temperature in the summer, it might be from 70°F to 95°F; or if it is a catalyst activity coefficient, it might be from 0.50 to 0.85. The disturbance value is expected to wander within those extremes. Using the range, R, as

$$R = HIGH - LOW \tag{3-14}$$

and the standard deviation, σ, as approximately 1/5 of the range, then

$$\sigma_d = \frac{R}{5}\sqrt{\frac{2-\lambda}{\lambda}} \tag{3-15}$$

In summary, to generate stochastic inputs for dynamic simulators, choose a time constant for the persistence of events and a range of the disturbance variable. Use Equation 3-12 to calculate λ and then Equation 3-15 to calculate σ_d. Then, at each simulation time interval, use Equation 3-11 to determine the perturbation to a variable.

Figure 3-1 illustrates one 400 min realization of a stochastic variable calculated above. The time constant is 40 min, the nominal input value is 20, and the range is 3 units. In this one illustration, notice that the variable averages about its nominal value of 20 and that the difference between the high and low values is nearly 3. From a period between 150 and 275, the value is below the average, indicating a persistence of 275 – 150 = 125 min, about three times the 40 min time constant. Some persistence values are shorter, some longer.

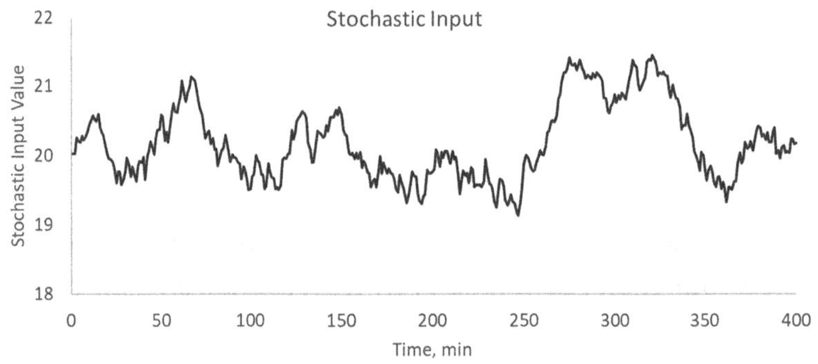

Figure 3-1. Illustration of generating a stochastic influence for a process.

This approach modeled a drifting influence as a first-order response to a Gaussian distributed driver. One could model drifts as higher order or driven by alternate forcing function distributions. However, I believe this adequately mimics how I consider randomly driven drifts in natural disturbances to respond and provides a relatively simple method to compute and assign values to the coefficients.

3.6 Simulating Faults

Faults could be related to many types of instrument failures, process issues, and data transmission errors. These could be spurious signals, progressive degradation, or step events. When demonstrating an automation technique for robustness to faults, be sure to simulate the fault effect in a reasonable and defensible manner.

3.7 Discontinuities

The simulator might contain IF-THEN-ELSE logical expressions. For instance, if the flow is laminar, use one pressure drop relation; if it is turbulent, use another. Or, if the process of heating a liquid transitions to nucleate boiling and then to film boiling, change the heat transfer constitutive relations for each of the mechanisms. Such changes in the models lead to either slope or value discontinuities in the modeled output. You may want to consider smooth (weighted average) model changes from one behavior to another within the transition region.

The numerical time increment leads to the fine discretization of the process response, which may appear as striations on a surface that an optimization is considering. Striations would lead an optimizer to the minimum within the "groove" and not permit it to jump out of that groove to identify the global optimum. I recommend using a multiplayer optimizer (one with many simultaneous trial solutions such as Leapfrogging or Particle Swarm) to have a higher probability of locating the global optimum.

3.8 Simulating Stiction

Another environmental affliction, stiction, is an aspect of the valve stem position in many flow control valves with air actuators. Valve stems enter the valve body through a hole filled with packing to prevent process fluid from leaking around the stem. If the packing is tight, this causes static friction to resist actuator pressure in moving the stem (called grabbing). Often, grabbing is exacerbated by corrosion on the stem or defects in any of the moving parts of the stem or actuator. When the actuator air pressure overcomes the spring back-force and the static friction, the stem jumps to a new position. With stiction, the stem changes only if the force imbalance exceeds a threshold, and if so, the stem jumps to a random position between the current location and the ideal target.

A convenient way to simulate this is to model the target stem position (that desired by the controller), x_{target}, and the actual stem position, x_{actual}, separately. If the difference between the target and the actual is less than a threshold difference, the actual position does not change. However, if the difference exceeds the threshold, Δ, the actual jumps to a random spot near the target.

$$IF(|x_{target} - x_{actual}| < \Delta)\ THEN\ (x_{actual} = x_{actual}) \tag{3-16}$$

$$IF(|x_t - x_a| \geq \Delta)\ THEN\ (x_a := x_a + (1+r)(x_t - x_a)/2) \tag{3-17}$$

In Equation 3-17, the target and actual subscripts are replaced with t and a. The r represents a uniformly distributed random number in the 0 to 1 interval. The $(1 + r)/2$ functionality in this model permits the valve stem to jump at least halfway to the target before stopping randomly. The user would specify the threshold, Δ. The user could also make the Δ value dependent on the stem position (modeling the local impact of defects) or the jump-to range larger or smaller.

Ideally, in a single-input, single-output (SISO) application, stiction is evidenced by a square wave (alternating steps) in the CV and a corresponding sawtooth wave (alternating ramps) in the MV. The pattern might not be exactly regular, and the two traces may be shifted by a delay or tempered by a filter. The evidence may be more complicated in a multivariable process and with continually drifting disturbance effects.

Figure 3-2 reveals the impact of stiction. Here, two independent SISO controllers are used—proportional-integral-derivative (PID) for mixed fluid temperature and proportional-integral (PI) for total flow rate. The temperature controller sends the signal to the hot water valve, and the total flow rate controller manipulates the cold water valve. One characteristic of stiction is the cold flow rate response in the 80 to 100 s and 170 to 200 s periods. In graph (a), the signal to the cold valve (the lower trace) makes a steady ramp up and then down, but in graph (b), the actual cold flow rate (also the lower trace) holds a past value and then makes abrupt steps up or down each time the stem jumps to a new position.

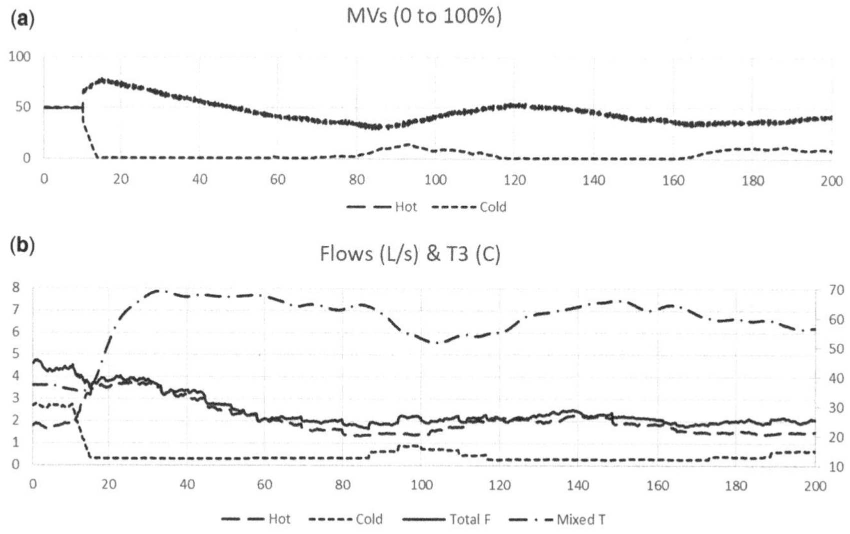

Figure 3-2. Stiction.

3.9 Simulating Lost Motion

Simulating lost motion simulates play or slop in mechanical linkage. A convenient way to simulate this is to model the target final element position (that desired by the controller), x_{target}, and the actual position, x_{actual}, separately. If the difference between the target and the actual is less than a threshold difference, the actual position does not change. However, if the difference exceeds the threshold, Δ, the actual follows the target with a difference of Δ.

$$IF(|x_{target} - x_{actual}| < \Delta)\ THEN\ (x_{actual} = x_{actual}) \tag{3-18}$$

$$IF(x_t - x_a \geq \Delta)\ THEN\ (x_a = x_t - \Delta) \tag{3-19}$$

$$IF(x_t - x_a \leq \Delta)\ THEN\ (x_a = x_t + \Delta) \tag{3-20}$$

In Equations 3-19 and 3-20, the target and actual subscripts are replaced with t and a. The user would specify the threshold, Δ. The user could also make the Δ value dependent on the final element position (modeling the local impact of the rotary arm angle).

3.10 Simulating Resolution

This could be termed *discretization* or *truncation*. Choose either the smallest interval for change, Δ, or the number of values, N, within a range, $R = y_{high} - y_{low}$. The relation is $\Delta = R/N$. Then, if y is the process value and $y_{display}$ is what will be transmitted or reported, calculate $y_{display}$ from y.

$$y_{display} = y_{low} + \Delta \bullet INT\left(\frac{y - y_{low}}{\Delta}\right) \tag{3-21}$$

Equation 3-21 truncates the y value to an integer. You could round it by using $INT\left(\frac{y - y_{low}}{\Delta} + 0.5\right)$.

3.11 Takeaway

- Make process simulators you intend to use to validate an automation technique with realistic, credible features.

Exercises

3.1 Show that Equation 3-1 generates Gaussian deviations with a mean of 0 and an expected standard deviation.

3.2 Generate your own environmental drifting disturbance, as in Figure 3-1, and show that it behaves as expected.

3.3 Combine noise and drift. Provide analysis to indicate that the simulation is as expected. This could simulate a feedforward signal on a noisy, wild flow measurement.

3.4 Feed an environmentally drifting disturbance into a model and see the impact on the model output. Show that the simulation results are as expected.

3.5 Add measurement noise to that modeled output. Show that the simulation results are as expected.

3.6 Explore stiction in a simulator. Show that the simulation results are as expected.

3.7 Explore lost motion in a simulator. Show that the simulation results are as expected.

3.8 Explore resolution in a simulator. Show that the simulation results are as expected.

4 Model Verification and Validation

4.1 Introduction

Verification means ensuring the simulator is true to the mental concepts it should mimic. The mental concepts for a model are converted to algebraic and calculus equations. These are then converted to a solution procedure—numerical methods to solve the transient equations and to provide numerical root-finding techniques when the relations cannot be explicitly represented. The procedure is then converted to executable code. In any of the stages, there could be mathematical or translational errors. In the executable code, "*" might erroneously be entered instead of "**" or "^," " + " might be used instead of "–," a variable may be misspelled such as "H0G" for "HOG," or a variable may not be properly initialized in a subroutine. In translating a concept to calculus, there could be an error in nested parentheses or in a mental jump through several algebraic moves to a result. A coefficient value might be entered with the decimal in the wrong place or with numbers transposed. There is a seemingly infinite number of ways to end up with an error.

Verifying a model means confirming there are no errors in the stages from concept to executable code. However, being verified does not mean the model is correct; the mental concept might be incorrect or incomplete.

Ideally, *validation* means ensuring that the simulator output matches the process. There are many reasons why this cannot happen. The process data may have noise and unmeasured disturbances that differentiate it from the deterministic simulator. The simulator might use 3.14159 for the value of π, which is not what Nature would use. The simulator will use convenient functionalities (e.g., fugacity coefficient values of unity); it will not incorporate all the nuances that Nature secretly uses.

Validation decides that the model is good enough for its intended purpose.

4.2 Verification

Double-checking all your mathematical procedures in the derivation and translation to code is important. However, not finding an error might mean that you are repeating the same error or, knowing the translation is easy, not critically checking it. Double-checking is good but not adequate. Here is a list of things to do to verify translating a concept into code.

1. List all variables used in translating the concept to calculus and algebra and explicitly define their meaning and dimensional units.
2. Double-check all derivations, or ask a friend to check them.
3. Look at each equation in each stage to ensure that each symbol has the correct meaning.
4. Look at each equation in each stage to ensure that the dimensional units are consistent.
5. Print a list of all the variables in the code to ensure that there are no misspellings.
6. Compare the simulator output to see if it matches what is expected.

    a. Enter extreme values as influences and determine whether the simulator responds properly to asymptotic conditions (zero time, infinite time, zero inflow, maximum inflow, very hot, zero ambient losses, etc.).

    b. Convert nonideal relations in the simulator to ideal relations and determine whether the simulator responds properly. You might need to look at internal variables to identify the impact, and you might compare your simulator results to those of other software.

    c. Cut the simulator time step in half. If this does not change the simulator results, the Δt is probably small enough.

    d. Cut the convergence criterion for root-finding steps in half. If this does not change the simulator results, the convergence threshold is probably small enough.

e. Verify that all variables in each subroutine are properly initialized. Display them on the first call and at each subsequent call.

f. Run the simulator for several cycles or return it to key conditions to see if the results are identical.

g. Have expectations for every intermediate variable and determine whether the simulator changes them as expected.

4.3 Validation

There are two general approaches to model validation. First, does the model behave as logically expected? Second, does the data support the model? This section is based on material in *Applied Engineering Statistics*, second edition, by R. R. Rhinehart and R. M. Bethea,[1] and *Nonlinear Regression Modeling for Engineering Applications: Modeling, Model Validation, and Enabling Design of Experiments*,[2] by R. R. Rhinehart.

There are two key concepts you must understand. First, your model is wrong. We can never derive a perfect model, one that truly represents the natural $y = f(x)$ phenomena. Nature always appears to be one step more complicated than our mental constructs of the phenomena. At best, we can hope to derive models that seem adequately consistent with the data considering the purpose of the model.

The second key concept is that Nature lies. The data has variability, often termed *measurement error*, the result of natural variability due to incomplete mixing, variable sample aging because of time variation in analysis, imprecision in chemical analysis, electronic or mechanically induced fluctuation, or other issues. Nature torments us with imprecise data, and our best recourse is to acquire enough data to reasonably estimate the location of the truth.

The concept for this section is that the model is imperfect, but it may be close enough to Nature's natural misrepresentation of itself to judge the model as useful for the intended purpose.

Validation should be both data-based and qualitative.

---

1. R. Russell Rhinehart and Robert M. Bethea, *Applied Engineering Statistics*, 2nd ed. (Boca Ratan, FL: CRC Press, 2022).
2. R. Russell Rhinehart, *Nonlinear Regression Modeling for Engineering Applications: Modeling, Model Validation, and Enabling Design of Experiments* (New York: John Wiley & Sons, 2016).

4.3.1 Logic-Based Validation Criteria

Logic-based validation is often called a *parametric* or *functional analysis*. It is a set of sanity checks to explore the question, "Does the model make sense with what we know or expect?" This validity testing decision is based on qualitative reasoning. Following are several guidelines for choosing which functional analysis to use for logic-based model validation:

- Take simulation time to its limits and examine the initial and final values. Are asymptotic values and their rates of change as expected? If the process is expected to approach an equilibrium condition, then at large time values, rates of change should be zero. However, the asymptotic rate of change might be constant if it is an integrating process.
- Take coefficient values to their extreme limits of 0, 1, or infinity and look at the asymptotic limits of the terms. Do they reduce to ideal conditions or functionalities? For example, if coefficients a and b in the Van der Waals equation of state are set to zero, then, satisfyingly, it returns the ideal gas law.
- Take variable values to extreme conditions (dilute or concentrated, hot or cold, high or low flow rate, short or long tube, early or long time) and look at the asymptotic limits of the model terms. Do they reduce to ideal conditions?
- Consider the sensitivities of the model output to either input variable or coefficient values. Does the model response change in a logically expected manner? Is the sensitivity as expected? This must be true throughout the ranges of all variables and coefficients. For instance, when mixing hot and cold water, as in Section 2.2.1, the mixed temperature should fall with the increasing flow rate of cold. The model derivative of T w.r.t. F_c should be negative. Further, in the limit of very large cold flow rates, the temperature should approach that of the cold water, and the derivative of T w.r.t. F_c should approach zero with extreme F_c values.
- Segregate data from disparate experimental conditions (Instrument 1 and 2, batch and continuous, high and low flow rates) and perform the regression on the segregated data. Are coefficient values independent of experimental conditions? If a coefficient is supposed to represent an inherent material property, then it should not change with experimental conditions. For example, reaction activation energy should not change with the flow rate through the reactor.
- Compare coefficient values to expectations. Are magnitudes and signs of coefficient values consistent with the expected values from related or similar situations? For example, hydrogen gas diffusivity should be similar to that of helium

gas. The overall heat transfer coefficient for one condenser should be similar to that for another. Distillation tray efficiency should be between 80% and 110% (internal reflux could exceed 100%). Is a coefficient value consistent with the trend in homologue materials? For example, because the specific heat of propane is larger than that of ethane, it would be expected that the specific heat of butane is even larger.

- Some coefficient values should change with experimental conditions. When phenomena are lumped into one coefficient, does it change as expected? For example, does increasing turbulence and tortuosity of the fluid path increase the dispersion coefficient? Does in-process transport delay increase with lower flow rates?
- Be your own devil's advocate. Try to find the weakness. You will tend to want to accept your model. You created it; it is your progeny. Therefore, it may be difficult for you to see its inadequacies.

4.3.2 Data-Based Validation Criteria and Statistical Tests

In contrast to the logical and asymptotic functional considerations discussed earlier, which are "fuzzy" because they are grounded in human expectations, the data-based statistical tests in this section are quantitative. However, considering the vagaries of experimental measurements, these quantitative measures must be statistically based, which also makes them fuzzy.

These tests must be appropriate for several data classifications: continuous-valued w.r.t. discrete or category, deterministic w.r.t. stochastic, and steady-state w.r.t. transient.

Comparing the model to data might involve formal statistical tests. However, when using first-principles models, we accept that perfection is unnecessary and are fairly sure that any critical statistical test will find imperfections and reject the model. Accordingly, a qualitative comparison is suggested here to evaluate whether the deviation between the model and the data is acceptable for the intended model use.

For steady-state deterministic models, we would like the data to reveal the following:

- The model outputs, y, should substantially go through the center of the experimental response data throughout the entire data range for each output variable. If there is only one model output, then after regression fitting the model to the data, the model likely goes through the center of the data. If not, suspect an error in the regression method. However, if there are several output predictions, some may show a consistent bias in an attempt to best fit all outputs with an inadequate model. If there is a bias, you have evidence that the model can be rejected.

- A residual is the difference between the model and the data. There should not be any trend in the residuals of any output variable w.r.t. any variable (input, output, chronological order, spatial order, or treatment). If the model is true to the data, residuals should be independently distributed (randomly above and below). If there is an unexpectedly long sequence of residuals of like sign, this is an indication of a model-data mismatch. Test residuals for each output w.r.t. every other variable (input, output, chronological order, special order, or treatment) because sorting the sequence of residuals by one variable may not reveal autocorrelation, but it might when sorted by another variable. If a test reveals autocorrelation in a particular region for a particular variable, this can be a clue as to where the model could be improved.

- The variability expectation from the propagation of uncertainty in the data model or from replicate trials should match the magnitude of the residuals. If the ratio is unexpectedly too small or too large, it suggests you do not understand the process of estimating variance. If the residuals are unusually small, you likely have too many degrees of freedom in the model (adjustable coefficients) relative to the number of data, and you can fit the model to noise. If the residuals are unusually large relative to the replicate variability, then it is likely the model is not a good fit for the data.

- Determine the 95% (or other) confidence interval on the predictions. If 95% of the data values (or about) are within the 95% confidence interval of the model prediction, and the outliers are not clustered in one region, then the model cannot be rejected.

Data will probably provide statistical evidence to reject the model. Although the model may not be true to the process, it may be fully functional for control. Seek perfection, but balance that with functional adequacy.

For steady-state stochastic models, the model's cumulative distribution function (CDF) should match the empirical CDF of the data. The deviations between the two distributions should be small, and the deviations should be independent—randomly above and below.

For a transient deterministic model to match the data, several aspects could be of concern. These include the final steady-state value, the rate of change of the model after an input change, the delay in the model response, the value of the modeled output after a desired time (before the steady state), or several other concerns. For any of these deterministic values that are important, use the four tests described earlier for the steady-state deterministic models.

In any case, the experiment design that generates data for model validation must generate data that critically and legitimately challenges the model. Here are three criteria:

1. There must be sufficient independent data values within the input variable range of interest. A desirable number would be 20 to 40 for each input variable. These should not be replicates but independent values throughout the entire range. Adding replicates will aid in estimating experimental variance, but replicates do not reveal the trend.
2. Further, there must be a large enough range on each input variable to ensure that the change in the output is large relative to the experimental variation. If not, the model trend will not be distinguishable from the noise. The range on each y must be much greater than the standard error of the deviations, $R_y \gg \sigma_e / \sqrt{n}$.
3. Finally, there should be no gaps in the data regions as viewed by any variable (input or output). If there are gaps, then no claims can be made about that region of the model.

4.4 Takeaways

- Make process simulators you intend to use to validate an automation technique with realistic, credible features, but do not strive for simulation perfection.
- Develop models that are adequate for the intended purpose.
- Qualitatively verify and validate the model and simulated application before implementing them.

Exercises

4.1 Use your favorite model and apply the logic-based criteria in Section 4.3.1.

4.2 Compare a deterministic version of your favorite model to data generated by a stochastic version of the same model per Chapter 3. Because the equations in the two models use the same deterministic equation, the random perturbations from the stochastic model should not cause the validation tests in Section 4.3.2 to reject the model.

4.3 Compare a deterministic version of your model to data generated by a stochastic version of a more rigorous version of the model. Because the equations of the two models are a bit different, the validation tests in Section 4.3.2 should reject the model.

4.4 Use model validation techniques to compare your favorite model to real-world data. Maybe the model cannot be rejected because the noise in the real world masks deviations. Maybe the comparison will identify a model weakness, but you may decide that the model is adequate for its purpose. Maybe the comparison will identify a model weakness, and you may decide that a model improvement is justified.

5 Control Issues

5.1 Introduction

If the process were linear, first-order, and unconstrained with measurable disturbances and a single controlled variable, control would be simple, and any number of controllers or methods to devise control would work. Unfortunately, processes within the chemical, paper, mining, materials, pharmaceutical, plastics, energy, and similar industries are not so simple.

Characteristics (difficulties) of processes, process control, and the manufacturing environment shape the methods that are practicable for the process industries. These characteristics of the chemical process industries (CPI) pose particular challenges and a unique relative importance of those problems compared to other control applications such as robotics, aerospace, and communications. Further, differences in (1) economic and risk considerations and (2) process standardization and equipment design between the process and other control applications restrict the permissible investment of sensing and control modeling. As a result, only a subset of control algorithms has found acceptance within the CPI.

There are many other advanced control approaches used outside the CPI. These include sliding mode control (an adaptive approach), fractional order (a nonlinear approach), state-space (or modern) control, and linear quadratic regulator (multivariable, linear, model-based, predictive approaches).

5.2 Elaboration on the Issues and Direction for Simulation Control Testing

Automation algorithms must be tested with relevant and credible simulations.

Section 1.9 presented a list of issues that shape what is practicable for process control. Here are some elaborations:

1. **Nonlinear process** – Include enough gain change so a linear controller tuned for one region does not work acceptably in another. Time constants and delays also change with variations in process throughputs; therefore, a controller tuned for one production rate will not work well at another rate. Simulating these process changes does not have to be excessive, just effective in troubling a linear controller. In the normal range of operation, process gain often has a 3:1 range, and time constants and delay can be scaled with the inverse of process throughput rate, a 5:1 range. Without process experience, researchers often see only one nominal operating state. I think it helps to relate a personal experience. I ask readers to consider the range of speeds at which they drive a car (from parking lots to alleys to curvy mountain roads to straight superhighways) and the implications for steering and speed control.

2. **Nonstationary process** – Properties or characteristics of the process and feed materials change over time. These could be due to heat exchanger fouling, loss of catalyst reactivity, or product purity. However, these may also result from the operational change of flow paths or switching pumps, reactors, and so on. Again, the extent of the change in simulator coefficients or the model should be enough to make a controller not work well under the new conditions. Values for factors representing feed composition, reactivity, and pressure losses can change in a 2:1 range. Consider a car going downhill or pulling a trailer uphill and the process gain impact on the automatic cruise control.

3. **Ill-behaved dynamics** – These include large dead times relative to other time constants or sampling intervals; processes that are integrating, open-loop unstable, or inverse-acting; and processes with disparate settling times for different variables.

4. **Disturbances** – Include unmeasured external disturbances that change over time. For example, an elementary view of distillation is that reflux and boil-up rates are the manipulated variables (MVs) used to control top and bottom composition. However, feed flow rate, composition, temperature, column heat losses, and reflux temperature are all continually changing and unmeasured disturbances. These

could be human-contrived events if, for instance, maintenance is underway. Perhaps a "friend" in the field wants to wake those in the control room by playing with a valve. I do not condone mistakes or practical jokes, but it is unrealistic to expect them not to occur just because they should not.

5. **Multivariable process** – Each MV move affects several controlled variables (CVs), requiring a coordinated plan for the several MV moves. There may be more MVs than CVs, the extra degree of freedom (DOF) situation, which requires an economic optimization of the best MV combination that meets the CV objectives. Alternately, there may be fewer MVs than CVs and a negative DOF, meaning optimization is required to limit deviations from set points. The DOF situation changes as production, product mix, raw material, process maintenance, and environmental factors (disturbances) shift active constraints. The economic weights for the DOF > 0 optimization change with economic factors affecting the business (waste penalty, raw material cost, energy cost, product price, etc.). The constraint violation and set-point violation concern values needed for the DOF < 0 optimization change with the same factors, but also with political factors (e.g., What is the boss's concern today? How soon is the labor contract to be renegotiated? How long has it been since a community or regulatory complaint?).

6. **Extra MVs** – Beyond square processes with an equal number of CVs and MVs, multiple-input, multiple-output (MIMO) studies also include extra MVs. Again, an elementary view of distillation is that reflux and boil-up rates are the MVs. However, column pressure can also be adjusted to control composition. It is an extra MV. There are desired resting values for the extra MVs, and control will switch to use the extra MVs when a primary MV is out of range.

7. **Extra CVs** – A system may be square (the same number of MVs and CVs), but one MV may encounter a limit or constraint of 0% or 100%. In this case, there are more CVs than remaining unconstrained MVs. What to do? Give up control of one CV, and if so, which? Use the residual MVs to best balance all CV deviations from set points, and if so, how does one assess the relative importance of each variable? Should one use steady-state or dynamic optimization? The answers will depend on the relative economic or safety impact of each variable.

8. **Auxiliary variables (AVs)** – For example, an AV may be monitored for safety, and if that process variable (PV) value gets out of range, an auxiliary controller may take over the MV from the primary controller. A classic example is that of a fired heater. As the inflow rate increases, inflow temperature decreases, or fouling increases, the firing rate must increase to keep the outflowing process

fluid at the set point. However, excessive firing can overheat the tubes, so an auxiliary tube temperature controller will be used to override the primary controller to reduce firing and prevent failure. As another example, in distillation, the differential pressure across the column is monitored as an indication of weeping or flooding. When this AV is out of range, normal control is overridden. How does either the normal or override controller operate when not selected? How is the transfer managed to prevent a bump in a transfer of control? External reset feedback is a common approach with override switches in proportional-integral-derivative (PID) control, preventing integral windup and bumpless transition.

9. **Interactive process** – Usually, each MV affects each CV, or at least some MVs affect several CVs. For instance, in mixing hot and cold water (to achieve a desired total flow rate and mixed fluid temperature), one might think the signal to the cold valve does not affect the hot water flow rate. But it does. If the cold water flow rate changes, the mixed flow rate changes, affecting the pressure drop on the mixed flow line and, consequently, back pressure on the hot flow rate, which changes the hot water flow rate.

10. **A PV is not an MV** – Unless cascade structures are used, the MVs are not the process conditions. The MVs are signals to the valves that influence the process. For example, a distillation composition controller might specify the reflux flow rate, but the flow rate can only be changed if a lower-level controller measures the flow rate and changes a valve stem position to regulate it. This adds multiple devices to the control structure. In a simulation study, include the signal transducers and final elements.

11. **Constraints** – Constraints include limits on the value and rate of change on the MVs and limits on the CVs and AVs. Action now can lead to a constraint violation in the future, so control should include the future impact of current actions on both CVs and AVs. Additionally, action on one MV can change the constrained value on another MV (e.g., the boil-up rate in a distillation column limits the feasible reflux range). Often control is simply a response to current and past conditions, such as the logic in a proportional-integral controller. The D action in PID has some anticipatory aspects, but it does not forecast the impact on the AVs. Constraints are often the key consideration in advanced process control installations.

12. **Noise** – Noise can be considered as independent sample-to-sample perturbations on the measurement from issues such as flow turbulence, imperfect mixing, mechanical vibration, or electronic interference. The distribution could be modeled as Gaussian, but often nonlinearities skew it. Many features create

perturbations persistence (mixing, thermal inventory, filtering), so the noise model might have autocorrelation. Include noise. If there is a derivative-equivalent aspect to your controller, the MV will be a joy to watch!

13. **Spurious signals** – Occasional extreme values of either a process influence or a measurement may happen because of a pulse (mechanical or electronic), or an event may appear because of a missed data transmission. This could be a one-sample event, or there could be a brief duration of a few samples. Events such as operator switching, compressor on-off, solenoid valves, water hammer, flow slugging, and compressor surge are all common sources. A median filter is commonly used to reject spurious signal values, but this delays measurement. Consider substituting extreme values for normal simulator values at random samplings.

14. **Sensor error** – Sensor error could be from calibration error or calibration drift (due to age, dust, humidity, temperature, etc.). Alternatively, it could be due to the actual failure of the device (sample line plugging, broken glass membrane on a pH probe, a blown fuse, etc.). Finally, a brief period of an erroneous signal may result from process maintenance. Include a systematic sensor error; do not pretend your controller can know the true PV value.

15. **Sensors** – Seeking to minimize costs, the number of sensors employed is usually the minimum essential number for control and analysis, and the sensors are often selected and located with a priority to cost, safety, and flexibility rather than the quality of control. Orifice flowmeters, for example, have a 3% to 7% measurement uncertainty, and we use flange taps because of maintenance convenience and safety override concerns about measurement accuracy. As long as the flowmeter is internally consistent, the numerical value of the flow rate (reflux, perhaps) is irrelevant to composition control. Often sensors measure the wrong thing because it is easy to "prove" to regulatory bodies by convenient calibration. Other issues include the following: there are not enough sensors to measure everything; sensors are often located in spots that lead to delay or lag; low-cost sensors often have high uncertainty; and sensors often measure a related but not primary value (e.g., pressure drop can infer viscosity, which infers polymer average molecular weight).

16. **Delay** – The delay's value is often variable and dependent on flow rates or analyzer duty. Any simulation should not pretend that the delay has a constant value; it is often inversely related to the process flow rate.

17. **Inverse action** – Here, the natural process response is initially in the wrong direction. The classic example is the boiler swell effect, and an inverse response

is also common in fluidized bed reactors. Arising from Laplace or Fourier transformed variable analysis, this is often termed *nonminimum phase behavior.* A Laplace transform might provide an efficient communication tool, but do not think that choosing a transformed model makes the process either linear or first-order. In my experience, inverse responses have occurred with nonlinear integrating behaviors.

18. **Integrating processes** – Tanks and other inventory-containing vessels have a fill or empty stage; they do not magically start at the level set point with inflow and outflow at a steady state. Any inventory accumulation could also be described as integrating, such as internal electronic heat generation in an insulated device, odor accumulation in a sealed room, impurity buildup in a process with a flow recycle, or the attrition loss of catalyst particles in a fluid bed. When the inventory threshold hits a limit, this triggers an event (e.g., turning on the computer fan or opening a purge vent) that changes the inventory. A simulation should include such on-off cycling for AV control because it affects the nominal control of the process. For instance, the level in a tank affects the discharge pressure, which can change the gain of a valve or cavitation override on a pump.

19. **Open-loop unstable** – This is characteristic of autocatalytic reactions. The faster it reacts, the hotter it gets, and the greater the initiation of other elements to react. Here, rates and attributes (consumption, inventory, temperature, or pressure) would seem to rise exponentially. Consider nuclear reactors, biological production, or exothermic chemical reactions. An explosion could also be characterized this way but on a very rapid time scale. However, the open-loop unstable characteristic is only valid with an infinite inventory. In reality, after a period of nearly exponential acceleration, the process settles as inventory is consumed. Using the open-loop unstable transfer function, $1/(\tau s - 1)$, is just a partial representation of the initial stage of process behavior, not the entire story.

20. **Batch and continuous** – Control during a batch reaction must accommodate the nonlinear and nonstationary issues as the batch evolves and its properties change. This would be similar to all the items mentioned earlier that are related to continuous flow processes. With batch, however, there is the additional aspect that feed material properties change from batch to batch, and the recipe must adapt from batch to batch. This involves more than tuning or modeling; it involves changing the batch recipe's models, set points, trigger values, and trigger times online.

21. **Individuality** – Replicated CPI processes are usually not the same. By contrast, every robot, car, camera, or airplane of the same series has the same behavior

(usually). Admittedly, some processes have standardized designs (air separations, fired boilers, water treatment, ion-exchange units, refrigeration units, etc.), and standardized process units would be shipped with a unit-specific control system. However, nearly every CPI plant has a unique design with unique behaviors. Even if originally built from a standard design, maintenance procedures, additions, upgrades, and replacements cause processes to evolve individually. This makes each control application unique. The model with one controller may need to be changed to apply to a "twin" application.

22. **Transitions** – Even continuous processes must transition from one product grade to another. The transition might be in product viscosity, polymer molecular weight, or chemical makeup. In the extreme, transitions for a continuous process include start-up and shutdown. The transition must happen as fast as possible to minimize cross-contaminated waste, but it must also be managed safely. Environmental, health, safety, and loss prevention (EHS&LP) are fundamental concerns. During transitions, the gain, time constants, and dead times might change in a 5:1 range.

23. **Start-up and shutdown transitions** – Batches must be filled and then emptied. Continuous operating processes must be started and shut down at the end of a campaign or when it is time for plant turnaround. These critical operations do not use the controller that is used during the normal batch processing operation. For example, during start-up, a distillation column typically operates with a batch charge, heating and phasing in pressure, level, and reflux control as needed, then operates in a total reflux manner until it is at the point that compositions reasonably match feed and product specifications. As a column is shut down, the collapsing vapor would create a vacuum that must be balanced by a controlled infusion of inert gas. These transitions usually require operator interaction, and not only do the controllers have to perform their tasks, but the switching operations must also make sense to operators to prevent errors.

24. **Initial capital cost** – The desire to minimize investment costs often drives design and equipment decisions. Such "rational" economic decisions often result in control-related undesirables, such as making CVs have a nonlinear and interacting response to MVs, limiting process information available to the control system, providing instruments and final control elements prone to faults, and confounding the measurement with inferential variables, delays, and noise.

25. **Cause-and-effect relations** – Depending on a viewpoint, one can substantially shift an input-output perspective. For example, on a primitive heat exchanger control, the signal from the product temperature controller to the steam valve

constitutes a single loop. In this case, steam pressure is a significant, unmeasurable disturbance to product temperature. However, in a cascade or ratio arrangement, the steam flow controller operates the valve and gets the flow rate set point from the primary temperature controller. In the cascade strategy, any flow rate fluctuation due to steam pressure is "immediately" corrected, eliminating that disturbance from the product temperature.

26. **Faults** – These take many forms, and the controller should be tested for its ability to handle the following:

    - Measurement error – Due to issues such as sensor failure, fouling, aging, and calibration drift
    - Final control element problems – Valve stiction, steam utility pressure, instrument air failure, capacity, and other problems
    - Process issues – Bypass, blockage, cavitation, choking, collapsed internals, broken mixer impeller, and so on
    - Control system issues – High traffic leads to missed or delayed transmission, central processing unit (CPU) overload, alarm/priority overload, and other issues
    - Calibration issues – Discrimination error dominates a signal, PV is beyond the calibrated range, bias or systematic error, and others

27. **Models** – Model development is expensive because processes are nonlinear, multivariable, and unique. Because processes are nonstationary, models are short-lived, require maintenance, or must be adaptive.

28. **Plant experience of the operator and process engineer** – Human issues include an understandable human-machine interface (HMI), process and control method complexity, easy and understandable process overview, and the training and education required to implement and maintain the controllers and associated aspects (HMI, models).

29. **Return on investment** – Control systems and algorithms must be economically justified. The essential minimal control is obvious to the naive process owner, who is just thinking about keeping processes at a set point. How does one justify the benefit and pay-out time for each additional level of control complexity? Typically, each significant control strategy advance leads to halving process variability. This permits operating set points that are closer to constraints or product specifications, less waste, or higher throughput, for example. Simulating various control strategies on the process (with characteristic upsets)

can reveal the improvements in variation and permit economic assessment of several benefits, such as increased throughput, improved uniformity, reduced off-spec material, and material and energy efficiency.

30. **Infrastructure** – The choice must be compatible with the legacy of field instrumentation, control systems, software, procedures, and training materials for the site.

31. **Bumpless transfer from manual (MAN) to automatic (AUTO)** – MAN mode is not an off status. In MAN mode, set points, controller output, model prediction, and model adjustment must be performed so that when the operator switches the controller to AUTO, no output change upsets the current conditions.

Because simple to implement problems will not include each of the aspects just listed, Controller testing will require addressing several benchmark challenge problems. Appendixes A through D provide four example simulators: hot and cold mixing, pH in a continuous stirred tank reactor (CSTR), a heat exchanger, and an automobile speed controller. I offer several on my website (*www.r3eda.com*), which are open code and free to use. I have used these for classroom teaching and control algorithm research.

5.3 Takeaway

- When developing or considering using a model-based controller, use the comprehensive issues in this chapter to test your controller or other automation algorithms.

Exercise

5.1 Consider a control application that is of interest to you. For each of the 31 items in Section 5.2, state how the issue could impact the application of interest.

6 Control Metrics Goodness

6.1 Introduction

There are many metrics for control quality (e.g., minimum variance in the controlled variable) and other desirability factors (e.g., computational efficiency and ease of tuning). Here, they are all collected as "goodness" metrics.

Usually, tuning is tested with set-point changes (servo mode) using deviations from the new set point as the measure of desirable controller performance (goodness). And usually, control goodness is based on observing the controlled variable (CV) trend in time. If overshoot and undershoot (e.g., too hot or too cold) material is blended downstream, then the integral of the error (IE) could be an appropriate measure of control goodness. "Error" is the controller actuating error, the deviation from the set point. In the IE method, "+" and "−" errors balance each other.

However, if either deviation is bad, the integral of the absolute value of the error (IAE) could be the right measure. For example, the customer may like a bit of extra purity, but the manufacturer may not like the extra energy it takes to make the higher purity. Here, either deviation is undesired.

But often, small deviations are inconsequential, and large ones are detectably bad. In this case, the integral of the squared error (ISE) is appropriate.

The integral of time-scaled absolute error (ITAE) and integral of time-scaled squared error (ITSE) penalize persisting oscillations (or any sort of deviation) from the set point by multiplying the actuating error by the time since the set-point change. These metrics seek to improve the settling time in servo mode.

All those measures quantify the impact of deviations from the set point throughout a settling period. However, an older, still popular criterion is the damping rate with quarter-amplitude damped (QAD) as the long-ago custom. In QAD, the controller is tuned so the second overshoot is one-fourth of the first. The overshoot is the beyond shoot. It could be either above or below the set point, depending on the direction of the activating set-point change. The overshoot is in the same direction as the set-point change.

However, many other attributes of control are important. Most processes operate for long periods without set-point changes, and the controller performs in the regulatory mode (countering undesired upsets) rather than in the servo mode (making set-point changes). So, the IE, IAE, and ISE responses to regulatory action are important.

Further, CV criteria for tuning often overlook the impact on the manipulated variable (MV). Exclusive use of the CV tuning criteria leads to aggressive controllers. However, because of safety, propagation of upsets in utility lines, thermal stresses, operator response to oscillating conditions, and robustness to process gain changes, operators usually prefer controllers to be more temperate and move the process in a fast enough but overdamped manner. So, a comprehensive control goodness assessment should include the MV "cost" to control the CV.

But there is more to consider. A good controller will be robust to faults (e.g., sensors) and process changes (piping reconfiguration). So, the testing should include such upsets, not just normal process disturbances.

Also, the controller must be able to balance action when there is a constraint, recover immediately when the constraint is removed, and take action that does not create a constraint violation in the future.

Finally, a good controller also provides human convenience, such as ease of understanding, tuning, maintenance, and transferability from MAN (manual mode) to AUTO (automatic mode) to cascade.

This chapter addresses all these issues.

6.2 CV Metrics in Regulatory Performance

Although we use the term *integral*, the actual metric is calculated numerically from sampled data, not continuum values, and over a finite time period, not for infinite time. Here is a nominal definition for ISE and its calculation using the rectangle rule of integration

$$ISE = \int_{t=0}^{\infty} e^2\,dt \cong \int_{t=0}^{t_{long\ enough}} e^2\,dt \cong \Delta t \sum_{i=1}^{N} e_i^2 \tag{6-1}$$

where $t_{long\ enough}$ represents a period long enough to encounter all the upsets that must be countered. t is time, e is the actuating error (the difference between the set point and CV), Δt is a small time interval (conveniently, the control sampling interval), $N = t_{long\ enough}/\Delta t$ is the number of samplings in the $t_{long\ enough}$ period, and i is the sample index.

Because ISE will progressively rise in time, it should be normalized (*NISE*) by the time duration

$$NISE \cong \frac{\Delta t}{t_{long\ enough}} \sum_{i=1}^{N} e_i^2 = \frac{1}{N} \sum_{i=1}^{N} e_i^2 \tag{6-2}$$

ISE is my favorite of all these measures of CV response because, when normalized by time, it represents the process variance during regulatory periods. Most of the time, processes are not being changed; they are regulated and kept at the same set point. Process variance, which determines how close the set point can be to product specifications, would be an excellent choice for assessing control goodness. This expansion shows the relation to σ and ISE normalized by collection time, $t_{long\ enough}$.

$$\begin{aligned} \sigma^2 &= \frac{1}{N-1}\sum_{i=1}^{N}(x_i - \bar{x})^2 = \frac{1}{N-1}\sum_{i=1}^{N}(x_i - x_{SP})^2 = \frac{1}{N-1}\sum_{i=1}^{N} e_i^2 \cong \\ &\cong \frac{1}{N}\sum_{i=1}^{N} e_i^2 = \frac{\Delta t}{t_{long\ enough}}\sum_{i=1}^{N} e_i^2 = ISE / t_{long\ enough} = NISE \end{aligned} \tag{6-3}$$

If the product specification is x_{Spec}, the CV variation is Gaussian with a standard deviation of σ, and the desire is to violate the specification only 5% of the time, then the set point must be

$$x_{SP} = x_{Spec} - 1.645\sigma = x_{Spec} - 1.645\sqrt{NISE} \tag{6-4}$$

This is a good measure of whether the process is noisy or noiseless during a change or stationary period.

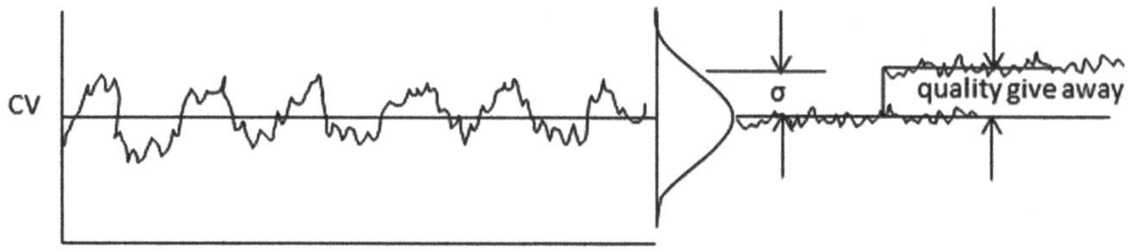

Figure 6-1. The relation to process variance and quality giveaway. To the left, is the current CV at steady state, and the distribution, indicating standard deviation. If the variability can be reduced (middle period), then the set point can be shifted closer to the constraint (right).

If the controller reduces variance during periods of no set-point change, you can move the set point closer to the specification and save expenses (see Figure 6-1).

But, in this regulatory case, testing controller tuning goodness would not be in response to set-point changes. To test in a regulatory situation, choose the controller coefficient values and collect data in an extended regulatory period long enough to experience the control response to the range of vagaries the process might experience.

Although ISE is my default choice, a particular application situation may make one of the other metrics (IE, IAE, or ITS) more appropriate.

Normalized IAE (NIAE), for instance, would be

$$NIAE \cong \int_{t=0}^{t_{long\ enough}} |e|\, dt \cong \frac{1}{N}\sum_{i=1}^{N} |e_i| \qquad (6\text{-}5)$$

NIAE is useful when all errors are bad. For example, if the product is not pure enough, the customer complains. If it is too pure, the manufacturing costs are high.

$NIE = \frac{1}{N}\sum e = \frac{1}{t}\int edt$ is the normalized integral of the error, which is useful when " + " errors are compensable for "–" errors due to downstream mixing.

$NITAE = \frac{1}{N}\sum i\,|e|\,\Delta t = \frac{1}{t}\int t\,|e|\,dt$ is the time-weighted integral of the absolute value of the error, which is useful when all errors are bad, and it places higher importance on persisting errors.

If the set point is close to the specification limit, the measured value will occasionally violate the limit. The extent of the violation could be a count of instances when the CV is out of range, the sum of the quantity of material out of range, or the quantity weighted by the degree of violation. My preference is the weighted quantity

$$P = \sum \Delta t v_i F_i \qquad (6\text{-}6)$$

where P is the simulator penalty when the violation is v, and the flow rate of the material is F. $v = 0$ unless the CV exceeds the specification, then $v = Abs(CV - spec)$. The sum should be over a long enough simulated time duration to effectively encounter the population of all disturbance realizations.

A good controller will have a penalty of zero, but it can have a significant penalty value if the set point is set at the specification.

Even a bad controller can have a penalty of zero if the set point is far enough from the limit. Accordingly, it is not just the off-spec penalty that is important, but also the cost of keeping a set point from the spec, termed *quality giveaway*.

Understand your process and its context, then choose the right CV metric or set of metrics.

6.3 CV Metrics in Ideal Servo Performance

Classic criteria for measuring control goodness include the CV response to a step-and-hold change of the set point, starting from an initial steady state and ending at a new steady state when the CV settles to the set point. This also requires the CV to be noiseless. If the CV is not noiseless, it will never settle to the set point but will continually fluctuate about the set point. Figure 6-2 illustrates such a classic situation. The set point makes a step, and the CV first rises above the set point with A representing the magnitude of the first beyond shoot. The beyond shoot is an overshoot when the set point increases, but it is a below shoot when the set point drops. The CV is illustrated as a temperature response in Fahrenheit, which is also the unit on A. B is the magnitude of the second beyond shoot. The B/A ratio is the damping ratio.

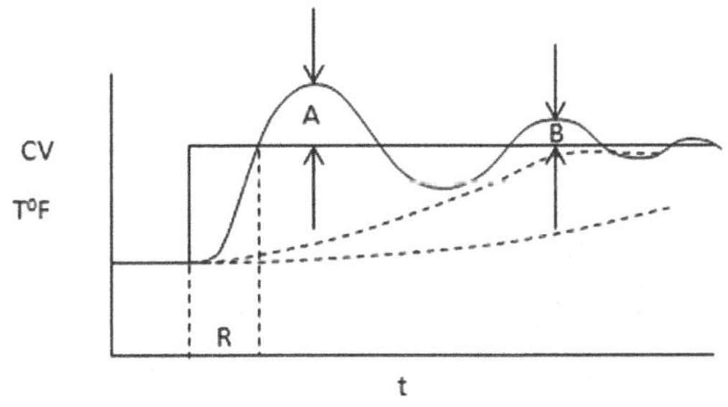

Figure 6-2. CV responses to servo action.

In the 1940s, a QAD ratio of $\frac{B}{A} = \frac{1}{4}$ was deemed as the best balance of overall performance and used as the target metric for controller tuning algorithms. Similar to IE, QAD blends over and under CV values to average at the set point. Making a step change in the set point is a convenience. However, in my experience, the persisting CV oscillation undesirably upsets downstream processes, and the aggressive MV action required for QAD upsets utilities other processes are using. Many people prefer an overdamped, nonoscillatory return to the set point, as illustrated by the two dashed lines in Figure 6-2.

Overrides to QAD include the following:

- **Safety** – Desire a quick response to prevent a problem
- **Stress** – Desire low MV travel or energy
- **Propagation of upsets** – Desire gentle changes
- **Quality** – Desire quick fixes to disturbances, not persistent oscillation
- **Constraints** – Amount and duration constraint violation, or amount of and duration of the steady-state offset

Rise time, illustrated as *R* in Figure 6-2, is the time for the CV to first rise to the set point. *R* has the units of time. Smaller is better for the CV response, but a smaller rise time leads to a more aggressive controller, increasing either the overshoot or the damping ratio and extending the persistence of oscillations.

The time to make three-fourths of the change is related to the rise time, but this could be the time to reach 50%, 63.2%, or 86.5% of the change or any number of targets. The 63.2% and 86.5% targets are the portions for a first-order process to travel one and two time constants. The 50% target is easy to read, but the process still has a long way to go. The three-fourths time is also easy to read and reports when the process has made substantial progress toward the final value. These are user choices, and as long as the choices are consistent, they are good measures of the speed at which the controller can move the process.

Settling time, ideally, is the time after the set-point change for the CV to settle to the new steady state. Mathematically, it would be infinity. At some point, the oscillations (or asymptotic response if overdamped) would be close enough to the set point that the deviation is no longer visible. But that depends on the pixel size of the display. So, conventionally, the settling time is the time for the process to settle within some specified distance from the set point that does not create concern. Perhaps it is within ±5% of the set-point change. A shorter settling time is preferred. However, the characterization depends on the user's choice of the distance. So, although this is a meaningful attribute, it is not independent of the user's choice.

IE metrics are also used in the servo performance of noiseless simulations. $IE = \sum e\Delta t = \int edt$ is the integral of the error, which is useful when "+" errors compensate for "−" errors due to downstream mixing. $IAE = \sum |e|\Delta t = \int |e|dt$ is the integral of the absolute value of the error, which is useful when all errors are bad. $ITAE = \sum i\Delta t|e|\Delta t = \int t|e|dt$ is the time-weighted integral of the absolute value of the error, which is useful when all errors are bad and persisting errors are considered to be worse. $ISE = \sum e^2\Delta t = \int e^2 dt$ is the integral of the squared error, which is useful when all errors are bad, but when small ones are not noticeable and when big ones are really bad.

In a servo response for an ideal process (noiseless and undisturbed process, which settles at the set point), the error approaches zero as time persists, and there is no need to normalize the integral measures. However, what process is noiseless and free from disturbances?

Although we call the aforementioned CV assessment metrics "integral," they are truly evaluated digitally as the sum. Calculus integration is implemented using the rectangle rule of numerical integration.

> Note: My un-normalized CV metrics have the Δt included in the sum, but Δt could be excluded. The Δt multiplier is just a constant that would not change the relative values of the ISE or similar metrics if you compare controller types or tuning values.

In a real process, noise and disturbances continually perturb the CV, and the integral metric collection must be stopped when the continuing contribution is due to the noise, not the set-point response. This is a subjective human judgment, confounding the truth that might be represented in the seemingly complex metric.

6.4 MV Metrics for Ideal Servo Response

Control goodness should also consider the impact on the MV, also termed the *controller output* (CO). In my opinion, exclusively using the metrics of CV performance to tune a controller often results in a controller that aggressively moves the valves. This undesirably and excessively upsets utility streams, stresses equipment, and propagates upsets downstream. One measure of MV action is energy, the sum of squared incremental changes in the MV at each sampling. Another is travel, the absolute value of the incremental MV changes since the last adjustment. In response to a step in the set point from an initial steady state, metrics for the servo responses of a deterministic process (noiseless and disturbance-free) are:

$$MV_{Energy} = \sum (MV_i - MV_{i-1})^2 = \sum \Delta MV_i^2 \tag{6-7}$$

$$MV_{Travel} = \sum \left| MV_i - MV_{i-1} \right| = \sum \left| \Delta MV_i \right| \tag{6-8}$$

The sums would start at the point of the set-point change and continue until the ideally noiseless and undisturbed process settles at the set point and there are no MV moves.

Note: If the MV must change from 50% to 60% but does it by jumping to 80%, then 55%, then 75%, and then settling at 60%, it traveled 95% just to make the 10% change.

6.5 MV Metrics in Regulatory Response

Paralleling the CV categorization, in the regulatory mode, normalized energy and travel are

$$NMV_{Energy} = \frac{1}{N} \sum (MV_i - MV_{i-1})^2 = \frac{1}{N} \sum \Delta MV_i^2 \tag{6-9}$$

$$NMV_{Travel} = \frac{1}{N} \sum \left| MV_i - MV_{i-1} \right| = \frac{1}{N} \sum \left| \Delta MV_i \right| \tag{6-10}$$

The sums would span a long enough duration to collect control responses to all aspects of disturbances and noise.

6.6 Qualitative Measures of Control Goodness

Although CV and MV issues are important, evaluating controller performance also requires assessing qualitative factors such as development, tuning, and maintenance. Control and all ancillary automation techniques are applied within the human and economic context of industry. These context aspects also shape desirability metrics such as those listed here:

1. Operator/technician/engineer time and skill level are the most important criteria. A tuning procedure should be fast, robust, and simple (no complicated calculations).
2. A tuning procedure should not be invasive. It should not upset the process.
3. Tuning must accommodate what the controller might encounter, not just what the process is at the instant it is being tuned. Processes are nonlinear, meaning the gain, time constants, and delays are not constant but depend on continually changing operating conditions. A tuning procedure should permit the tuner to balance CV and MV issues in the situation context over the likely future conditions.

4. Safety and reliability are important. Something bad happens when an element in the control system fails. The more complex or larger the number of elements in the system, the greater the probability that one fails. A key metric of a control system is its reliability, the probability that it operates as desired. Another is the design to mitigate any consequence of a failure, such as fail-closed or fail-open valves. Auxiliary control systems, overrides, and switches are used to take over when a process variable exceeds a limit. Usually, fancy supervisory control does not directly send the signal to a valve but creates a set point for a lower-level flow or temperature controller. Then, when the complicated, model-based, constraint-avoiding controller hits a snag, the lower-level proportional-integral (PI) controllers sustain the process at the last feasible set points. Do not think your design can be isolated from reliability, safety, or override issues.

5. Transfers must be bumpless. Controllers start in MAN mode and are placed in AUTO mode when the operator feels the system is stable. Often, operators switch from AUTO to MAN to make their own corrections, then switch back to AUTO when comfortable. *Bumpless transfer* means that the controller output and set point do not change to alternate values in the transition. The MAN mode is not the OFF mode. In MAN, the controller model, set point, and model bias should be updated. Consider the cruise controller on your car. When you get up to speed and press the cruise control button, you do not want the MV to go to zero or its last used value. As an additional example of the need for bumpless operation, when tuning is adjusted, you only want new coefficient values to affect new actions, not scale the history. Consider a controller with an integral multiplied by the gain and divided by the time constant. If the controller accumulates the integral $I_i = I_{i-1} + \Delta t \cdot e_i$ and then computes action by $u_i = K(e_i + I_i/\tau)$, which is how the PI equation and block diagram seem to indicate the code structure, then changing a coefficient value rescales the integral history, which makes a bump in the controller output. For bumpless tuning in the AUTO mode, the integral should be calculated as $B_i = B_{i-1} + K\Delta t e_i / \tau$; then compute the action by $u_i = Ke_i + B_i$. The control action is identical under normal operation. To differentiate the approaches to computation, I renamed the scaled integral as controller bias, B. Test your controller to ensure it provides bumpless operation for diverse events.

6. Simplicity is important. Processes and controllers in the chemical process industries (CPIs) are managed by operators who might have an associate degree; if they do, it could be a degree that is not technical. Tuning and diagnosing abnormal situations must be easy—at a non-college-bound, high school, infrequently refreshed skill level.

7. Costs are important. The CPI cannot justify the expense of top-performance computers like aerospace, military, critical infrastructure utilities, or healthcare organizations. Each added element in a control system incurs an initial installed cost and an annual maintenance expense. Fancy algorithms increase both the initial costs and annual expenses. There are also the process costs. A more expensive control system could be justified if it lowers the cost of the process or expenses, such as process intensification designs to integrate heat or recycle. Costs of operation include penalties for specification violations and the quality giveaway when a set point is safely far from a specification. The cost of a controller cannot be evaluated in isolation from the process cost.

8. Another important metric related to the CVs is the measure of a violation of a constraint or specification. pH, for instance, has associated limits for discharge, usually between 6 and 9. For practical considerations, a controller that can keep the pH between 6.99 and 7.01 is no better than one that keeps the pH between 6.5 and 8. Both meet specifications. Although a normalized ISE (NISE) of 0 beats an NISE of 5, the "improvement" may be purely academic and inconsequential to the customer or society if neither controller causes discharge violations. Metrics that assess the CV violation could be related to the quantity (material or time) and degree of violation of a limit (product, regulatory, or safety). If the material is out of specification, it must be recycled, held, or converted to waste, or the product must be downgraded. If safety or regulatory limits are exceeded, an undesired event or fine is likely. In any case, there is a cost penalty, and the override action must handle the material or control the process.

9. Often overlooked in assessing the desirability of a controller are the operator and engineering requirements to implement, initialize, and maintain a controller. Process engineers are typically entry-level BS graduates. They might set up a control strategy, and the design and modeling required must be consistent with that skill level. However, the day-shift engineer is off duty for three-fourths of the process operation time and does not remain at the control panel when on duty. The shift-to-shift on-duty operators usually maintain control. Some might have an associate degree, and several will likely have attended training courses on proportional-integral-derivative (PID) control. Although they may have extensive and valuable process experience, their technical/mathematical skill level is typically at a low high-school level. Their solution to nearly any perceived problem with a controller is to place it in MAN mode and wait for the responsible person to diagnose and fix it. The CPI is generally averse to anything complicated, and the KISS (keep it simple and safe) principle guides choices. The simplicity of the control

concept, obtaining models, tuning, and bumpless MAN-AUTO transitions is essential for this environment. Accordingly, a controller attribute that is often overlooked in science or engineering articles or textbooks about control is simplicity, related to every aspect of control strategy development, implementation, and operation.

10. Another metric to consider is the cost of the system, the process, and the control elements. This includes initial installation capital and annual maintenance expenses, both of which are effectively proportional to the number of devices. Installing and maintaining the process and the devices in the control system are dominant issues. Measurements are not free. Advanced control strategies may require extensive process testing (with associated process and material costs) and front-end engineering (setup costs). Be sure to consider these. In addition, operating set points must be safely away from a limit or specification to prevent violations and operating at higher energy or lower capacity costs to achieve higher purity. Any specification violation generates waste, material to be recycled, or a hazard, each of which has a cost impact. Placing set points safely far from specifications prevents violations, but this means that, on average, the process is doing more than needed (impurity is lower, weight is higher, viscosity is higher, etc.). The quality is higher on average but given away. This quality giveaway increases manufacturing costs. Consider all economic indicators.

11. Include constraint handling and recovery when the constraint is no longer active.

12. Include optimizing MV use to minimize costs when there are extra MVs.

13. Include miscellaneous benefits (process knowledge validation and dissemination, personnel training, process diagnosis and health monitoring, predictive maintenance, and politics).

6.7 Multiple-Input, Multiple-Output

Quality metrics for each CV and MV must be assessed for a multiple-input, multiple-output (MIMO) process. It would be convenient to have an overall CV and overall MV performance metric. In my hot and cold water mixing simulator in Appendix B, I used equal concern (EC) factors to combine the impact. For example, a 5°C deviation in the temperature set point could be considered as having the same concern as a 1 L/s deviation in flow rate. So, the combined CV metric would be $\frac{nISE_1}{EC_1^2}+\frac{nISE_2}{EC_2^2}=\sum\frac{nISE_i}{EC_i^2}$. Because the MVs range from 0% to 100%, it is likely that the weighting on the *nTravel* is identical for the two MVs. So, the overall MV metric would be $\sum nTravel_i$. However, if changing one utility has an adverse impact on other processes, there might also be equal concern weighting on the MVs, $\sum \frac{nTravel_i}{EC_i}$.

Note: The EC value is squared in the denominators of the CV metric but not in the MV metric to match the numerator dimensions.

Usually, greater MV work will improve the CV closeness to the set point. However, excessive MV work upsets utilities, creates operator discomfort, and hurts process efficiency. The process owner seeks a balance, minimizing both *NISE* and *NTravel*, but without excess in either. One could weight, *w*, the importance of the two and combine them, $w_{CV} \sum \frac{NISE_i}{EC_i^2} + w_{MV} \sum \frac{NTravel_i}{EC_i}$. Or create any number of joint combinations, such as a product, $\left(\sum \frac{NISE_i}{EC_i^2}\right)\left(\sum \frac{NTravel_i}{EC_i}\right)$. One could also use a Pareto analysis: Generate data for $\left(\sum \frac{NISE_i}{EC_i^2}\right)$ w.r.t. $\left(\sum \frac{NTravel_i}{EC_i}\right)$ for multiple experimental conditions and compare controllers on a dominance basis.

If a MIMO system is "square," it has the same number of MVs as CVs, and the set of MVs effectively controls the CVs. There are no constraints. However, if an MV hits a constraint (perhaps 100%), there are fewer MVs than needed. The constraint could be due to another auxiliary variable (AuxV), not one of the CVs hitting a constraint. For instance, you might be driving up a mountain and desiring to maintain speed, but if the engine overheats, you must downshift and reduce speed. The AuxV (perhaps engine temperature) overrides control of the primary CV (car speed). When there is a constraint, one response is to give up control of one CV and let the free MVs keep the other CVs at their set points. But which should be paired with which, and what action should be taken if a second variable hits a constraint? An alternate constraint-handling philosophy is letting the remaining free MVs take action to balance the constraint and set-point violations of all variables. Another possible action is temporarily shifting a set point, such as lowering the production rate.

In contrast, the MIMO system might have more MVs than CVs. As a distillation example, heat in the reboiler is required to boil the material to separate components. Power to the reboiler, for instance, the steam flow rate, would be one MV. However, it might be possible to use a heat recovery heat exchanger to preheat the column feed with a hot bottoms product. The temperature of the feed would be another MV that has the same impact as the reboiler steam flow rate. When there are extra MVs, the objective is to set them to meet set points at the lowest operating cost.

All in all, control goodness should include the controller's ability to handle constraints and economically optimize when there are extra MVs.

6.8 Control Perspectives Goodness

My perspective on these metrics is from my experience in the process industries. Characteristically, processes are nonlinear and relatively slow (distillation columns might have a 4-hour transient), and the operational staff will have substantial concerns related to safety. The best criterion for judging tuning goodness depends on the application.

Usually, controller tuning is tested with set-point changes (servo mode) using deviations from the new set point as the measure of goodness. However, regulatory mode performance is probably more meaningful to quality and economics.

To test in a regulatory situation, choose the controller coefficient values and collect data in an extended regulatory period long enough to experience the control response to the range of vagaries the process might experience. Then, tweak a parameter value and observe the change in ISE and travel (or the appropriate metrics) from a similar period. Let the results guide sequential parameter adjustment and seek to minimize the quality metrics. (This is often termed *evolutionary optimization*. It evolves controller coefficient values in time to optimize the ISE, but the heuristic rules for tweaking are not related to contemporary optimization algorithms by similar names.) Such an approach is predicated on sequential regulatory periods expressing similar upsets that must be regulated. If a spurious event happens in one period, the optimizer cannot use data from that period as a rational performance compared to other periods.

Although tuning for regulatory response is more meaningful, tuning for servo response is much more convenient.

A key characteristic of processes is that they are nonlinear, which means several process characteristics impact tuning. The gain, time constants, and delays are not constant but change with operating conditions. A simple nonlinear process example would be mixing hot and cold water in a stirred tank with a temperature sensor in the downstream line. Equation 6-11 represents the mixing dynamics; Equation 6-12 represents the transport-delayed temperature. The controlled variable would be T, and the MV can be considered the hot water inflow.

$$\frac{V_T}{F_h + F_c}\frac{dT_T}{dt} + T_T = \frac{T_h F_h + T_c F_c}{F_h + F_c} \tag{6-11}$$

$$T(t) = T_T(t - \theta), \quad \theta = V_P/(F_h + F_c) \tag{6-12}$$

The time constant in the ordinary differential equation (ODE) depends on the two flow rates and the tank volume. The gain in the ODE depends on the two flow rates and the inflow temperatures. The delay in the measured T depends on the two flow rates. In process operation, the gain, time constant, and delays change over time in response to operating conditions and disturbances.

Processes continually change in response to variations in production rates, raw material properties, in-tank inventory, and other conditions. Tuning must accommodate conditions the controller might encounter (e.g., the range of inflow rates and temperatures), not just the process conditions at the instant it is being tuned. Similarly, controller performance assessment must reveal the global performance, not just that at one point. Perhaps the worst CV and MV metric values over the range of operating conditions should be reported.

Tuning to minimize CV performance metrics leads to undesirably aggressive MV action. Because of safety, propagation of upsets in utility lines, thermal stresses, operator response to oscillating conditions, and robustness to process gain changes, operators usually prefer controllers to be more temperate and move the process in a fast enough but overdamped manner. This balance is usually subjective. Seeking a controller that rapidly moves the CV to the set point with very little overshoot seems to be desired.

Cascade and ratio are common CPI control strategies in which the supervisory (primary) controller changes the secondary controller set point. One might consider that the set point following response of a secondary controller can be defined by observing a step-and-hold change in the set point. However, the reality is that the set point to a secondary controller is the MV from the primary controller, which changes continuously in small increments. Step-and-hold set point responses are important for visualization and convenient for simplicity in analysis. Still, because this may happen infrequently, it is much less than half of the story and should not dominate the analysis.

6.9 Takeaways

Any controller tests should be grounded in an assessment of all the important factors: (1) CV performance, (2) MV performance, (3) constraint handling, (4) cost (devices and people), (5) quality giveaway and penalty for specification violations, (6) economic optimization with extra MVs, (7) robustness to unexpected upsets, and (8) any of the more qualitative issues outlined in Section 6.6 that are relevant process-specific characteristics. Also, each should be considered over the entire range of operating conditions (including constraints) and the vagaries Nature uses to taunt the controller.

Exercises

6.1 Explain the derivation indicated by Equation 6-3.

6.2 Use the concept in Figure 6-1 to determine how a purity set point would change if a new control strategy could halve the σ of the controlled variable (CV).

6.3 Implement any CV or manipulated variable (MV) metrics on a control simulation. Show how CV and MV metrics are related to controller aggressiveness if changed.

6.4 Develop a rubric to rate controllers using the criteria 1, 2, and 3, or 6 and 9 in Section 6.6.

7

First- and Second-Order Plus Dead Time Models

7.1 Introduction

Classically, the analysis of control techniques and controller synthesis (designing both feedforward and feedback controllers) was based on first-order plus dead time (FOPDT) models. More recently, second-order plus dead time (SOPDT) models have become implementable. These are linear models, usually expressed as Laplace transfer functions. Although one could locally linearize first-principles models, the classic approach to generating FOPDT and SOPDT models is empirically by step-testing the process and observing the response.

An FOPDT model is a simple approximation of a process variable's dynamic response (the transient or time response) to an influence. It is alternately termed *first-order lag plus dead time* (FOLPDT), or *dead time,* and the *D* may be replaced with *delay,* changing the acronym to FOLPD. The FOPDT model is often a reasonable approximation to process behavior and has demonstrated utility as a basis for developing controller tuning rules, structuring decouplers and feedforward control algorithms, and communicating essential process attributes; it is also a computationally simple surrogate model in simulations for training and optimization.

FOPDT and SOPDT models are designed to match open-loop stable process behaviors. If the process is purely integrating, open-loop unstable, or inverse-acting, alternate model types should be used.

There is no claim that the FOPDT model is a true representation. The process is likely higher-order and nonlinear. However, an FOPDT model is often a practicable representation for local (linear) control, balancing multiple aspects of utility.

7.2 Things to Accept

FOPDT models are not the process. Even if we term them as process models and use the subscript p on model coefficients, they are not. They are ideal approximations of reality. For example, if the process is high-order without any delay, the FOPDT model (from a one-fourth/three-fourths parametric fit, which has the model match the process at the initial and final steady-state values and when the process has moved one-fourth and three-fourths from the initial to the final value) pretends that it has a first-order response after a pure delay. See Figure 7-1. The ideal FOPDT model is not the process.

Empirical testing includes process nonlinearities, measurement noise, valve stiction, and continuing unmeasured disturbances. Even if a process was truly FOPDT, getting coefficient values from empirical testing would not return the true values.

The process behavior changes with the operating level and due to continual changes in equipment, environmental influences, and raw material. The FOPDT models must change over time.

The method you use to get the τ_p and θ_p values (one point, two points, different two points, or regression) returns values that are different from those produced by other methods.

Although imperfect, FOPDT models are usually good enough for the control application. They provide a balance of sufficiency with perfection. Continuing that acceptable imperfection balance, use first-principles approaches to determine FOPDT model coefficient values as described in Section 2.2.3.

7.3 Reaction Curve Approach

In FOPDT modeling, typically, we consider that the influence has remained constant in the recent past and that the process variable (PV) has achieved a steady value. Then, we consider that the influence makes a step change and holds that new value until the PV reaches its new steady state. The dead time concept represents the time duration after the influence changes for which the PV does not change. It is like a transport delay in a pipeline with plug flow or a laboratory analysis time.

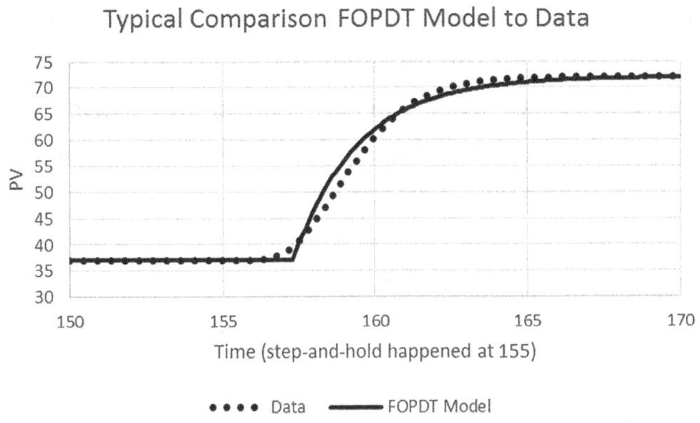

Figure 7-1. Reaction curve concept.

Figure 7-1 compares the FOPDT model to the data. For clarity, the data is ideally noiseless, S-shaped, and indicated by the dots. The step-and-hold in the influence happened at the time of 155. However, the high-order process does not begin to reveal a response until about 156, when it starts to rise slowly, achieving its fastest change rate at about 159. For the best fit to the data, the FOPDT model (solid curve) has a longer delay and does not begin to make a change until a time of about 157. Because it is a single lag, its fastest rate of change is when it starts to respond. The model has a delay longer than the process and must change rapidly to catch up to the process. The model rises above the process in the 158 to 162 period, then relaxes to the final steady-state value slightly more slowly. This model best balances the "+" and "–" deviations from the data.

Again, the model is not the process. It is the best fit of the model concept (a reduced order idealization) to the process data.

7.4 FOPDT Model

Mathematically, the FOPDT model could be stated as an ordinary differential equation

$$\tau_m \frac{d\tilde{y}'(t')}{dt} + \tilde{y}'(t') = K_m u'(t' - \theta_m), \qquad \tilde{y}'(t'=0)=0 \tag{7-1}$$

The model gain, K_m, is the multiplier for the influence change that determines the new steady-state value for the PV. The FOPDT model pretends that the process starts with an initial steady state, and the PV does not change until after the delay. Once the delay, θ_m, duration has passed, the PV follows a first-order exponential trajectory to the final steady-state value. The FOPDT time constant, τ_m, indicates how fast the PV moves toward the new value.

Contrasting some conventions, I have used the subscript m for *model*, not the subscript p for *process*, to acknowledge that the model is not the process. Because FOPDT models are intended to represent the process, classically, the subscript p is used instead of m as in Equation 7-1. However, because the model is not the same as the process, as illustrated in Figure 7-1, I am using the subscript m.

I have explicitly placed a squiggle hat over the model response variable, $\tilde{y}'(t')$, to indicate that it is the model, not the process. I also used the prime mark to indicate that the model influence, u', response, y', and time, t', are each a deviation from the initial steady conditions and the time for the step-and-hold influence. In Figure 7-1, the change happens at 155. Although $t = 155$, at that instant $t' = 0$. Similarly, the initial process value is $y = 37$, but the deviation value is $y' = 0$.

The Laplace representation is

$$\hat{y} = \frac{K_m e^{-\theta_m s}}{(\tau_m s + 1)} \hat{u} \tag{7-2}$$

which does not provide any different or unique information from Equation 7-1. The denominator exponent on the s operator, the lag, is 1, indicating that the model is first-order. The model gain, time constant, and dead time are each revealed. The response is y and the forcing function is u. The hats over the $\hat{y}$ and $\hat{u}$ variables mean that these are the Laplace transformed versions, not the time-domain versions. By default, implicit in Equation 7-2, the derivation of the Laplace transform starts with an initial steady state, and the values represent deviations from that initial steady state.

Although the concept for the model is a response to a step-and-hold influence from an initial steady state, and although this makes for convenient analytical solutions, it is a generic model, and the input pattern is not restricted to simple steps or ramps when the model is solved with numerical methods.

7.5 Determining Model Coefficient Values

The classic textbook method for generating FOPDT models is the reaction curve technique, a precomputer era technique. It is simple to understand and implement. Here is a summary: Start at a steady state and then make a step-and-hold in the MV. Observe the controlled variable (CV) and wait until it reaches a steady state. The gain is the difference between the final and the initial CV steady-state values divided by the value of the input step change. To determine the delay and time constant values, draw a tangent line to the point on the data with the highest rate of change (the inflection point) and extrapolate the tangent line to the initial and final steady-state values. The difference between the time of the step input and the tangent intersection of the initial steady-state value is a reasonable estimate of the model dead time. The time difference between the tangent intersections of the final and initial steady-state values is a reasonable estimate of the model time constant.

However, I believe reaction curve techniques do not express best practices in the computer era. If the process is nonlinear, the single step shows the model coefficient values on one side of the nonlinearity, not the best representation of the initial value. The step creates a one-sided deviation from the original desired value. And if the process is subject to disturbances and noise, the model coefficient values from a single step are confounded. A better approach for generating a good model is to make the input pattern up-down-down-up after reaching a steady state between each MV step. Then, average the model coefficients from the four steps to temper noise, disturbances, and nonlinearity. The second down step provides CV deviations that counter the initial up step.

Unfortunately, this better technique for model validity might take a long time. And the time-related cost might not be balanced by the model improvement.

However, a crude approximation FOPDT model is often all that is needed. In such cases, a one-step reaction curve technique can be a simple and fast method to get a good enough model.

The steps must be large enough to make a noticeable change in the response. If the change is small relative to normal noise and drifts, the FOPDT model coefficients will have a large uncertainty.

Once the response curves are completed, the model coefficients are calculated from a few points on the response curve. There are multiple twists on the method. A parametric approach seeks to match the initial and final model steady states to the reaction curve and then use two points in between to determine the time constant and dead time values. This is easily derived from the analytical solution to Equation 7-1 for $t' \geq \theta_m$.

$$\tilde{y}'(t'-\theta_m) = K_m u'\left[1 - e^{-(t'-\theta_m)/\tau_m}\right] \tag{7-3}$$

Here, $K_m u' = \tilde{y}'_{SS}$, and with a bit of rearrangement

$$t' = \theta_m - \tau_m \ln\left(\frac{\tilde{y}'_{SS} - \tilde{y}'}{\tilde{y}'_{SS}}\right) \tag{7-4}$$

When y has made one-fourth of its total change, $\left(\frac{\tilde{y}'_{SS}-\tilde{y}'}{\tilde{y}'_{SS}}\right)=1-0.25=0.75$, read the time at point $t'_{0.25}$. And when y has made three-fourths of its total change, $\left(\frac{\tilde{y}'_{SS}-\tilde{y}'}{\tilde{y}'_{SS}}\right)=1-0.75=0.25$, read the time at point $t'_{0.75}$. Then, the two versions of Equation 7-4 are linear in the two unknowns, θ_m and τ_m. The solutions provide

$$\tau_m=\frac{t'_{0.75}-t'_{0.25}}{-\ln(1/3)}\cong 0.91\left(t'_{0.75}-t'_{0.25}\right) \quad (7\text{-}5)$$

$$\theta_m=t'_{0.25}+\tau_m\ln(0.75)\cong t'_{0.25}-0.29\tau_m \quad (7\text{-}6)$$

"Eyeballing" the total change in the CV, then visually interval halving to get the 50%, then interval halving each half to get the one-fourth and three-fourths points is relatively easy. Reading the one-fourth and three-fourths times is also easy. In my experience, this parametric fit is easy and generates better models than the classic slope extrapolation method for nonlinear processes.

This four-MV-step approach has a few drawbacks.

- It requires operator attention for an extended time to wait for four steady-state periods.
- It may create process deviations that impact downstream quality.
- It requires the human to interpret the signal to choose when a steady state has happened to provide data for mathematical analysis.
- It only uses a small part of the data generated, and uncontrolled disturbances can substantially confound it.

7.6 The Skyline Approach and Least-Squares Regression

7.6.1 Skyline Input

In the computer era, by contrast, nonlinear least-squares regression is simple to implement; and a skyline input function has advantages in operational duration, magnitude of upsets, and number of excitations over classical methods. A skyline pattern in the controller output could look like the example in Figure 7-2.

To generate a skyline input, step the MV to random values within a desired range and hold for a random time interval between 10% and 90% of the process open-loop setting

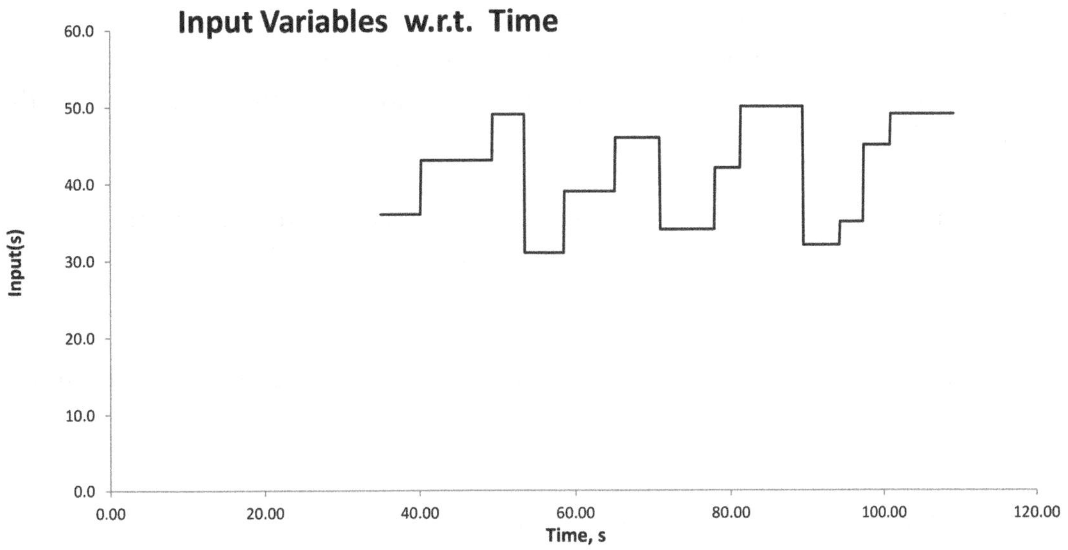

Figure 7-2. Example of a skyline input.

time. Repeat the steps about 10 times. The range of Umin to Umax should be large enough to create significant deviations in the CV relative to noise and disturbances and small enough to keep the response within the local range of CV values for which the model will be used.

Here is a sample of code with "r" representing a uniformly distributed random number $0 < r < 1$, and ST is the estimate of the open-loop settling time of the process. Feel free to adjust the changepoint interval or revise the "Timer >= changepoint" antecedent to just ">."

```
IF Timer >= changepoint THEN
        changepoint = changepoint + (0.1 + 0.8 * r ) * ST
        U = Umin + r * (Umax – Umin)
END IF
```

"Timer" is the computer time; and, not shown, the "changepoint" should be initialized by the Timer. With this method, you do not have to be present to watch and choose when to make the next input change after detecting that the process has achieved a steady state. The time duration and CV deviation away from the nominal value is much less w.r.t the standard step-and-hold method of up-down-down-up steps (which requires five settling times if starting from an initial steady state). Within the same duration, the skyline input provides a much greater richness of signal/influence to minimize confounding influences of noise and disturbances.

During the trial period, external disturbances may make the CV drift away from the nominal operating conditions. If so, the skyline MV values could be biased by a value to keep the CV within the desired operating range. I have used fuzzy logic to guide the progressive bias of the U range.

The skyline pattern does not have to be step-and-holds. It could include any pattern, like a sawtooth, in which the MV jumps halfway to the new value and then ramps to the target by the end of the interval.

Collect the MV(t) and CV(t) data for subsequent regression (fitting the model to the data).

7.6.2 Skyline and Least-Squares Regression

The nonlinear regression method seeks to fit the model to all data points, not just the several selected points in a classic reaction curve fit, enabling it to reject noise and disturbances better.

The skyline and least-squares regression approaches do not require operator attention or judgment, which lessens the possibility of operator error or bias. The skyline input has many ups and downs; accordingly, this tempers the influence of environmental drifts on confounding the CV response to the MV. The skyline and regression method does not extend the period of off-nominal production; each "+" or "–" period is shorter, creating less objection by a quality manager. The skyline and regression method does not require an initial steady state, and the entire test period takes less time.

For a nonlinear process, gain, delay, and time constant change with the MV and CV. The FOPDT model is linear and may not provide a great match to the process over a wide operating range. Just because the optimizer converges on the best model, it does not mean that the model actually fits the data. So, after the regression, see if the fit is satisfactory for your model use purposes. Figure 7-3 reveals the data and best FOPDT model from the skyline MV input sequence above for a pilot-scale process flow rate response to controller output. The model is the solid line, and the data are the dots. Note that early-time data is usually below the model, but late-time data is usually above. Perhaps some drifting influence was affecting the data. Also, note the kinks in the model at times after 80 and 100. These are expected responses to small changes in the MV at those times, but these are not expressed in the data. Perhaps valve stiction prevented the valve from moving even though the MV made changes. Finally, this data is not expected to be a linear response. Although the linear FOPDT

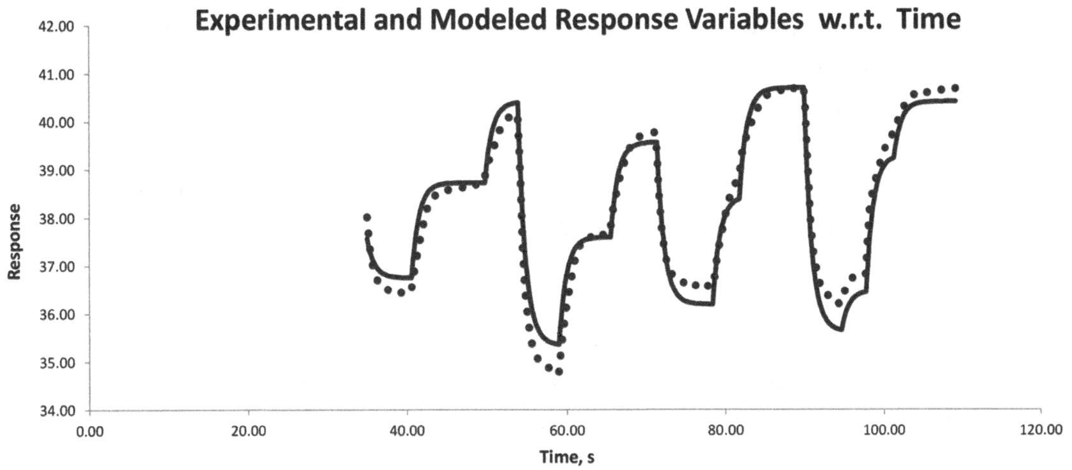

Figure 7-3. Experimental and model response.

model does not perfectly match the data, it is a good representation of the process dynamics.

A common presumption is that the process starts from an initial steady state. However, in my experience, a real process usually does not. If it starts at an initial steady state, then $\tilde{u}_{base}$ and $\tilde{y}_{base}$ are the initial values and $\tilde{y}'_{initial} = 0$. The user can specify those values. But, if the process does not start at an initial steady state, you cannot presume to know those values.

It appears that there are six model coefficients: $\tilde{\tau}, \tilde{K}, \tilde{\theta}, \tilde{y}_{initial}, \tilde{y}_{base}, \tilde{u}_{base}$. However, there are only five independent model coefficients. There is a steady-state relation between y and u. The FOPDT model indicates that at steady state (SS): $\tilde{y}'_{SS} = \tilde{K}\tilde{u}'_{SS}$. Converting the steady-state deviation variables to CV and MV reveals the relationship between $\tilde{K}$, $\tilde{y}_{base}$ and $\tilde{u}_{base}$

$$\tilde{y}_{base} - \tilde{K}\tilde{u}_{base} = y_{SS} - \tilde{K}u_{SS} = c \tag{7-7}$$

where c is some value defined by the process and the modeled process gain (not deviation variables). Ideally, for a linear process with both y and u measured on an absolute scale (as opposed to a deviation scale), $c = 0$, but this is not necessarily true for either a nonlinear process or one for which $u = 0$ does not mean $y = 0$. For instance, in heating, a heater power value of zero might lead to an ambient temperature of 63°F, or on an alternate scale, it might be 17°C. You could determine the c value by holding the MV at the u_{SS} value and waiting until the steady state to find the y_{SS} value. But it might take forever to determine a steady-state period. Fortunately, you do not need to do so. Note that Equation 7-7 places a constraint on the three model coefficients

$\tilde{K}, \tilde{y}_{base}$, and $\tilde{u}_{base}$. This enables you to choose the value of one, reducing the number of adjustable model coefficients (in optimization termed the *decision variables* [DVs]) to five: $\tilde{\tau}, \tilde{K}, \tilde{\theta}, \tilde{y}_{initial}$, and $\tilde{y}_{base}$.

My preference is to choose the $\tilde{u}_{base}$ value, and to choose it as the nominal u-value about which the skyline function is centered, the average of u_{max} and u_{min}. Alternatively, you could choose the initial u-value as $\tilde{u}_{base}$.

After generating the skyline MV input, use regression to adjust the five model coefficients to make the model best match the measured y response using the conventional sum of squared y deviation as the optimization objective function (OF, alternately termed the *optimization target* or *cost function*).

The traditional optimization statement is

$$\min_{\{\tilde{\tau}, \tilde{K}, \tilde{\theta}, \tilde{y}_{initial}, \tilde{y}_{base}\}} J = \sum_{i=1}^{N} (y_i - \tilde{y}_i)^2 \tag{7-8}$$

Note that the regression occurs at discrete time-sampling points; it is not continuous for all time. In the optimization statement of Equation 7-8, the index i is the time counter with $t = i\Delta t$. Note also that although the model delay, $\tilde{\theta}$, could have a continuum of values, when time is discretized into Δt intervals, the delay must have an integer number of Δt intervals. So, the delay becomes $\tilde{n} = INT\left(\frac{\tilde{\theta}}{\Delta t} + 0.5\right)$. Because of the delay, this is a nonlinear regression. Because the delay is constrained to be integer multiples of the sample interval, the optimization is termed as a mixed integer type (integer and continuum)—mixed integer, nonlinear, and constrained.

To calculate the optimization OF value, J, for each trial solution of the DVs, the code could be written as

```
Ndelay = INT(thetamodel/dt + 0.5)                    'discretized model delay
clambda = Exp(-dt / Taumodel)                        'model coefficient
lambda = 1 – clambda                                 'complement to model coefficient
ymodeldev = Yinitial – Ybase                         'initial deviation variable value for model
SSD = 0                                              'initialize the Sum of Squared Deviations
For DataNumber = Ndelay + 1 To Ndata                        'start with first value after the delay
        influence = udata(DataNumber - Ndelay) – Ubase                   'calculate influence
        ymodeldev = lambda * Kmodel * influence + clambda * ymodeldev           'model
        ymodel(DataNumber) = ymodeldev + Ybase                 'convert to CV
        SSD = SSD + (ymodel(DataNumber) – ydata(DataNumber))^2        'sum d^2
Next DataNumber
```

7.7 Empirical SOPDT Modeling

7.7.1 Model

The SOPDT model is also in deviation variables. This set of equations represents one first-order process feeding a second first-order process. It is not interactive and does not permit oscillatory behavior. Although feedback interaction is commonly encountered in electronic processes, this series structure is most common in chemical processes. However, one could easily represent this in one second-order ordinary differential equation.

$$\tilde{\tau}_1 \frac{d\tilde{y}_1'(t)}{dt} + \tilde{y}_1'(t) = \tilde{K}u'\left(t - \tilde{\theta}\right) \tag{7-9}$$

$$\tilde{\tau}_2 \frac{d\tilde{y}_2'(t)}{dt} + \tilde{y}_2'(t) = \tilde{y}_1'(t) \tag{7-10}$$

where

$\tilde{y}_1'(t) = \tilde{y}_i(t) - \tilde{y}_{base}$

$i = \{1, 2\}$ to indicate the first or second lag. The second lagged variable is the measurement.

$u'(t) = \tilde{u}(t) - \tilde{u}_{base}$

t = time after the user-defined starting point

The initial condition for the second lag is the initial measurement, and for the first lag is an initial value.

$$\tilde{y}_2'(t=0) = y_{data}(t=0) - \tilde{y}_{base} \tag{7-11}$$

$$\tilde{y}_1'(t=0) = \tilde{y}_{1,initial} - \tilde{y}_{base} \tag{7-12}$$

A common presumption is that the process starts from an initial steady state. However, in my experience, a real process usually does not. If it starts at an initial steady state, then $\tilde{u}_{base}$ and $\tilde{y}_{base}$ are the initial values and $\tilde{y}_{1initial}' = \tilde{y}_{2initial}' = 0$. The user can specify those values, but if the process does not start at an initial steady state, you cannot presume to know those values.

It appears that there are eight model coefficients: $\tilde{\tau}_1, \tilde{\tau}_2, \tilde{K}, \tilde{\theta}, \tilde{y}_{1,initial}, \tilde{y}_{2,initial}, \tilde{y}_{base}$, and $\tilde{u}_{base}$. However, there are only seven independent model coefficients. Again, there is a steady-state relation between y and u. The model indicates that at a steady state: $\tilde{y}_{SS}' = \tilde{K}\tilde{u}_{SS}'$.

Converting the steady-state deviation variables (SS) to CV and MV reveals the relationship between $\tilde{K}$, $\tilde{y}_{base}$, and $\tilde{u}_{base}$

$$\tilde{y}_{base} - \tilde{K}\tilde{u}_{base} = y_{SS} - \tilde{K}u_{SS} = c \tag{7-13}$$

where c is some value defined by the process and the modeled process gain. You could find the c value by holding the MV at the u_{SS} value and waiting until steady state to determine the y_{SS} value. But it might take forever to identify a steady-state period. Fortunately, you do not need to do so. Realizing that Equation 7-13 places a constraint on the three model coefficients $\tilde{K}, \tilde{y}_{base}$, and $\tilde{u}_{base}$, permits you to choose the value of one, reducing the number of DVs from eight to seven: $\tilde{\tau}_1, \tilde{\tau}_2, \tilde{K}, \tilde{\theta}, \tilde{y}_{1,initial}, \tilde{y}_{2,initial}$, and $\tilde{y}_{base}$.

My preference is to choose the $\tilde{u}_{base}$ value, and to choose it as the nominal u-value about which the skyline function is centered, the average of u_{max} and u_{min}. Alternatively, you could choose initial u-value.

There are two initial values for the state variables, $\tilde{y}_{1,initial}$, and $\tilde{y}_{2,initial}$. I have chosen to set $\tilde{y}_{2,initial} = y_{data(t=0)}$. This reduces the number of regression variables from seven to six: $\tilde{\tau}_1$, $\tilde{\tau}_2$, $\tilde{K}, \tilde{\theta}, \tilde{y}_{1,initial}$, and $\tilde{y}_{base}$.

After generating the skyline MV input, use regression to adjust the six model coefficients to make the model best match the measured CV response using the sum of squared deviation as the OF.

7.7.2 Empirical SOPDT Modeling – Regression

The traditional vertical least-squares optimization statement is

$$\min_{\{\tilde{\tau}_1, \tilde{\tau}_2, \tilde{K}, \tilde{\theta}, \tilde{y}_{1,initial}, \tilde{y}_{base}\}} J = \sum_{i=1}^{N} \left(y_i - \tilde{y}_i\right)^2 \tag{7-14}$$

Note that the regression occurs at discrete time-sampling points; it is not continuous for all time. In the optimization statement of Equation 7-14, index i is the time counter with $t = i\Delta t$. Although the model delay, $\tilde{\theta}$, could have a continuum of values, when time is discretized into Δt intervals, the delay must have an integer number of Δt intervals. So, the delay becomes $\tilde{n} = INT\left(\frac{\tilde{\theta}}{\Delta t} + 0.5\right)$. Because of the delay, this is a nonlinear regression. Because the delay is constrained to be integer multiples of the sample interval, the optimization is a mixed integer (integer and continuum) optimization classified as mixed integer, nonlinear, and constrained.

To calculate the optimization OF value, *J*, for each trial solution of the DVs, the code could be

```
Ndelay = INT(thetamodel/dt + 0.5)                'discretized model delay
clambda1 = Exp(-dt / Taumodel1)                  'model lag coefficient for y1
lambda1 = 1 - clambda1
clambda2 = Exp(-dt / Taumodel2)                  'model lag coefficient for y2
lambda2 = 1 - clambda2
ymodeldev2 = ydata(Ndelay) - Ybase                        'start y2 at first measurement
ymodeldev1 = yInitial - Ybase                    'initialize y1, not necessarily at steady state
SSD = 0
For DataNumber = Ndelay + 1 To Ndata
    influence = udata(1, DataNumber - Ndelay) - Ubase
    ymodeldev2 = lambda2 * ymodeldev1 + clambda2 * ymodeldev2           'second lag first
    ymodeldev1 = lambda1 * Kmodel * influence + clambda1 * ymodeldev1   'first lag after
    ymodel(DataNumber) = ymodeldev2 + Ybase
    SSD = SSD + (ymodel(DataNumber) – ydata(DataNumber))^2  'sum d^2
Next DataNumber
```

7.8 Empirical Modeling and Skyline Response Comments

This regression method seeks to fit the model to all data points (after the delay, actually N – Ndelay number of data points). By contrast, traditional reaction curve or parametric approaches fit the model to several selected data points (initial, final, and two intermediate points) and presume that the process starts at an initial steady state. So, this many-data approach better rejects noise and disturbances over classical reaction curve or parametric approaches.

The skyline and regression method does not require operator attention and judgment when a steady state has been achieved, which lessens the possibility of operator error or bias.

While running the process tests, the skyline method keeps the CV closer to the target value because it changes the MV to a new value before a steady state. It also takes less time and provides many more than just a few influences.

The skyline input has many ups and downs, which temper the influences environmental drifts have on confounding the CV response to the MV.

Because of these advantages, a choice in generating an empirical model could be the skyline MV influence and the regression approach to determining model coefficient values, not the old-school (precomputer) best practice "reaction curve" approach that still dominates the direction given in textbooks.

Using the classical reaction curve or a parametric approach to get an FOPDT is relatively simple, and either can generate fully functional models for control. However, determining the two time constant and delay values for an SOPDT model is not nearly as easy. The equations are more complicated, and the error propagation due to noise is larger. However, this is not a problem with the skyline and regression approach to get an SOPDT model.

In the classic reaction curve methods, the up-down-down-up response should start with an initial steady state, make equal step changes in the MV, and wait for the CV to reach a steady state before making the next MV step. The first up-and-down changes should return the CV to the initial location. If they do not, something else is affecting the process. You should then connect the initial and final steady-state CV values as the baseline and use the differences between the CV and the baseline as the measure of the CV response to the MV. The second down-and-up is to push the process to the other side of the base value so you can see if there is any nonlinearity impact and to provide a CV balance to the first up-down upset. Again, the CV should return to the initial location.

The steps should be large enough to reveal the CV change clearly but small enough so the CV remains in the local region. Usually, ±10% changes are made in the MV, such as starting at 25%, then going to 35%, then 25%, then 15%, then 25%. The skyline range should be similar.

If the process is linear, in the classical up-down-down-up method, the steps up and down should each provide an identical response. The two steps up should be identical. The two steps down should be identical. If the process is linear, when you flip the down responses (turn the page over and look at them through the paper), they should be exactly the same as the up responses. However, Nature is not required to comply with human convenience. Just because we choose to use an FOPDT or SOPDT model does not mean that the process is linear, with stationary coefficient values, and either FOPDT or SOPDT.

Real processes have noise and disturbances. When doing classic step tests, this makes it difficult to know when the process transient from the MV change is over and when the operator can implement the new MV change. This also makes it difficult to determine the ΔCV at steady state for the process gain calculation. This procedure requires starting at an initial steady state and waiting for four additional steady-state points, which makes the test period last about $5 * 5 = 25$ time constants—a long period. During the test period, the process is pushed off the nominal CV value for about $2 * 5 = 10$ time constants and then pushed off the nominal in the other direction for another 10 time constants—a long time for the quality manager to accept. After you experience these difficulties, you can see that the classic reaction curve techniques are not as good as using the skyline and regression methods to get your model.

By contrast, the skyline and regression method has many steps, not just one or a few, providing greater noise and disturbances temperance than classic reaction curve modeling techniques. The skyline and regression method does not extend the period of off-nominal production. Each "+" or "−" period is shorter, creating less objection by a quality manager. The skyline and regression method does not require an initial steady state, and the entire test period takes less time.

In models for a nonlinear process, gain, delay, and time constant change with the MV and CV. This FOPDT or SOPDT model is linear and may not provide a great match to the process over a wide operating range. Just because the optimizer consistently converges on the same model does not mean that the model actually fits the data. As a similar concept, one can seek to fit a straight line through curved data. There is a best line, a line that minimizes the sum of squared deviations (SSDs) between the model and data. However, the fact that there is a best straight line does not mean that it adequately corresponds to the data. So, after the regression, see if the fit is satisfactory for your model use purposes.

See Section 2.2.3, "Notes About Model Coefficients," to see how to use first-principles models to estimate the FOPDT model coefficient values.

7.9 Takeaways

- There are many valid approaches to generate FOPDT or SOPDT models of a process that is open-loop and stable. Several are presented here, along with a discussion of the experimental effort to generate models that are not corrupted by noise, disturbances, or nonlinearity.
- The model is not the process, and depending on the approach used to generate model coefficients, the approximating models will be slightly different from each other.

I have posted my Excel/VBA file for generating FOPDT models on my website, *www.r3eda.com*. It also offers a user guide to generate skyline data, use the software, and explain the nonlinear regression procedure. The VBA code is open. My entire academic career has been devoted to enabling students with best practices in engineering. I am using the website to extend beyond the classroom. I hope you find it useful.

Exercise

7.1 Generate process-controlled variable responses to a skyline manipulated variable (MV) input. Then, access the first-order plus dead time (FOPDT) regression software from *www.r3eda.com* and use it to generate the best FOPDT model.

8 Ill-Behaved Dynamics

8.1 Introduction

More than simply nonlinearity (changing gain) and nonstationary dynamics (changing dead time and time constants), some processes have dynamic trends that cause problems for controllers. If the process model is first-order, then control is relatively simple. Simple generic model control—steady-state or one-step ahead, model-based control (MBC) controllers—may not always be effective due to dead time or high-order, integrating, open-loop unstable, and inverse-acting processes.

8.2 Ill-Behaved Dynamics

The ill-behaved dynamics include dead time and high-order, integrating, open-loop unstable, and inverse-acting processes.

8.2.1 Dead Time

Dead time is a delay in the process model. The delay could be due to any number of process features, including the following:

- Sampling delay (time to transport material to a lab and then analyze it)
- Interval in control action
- Transport time of in-process material
- The apparent dead time when using a simple model for a high-order process

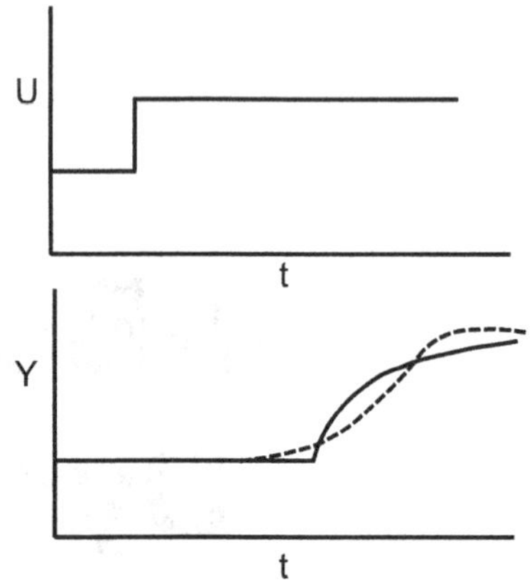

Figure 8-1. Illustration of dead time.

Dead time is illustrated in Figure 8-1. The upper graph shows a step-and-hold in a process input, and the lower graph shows the response. The dashed line is the process, and the solid line is a first-order plus dead time (FOPDT) model. Note that the step input occurs near the left of the graph, but the dashed response does not start changing until about the middle of the graph. Moreover, the best fit of the FOPDT model to the process adds further delay to the model.

In Laplace notation, the model for dead time is

$$\hat{y} = Ke^{-\theta s}\hat{u} \tag{8-1}$$

Here θ is the delay with the units of time, and K represents the process gain. y is the response variable, and u is the influence variable, but $\hat{y}$ and $\hat{u}$ are the Laplace transformed versions of the deviation of the variables, usually from an initial steady state.

There are two problems for control. The first is when the controller takes action and nothing happens. It takes more action, and nothing happens. The controller continues to take action and nothing happens. Suddenly, all the past actions begin to be expressed, and the process overcorrects.

The second problem is that if the model has a delay, it is impossible to take action that makes the modeled controlled variable (CV) immediately move toward the set point.

The traditional model-based solution is to take action for a post-delay response, not an immediate one.

8.2.2 High-Order Process

High-order, open-loop, stable processes can be modeled as a sequence of lags, as illustrated in Figure 8-2. Again, the upper figure is the controller output, and the lower figure indicates the process response. Following the step, the first response is a first-order lag, immediately moving toward the steady-state target but approaching it in an asymptotic manner. The output of each lag drives the next. Initially, the response of the first lag is small, so the driver for the second lag is small, and the initial rate of change of the second lag is small. As a result, the early-time drivers for subsequent lags are smaller and smaller, and each lag seems to wait before beginning to move.

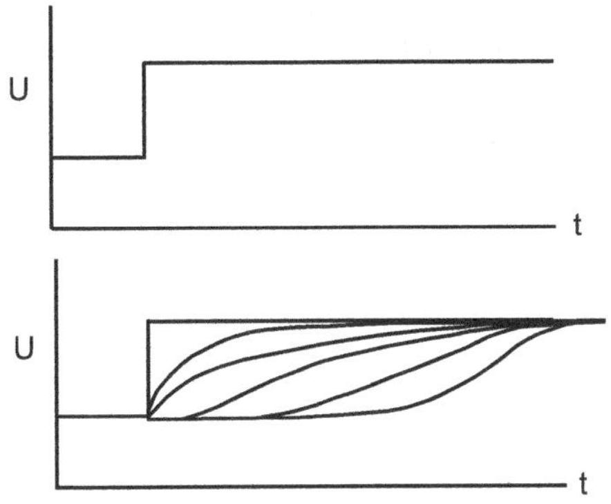

Figure 8-2. Illustration of a high-order process.

An example would be composition on trays in a distillation column. There are many inventory locations in sequence, and one tray cannot respond until the one before it can feed its changed material.

A Laplace transform representation of the high-order process starts with a first-order model of each stage.

$$\hat{y}_1 = \frac{K_1}{(\tau_1 s + 1)} \hat{u} \quad (8\text{-}2a)$$

$$\hat{y}_2 = \frac{K_2}{(\tau_2 s + 1)} \hat{y}_1 \quad (8\text{-}2b)$$

$$\hat{y}_3 = \frac{K_2}{(\tau_3 s + 1)} \hat{y}_2 \quad (8\text{-}2c)$$

Substituting successive solutions and representing n stages

$$\hat{y}_n = \frac{K}{\prod_i^n (\tau_i s + 1)} \hat{u} \quad (8\text{-}3)$$

where $K = \prod_{i}^{n} K_i$ represents the overall process gain, y_n is the response variable, and u is the influence variable, but $\hat{y}$ and $\hat{u}$ are the Laplace transformed versions of the deviation of the variables, usually from an initial steady state.

Like dead time, the high-order process takes a significant amount of time to begin expressing the results, causing the same problems for a controller as dead time.

8.2.3 Integrating Process

In an integrating process, the inventory quantity does not affect inflow, outflow, or generation. Classic examples are filling a tank or a sack, adding salt to a pot of soup, or adding heat to an insulated object. If the controller can only add material, once too much material has been added, it cannot be removed.

The Laplace model could be presented as

$$\hat{y} = \frac{K}{\tau s}\hat{u} \tag{8-4}$$

where K is the overall process gain and τ is the time constant, but because these are a ratio, only the quotient is needed. This might be expressed as $\hat{y} = \frac{K'}{s}\hat{u}$ or $\hat{y} = \frac{1}{\tau' s}\hat{u}$, depending on personal preferences. y is the response variable and u is the influence variable, but $\hat{y}$ and $\hat{u}$ are the Laplace transformed versions of the deviation of the variables, usually from an initial steady state.

8.2.4 Open-Loop Unstable Process

In unstable processes, the greater the inventory, the faster the inflow, outflow, or internal generation increases inventory. An example is an exothermic chemical reaction, in which a reaction generates heat, making the reaction rate faster. Another example is microbe growth. One cell splits into two, then two become four, then four become eight, and so on. Another example is a nuclear reactor. A thermal neutron (relatively low energy) is captured by a uranium nucleus causing it to fission and release four high-energy neutrons. One escapes from the reactor, two are captured by the control rods, and the fourth is slowed by bumping into hydrogen nuclei in the water. It becomes a thermal neutron that is captured by a uranium nucleus. If the control rods are partly removed, then 1.1 neutrons are left to cause 1.1 atoms to fission, generating $(1.1)^2 = 1.21$ thermal neutrons, then generating $(1.1)^3 = 1.331$, and compounding.

Forever compounding is an ideal concept. Eventually, fuel or food is depleted, and growth stops or goes to extinction. As inventory growth increases, other input-output mechanisms also happen. However, in the early time of the effect, the compounding seems unbounded and models can represent that.

The Laplace transform of a first-order unstable process is

$$\hat{y} = \frac{K}{(\tau s - 1)}\hat{u} \tag{8-5}$$

Note the negative sign in the denominator. Again, y is the response variable and u is the influence variable, but $\hat{y}$ and $\hat{u}$ are the Laplace transformed versions of the deviation of the variables, usually from an initial steady state.

8.2.5 Inverse-Acting Process

In an inverse-acting process, in response to a step-and-hold in the input, the process moves in one direction and then reverses sign and moves in the other direction, going beyond the original value. Some use the term *nonminimum phase*. This happens when there are two (or more) mechanisms for the input to affect the output; they act in opposite directions, and the one with a smaller time constant also has a smaller gain magnitude. A classic example is the swell effect in a boiler, illustrated in Figure 8-3.

The boiler illustrated in Figure 8-3 shows the fire, boiling bubbles in the water, and vapor leaving the liquid to enter the steam main. Initially, the rate of steam removal

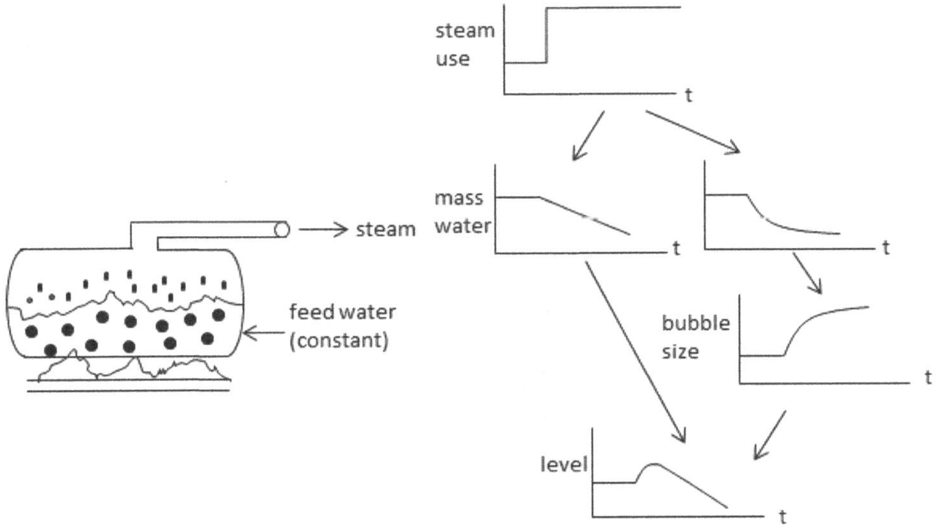

Figure 8-3. Boiler swell effect example.

is balanced by the rate feedwater is entering, and the water level in the boiler remains the same. Also, the firing rate is matched to meet the steam demand. On the right-hand side of Figure 8-3, the top sketch shows an increase in steam demand. This has two effects indicated by the sketches below it. It removes water from the boiler, lowering the mass of water in the boiler, and because steam is being removed faster than it is being generated, it reduces the pressure in the boiler. Reducing pressure in the boiler increases the bubble size, causing the water level to swell. These are the two effects in opposite directions. The reduced pressure causes the liquid level to rise, and this happens rapidly. The inventory imbalance causes the liquid inventory to fall, but although it will eventually have a larger effect than the bubble swell effect, it happens at a slower rate. The lower sketch in Figure 8-3 shows how the two mechanisms affect the water level. First, there is a rise; then, there is a progressive fall.

If the feedwater level controller is a simple device observing the water level and increasing or decreasing the feedwater flow rate to regulate the level, the controller will see the level rise and reduce the feedwater flow rate. This, of course, is the wrong action. A traditional solution uses three-element control using the steam demand to feedforward a trim to the water level feedback controller.

As another example, consider a fluidized bed coal gasification reactor, illustrated in Figure 8-4. Crushed coal pellets are fed into the top of the pressurized reactor. Upward-flowing oxygen (O_2) and steam fluidize the particles in the reactor. The oxygen partly burns the coal and forms carbon monoxide (CO), which the steam converts into carbon dioxide (CO_2) and hydrogen (H_2). The heat in the reactor also causes volatile compounds in the coal to become gas products. The desirable products are heat, CO, H_2, and volatiles such as methane (CH_4). There are several inverse mechanisms. Increasing the coal feed rate will initially increase the fluidized bed level and gas production

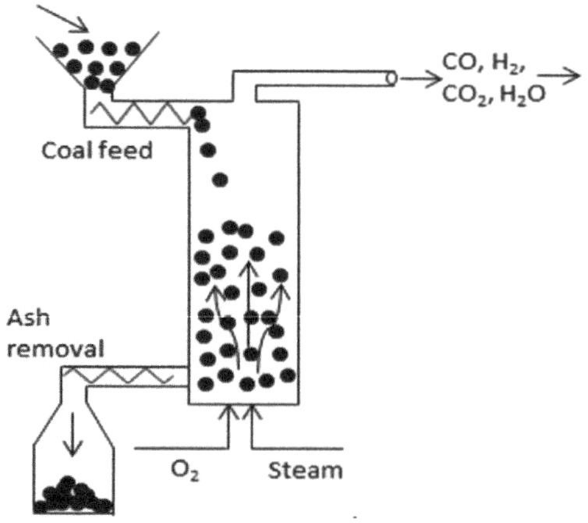

Figure 8-4. Coal gasifier example.

from the volatilization of the coal. It also cools the reactor, subsequently lowering the gas density (bed height) and gas production rate. Similarly, increasing the O_2 feed rate initially expands the bed (increasing level), accelerates carbon conversion, and increases the gas production rate, but later leads to coal depletion in the reactor with the opposite impacts.

Figure 8-5 illustrates why this is a problem for control. The left-hand side shows a desired set-point change and the step-and-hold action required to effect that. This is an open-loop response with the controller in manual (MAN) mode. The smooth curve in the upper-left sketch represents a desired path to the set point, but the process (the lower noisy curve) has inverse action and first moves in the wrong direction, then reverses and eventually reaches the set point.

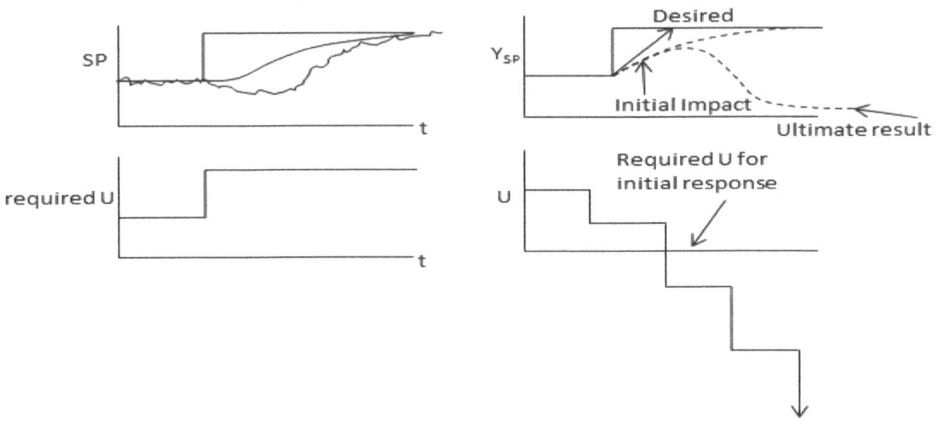

Figure 8-5. The control problem of inverse action.

The right-hand side of Figure 8-5 reveals the control-related problem. If the controller's objective is to start the process moving toward the set point, the controller must move the manipulated variable (MV) in the wrong direction. This initially moves the process in the right direction, but then the process reverses and gets even farther from the set point. Continued "correction" takes the controller output to a limit and in the wrong direction.

A simplified representation of this is a process with two mechanisms by which U affects X as illustrated in Figure 8-6 in Laplace notation, and the net effect is that the two independent mechanisms are additive.

Two modeled mechanisms are added to generate the process model.

$$\hat{X} = P_1\hat{U} + P_2\hat{U} \tag{8-6}$$

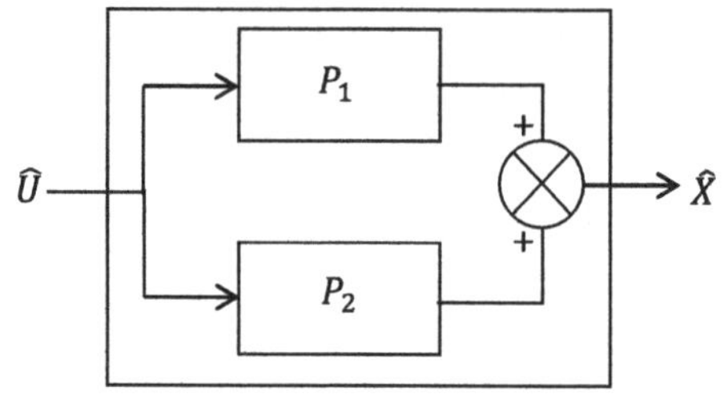

Figure 8-6. Dual path mechanism.

Assuming first-order processes for each mechanism and using K for the gains and τ for the time constants

$$P_1 = \frac{K_1}{\tau_1 s + 1}; \; P_2 = \frac{K_2}{\tau_2 s + 1} \tag{8-7a,b}$$

$$\hat{X} = \left(\frac{K_1}{\tau_1 s + 1} + \frac{K_2}{\tau_2 s + 1} \right) \hat{U} \tag{8-8}$$

Reducing to one fraction

$$\hat{X} = \frac{K_1 \tau_2 s + K_1 + K_2 \tau_1 s + K_2}{(\tau_1 s + 1)(\tau_2 s + 1)} \hat{U} \tag{8-9}$$

Regrouping terms

$$\hat{X} = \frac{(K_1 \tau_2 + K_2 \tau_1)s + (K_1 + K_2)}{(\tau_1 s + 1)(\tau_2 s + 1)} \hat{U} \tag{8-10}$$

Divide by $K_2 + K_1$ (if $K_2 \neq -K_1$)

And combining variables

$$K_P = K_1 + K_2 \tag{8-11}$$

$$\tau_3 = \frac{K_1 \tau_2 + K_2 \tau_1}{K_1 + K_2} \tag{8-12}$$

$$\hat{X} = K_P \frac{\tau_3 s + 1}{(\tau_1 s + 1)(\tau_2 s + 1)} \hat{U} \tag{8-13}$$

If τ_3 is negative, a positive change in U has a positive deviation value, but τ_3 makes it have the opposite sign of U. So $\frac{dU}{dt}$ and U have opposite influences on X. But the $\frac{dU}{dt}$ term eventually becomes zero, leaving only the U influence. If $\tau_3 \frac{dU}{dt}$ has a larger initial influence than U, X moves in one direction and then the other.

What makes τ_3 negative? See Equation 8-12. Note that τ_1 and τ_2 are both > 0 for a stable process. If K_1 and K_2 are both > 0, then $\tau_3 > 0$; or if K_1 and K_2 are both < 0, then $\tau_3 > 0$. To make $\tau_3 < 0$, K_1 and K_2 must have opposite signs, and the process with the larger gain magnitude must have the larger time constant.

8.3 Takeaway

If the process dynamics are well-behaved (first-orderish), then simple model-based controllers (generic model control with a steady-state and process-model-based control) will work. However, if the model has ill-behaved dynamics, more complicated controllers (predictive functional control [PFC] and horizon predictive control [HPC]) are required.

Exercises

8.1 Describe a few processes that have one of the ill-behaved dynamic behaviors represented in Sections 8.2.1 to 8.2.5.

8.2 Express the five ill-behaved dynamic behaviors in calculus, algebraic, or numerical notation. (This chapter used the Laplace notation.)

8.3 Create a computer process simulation to generate data that demonstrates each of the five ill-behaved problems. Explore the impact of simulator coefficient values.

Section 2

Simple Controllers – MISO and Unconstrained

9 Simple Internal Model Control – Unconstrained

9.1 Introduction

Internal model control (IMC) solves the process problem of ill-behaved dynamics, the response trend in time that gives simple controllers problems. Here, it is presented for linear, stationary models. Because an IMC is a linear model-based controller, the concept does not belong in this book about nonlinear control techniques; however, being linear, it is amenable to diverse analysis objectives that prove stability, zero steady-state offset, and such desirable control attributes. It also reveals techniques, such as biasing the set point to remove a steady-state offset, that support the structure of the nonlinear model-based controllers of this book. Because IMC is linear, this chapter is an introductory aside.[1]

9.2 IMC

Figure 9-1 shows the Laplace-transformed representation of the block diagram of the IMC strategy. The process is the block labeled *P*, with *u* as the input (the manipulated variable [MV] or the controller output [CO]) and *y* as the output (the controlled variable [CV]). The value of *u* changes at the beginning of a control interval and holds that

---

1. The landmark IMC publication is by C. E. Garcia and M. Morari. "Internal Model Control: A Unifying Review and Some New Results," *Industrial and Engineering Chemistry Process Design and Development* 21, no. 2 (1986): 308–323.

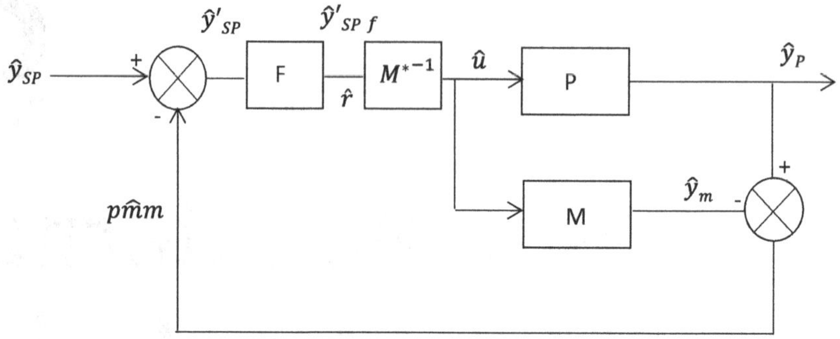

Figure 9-1. IMC structure.

value to the end of the interval. The controller process model is block M. Both P and M respond to the MV, and after each sampling interval, Δt, y_p (the process variable) is measured and y_m (the modeled value) is updated.

The difference between y_p and y_m represents the process-model mismatch, pmm. It could result from model error, unmeasured disturbance, a change in the process behavior, or noise. If the process value, y_p, is to be at the set point and u is the correct value to make this happen, then the model value, y_m, will not be at the set point because of the model mismatch. Because the controller action is based on the model, the set point for the model must be different from the process set point. A simple correction is to bias the set point for the model by the pmm. y'_{SP} is the biased set point for the model. $y'_{SP} = y_{SP} - pmm$. (This can be expressed in many ways, including $y'_{SP} = y_{SP} - y_p + y_m = e + y_m$.)

The block labeled F is a filter that has several purposes. If there is measurement noise, the *pmm* will be noisy, making the biased set point and the CO noisy. First-order filtering is commonly used to temper noise. In addition, if the set point makes a step change, the control action could have a bump or a spike, so filtering the biased set point provides noise reduction and a set point softening response. The filter output is the reference trajectory, r, the desired path for the model to move to the biased set point. However, if the model is high-order, it cannot follow a first-order path, so in IMC, the filter is of the same order as the modeled lag-lead difference. Then, the model can track such a higher-order path.

For simplicity, set all filter time constants to the same value. If this is a second-order model of the process, use

$$Filter = \frac{1}{(\tau_w s + 1)^2} \tag{9-1}$$

Finally, the block indicated by M^{*-1} represents the inverse of the model. Its function is to determine the u value that moves the model along the reference trajectory. However, if the model has a delay in it (in Laplace notation, $e^{-\theta s}$), making the model immediately track the reference trajectory is impossible. To solve this problem, remove the dead time from the model in this function (keep it in the M function), which effectively seeks to determine the u value that moves the model along the reference trajectory after the delay. This is similar to the action in a Smith predictor. The asterisk on the M indicates that there is no delay term, and the power of –1 represents the model inverse.

However, there is one more issue if the process has inverse action: seeking to invert its model and follow the reference trajectory will move the CO in the wrong direction (see Section 8.2.5). The model delay and inverse action are termed *noninvertible features.* Remove them both from the M^{*-1} function.

Figure 9-1 can be rearranged to appear like conventional control in which the actuating error enters the controller (within the dashed line), and the controller sends the MV signal to the process. To obtain the Figure 9-2 rearrangement, slide the $y_p - y_m$ junction in Figure 9-1 along the pmm line and through the $y_{SP} - pmm$ junction.

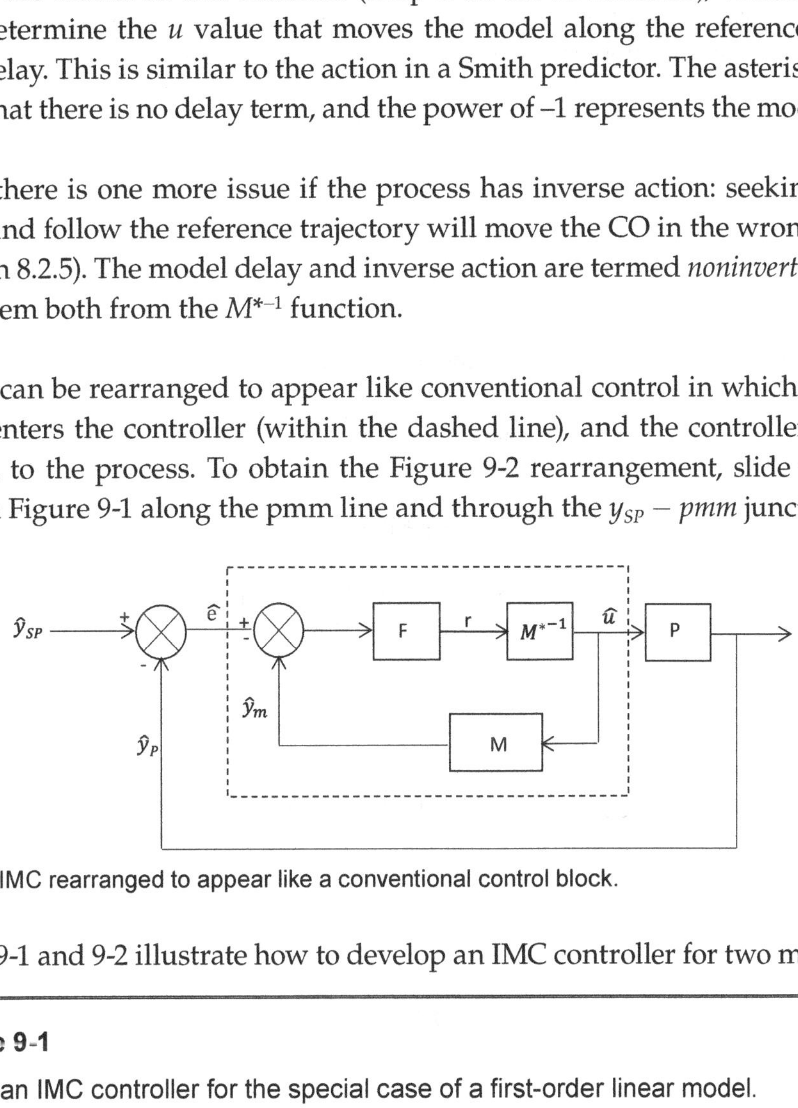

Figure 9-2. IMC rearranged to appear like a conventional control block.

Examples 9-1 and 9-2 illustrate how to develop an IMC controller for two model types.

Example 9-1

Develop an IMC controller for the special case of a first-order linear model.

$$\tau_m \frac{dY_m}{dt} + Y_m = K_m U, \qquad Y_m(t=0) = Y_{m,0} = K_m U_{initial}$$

Note that this model is in deviation variables showing deviation from an initial steady state. τ_m and K_m are the time constant and gain of the process model.

$$Y_{m,0} = K_m U_{initial}$$

Laplace-transformed: the model is a first-order lag with no lead.

$$\hat{Y}_m = \frac{K_m}{\tau_m s + 1}\hat{U}$$

So, the IMC filter will be first-order, with a filter time constant, τ_{want}, that represents the desired first-order trajectory for the response to a set-point change.

$$\hat{r} = \frac{1}{\tau_w s + 1}\hat{Y}'_{SP}$$

The biased set point will be

$$\hat{Y}'_{SP} = \hat{e} + \hat{Y}_m$$

There are no uninvertible parts to the model, so the inverted model will be

$$M^{*-1} = \frac{\tau_m s + 1}{K_m}$$

Using Figure 9-2, the controller block diagram is

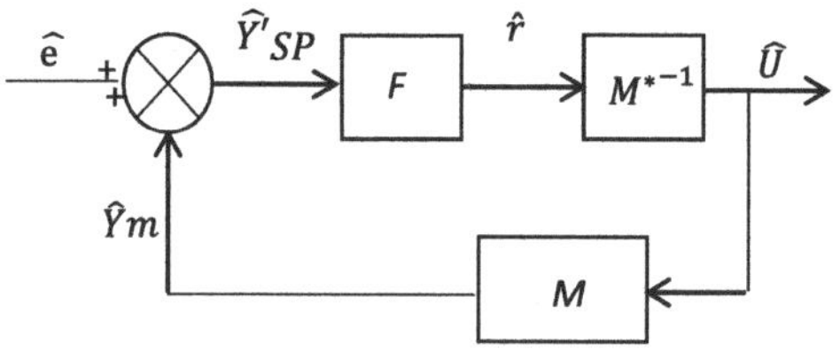

Composing the Laplace control rule from the block diagram

$$\hat{U} = M^{*-1}\hat{r} = M^{*-1}F\hat{Y}'_{SP} = M^{*-1}F(\hat{e} + \hat{Y}_m) = M^{*-1}F(\hat{e} + M\hat{U})$$

$\hat{U}$ is implicit (on both sides of the equation). Solving for $\hat{U}$ explicitly

$$(1 - M^{*-1}FM)\hat{U} = M^{*-1}F\hat{e}$$

$$\hat{U} = \frac{M^{*-1}F}{1 - M^{*-1}FM}\hat{e}$$

Inserting expressions for M^{*-1}, F, and M

$$\hat{U} = \frac{\dfrac{\tau_m s + 1}{K_m}\dfrac{1}{\tau_w s + 1}}{1 - \dfrac{\tau_m s + 1}{K_m}\dfrac{1}{\tau_w s + 1}\dfrac{K_m}{\tau_m s + 1}}\hat{e}$$

Rearranging to simplify

$$\hat{U} = \frac{1}{K_m}\frac{\tau_m s+1}{\tau_w s}\hat{e} = \frac{1}{K_m}\left(\frac{\tau_m}{\tau_w}+\frac{1}{\tau_w s}\right)\hat{e}$$

$$\hat{U} = \frac{1}{K_m}\frac{\tau_m}{\tau_w}\left(1+\frac{1}{\tau_m s}\right)\hat{e}$$

We discover that it is the familiar proportional-integral (PI) control rule where $K_c = \frac{1}{K_m}\frac{\tau_m}{\tau_w}$ and $\tau_i = \tau_m$.

The IMC structure becomes conventional PI control with a first-order model, with one user-chosen tuning factor, τ_w, to shape the trajectory and attenuate noise.

For users, it is satisfying to know that the linear form of the simple model-based control, internal model control (IMC), is a familiar, widely accepted, age-old approach.

Example 9-2

Develop an IMC controller for a second-order plus dead time linear model. Note that this model is in deviation variables—deviations from an initial steady state.

Laplace-transformed, the model is

$$\hat{Y}_m = \frac{K_m e^{-\theta_m s}}{(\tau_{m,1}s+1)(\tau_{m,2}s+1)}\hat{U}$$

So, the IMC filter will be second-order, with a filter time constant, τ_{want}, that represents the desired second-order trajectory for the response to a set-point change.

$$\hat{r} = \frac{1}{(\tau_w s+1)^2}\hat{Y}'_{SP}$$

The biased set point will be

$$\hat{Y}'_{SP} = \hat{e} + \hat{Y}_m$$

The dead time is not an invertible part of the model, so the inverted model will be

$$M^{*-1} = \frac{(\tau_{m,1}s+1)(\tau_{m,2}s+1)}{K_m}$$

The controller block diagram is

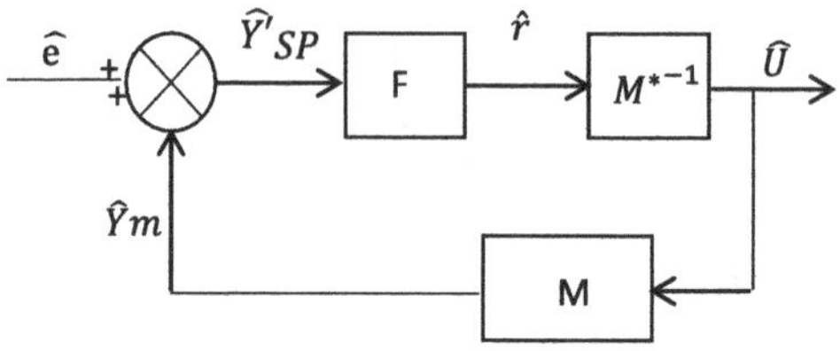

Rather than remaining in the Laplace domain, I will translate each function to its numerical solution. In the automatic (AUTO) mode

$$e = y_{SP} - y_p$$

$$y_{1md} = \left(\frac{\Delta t}{\tau_{1m}}\right) u_d + \left(1 - \frac{\Delta t}{\tau_w}\right) y_{1md,\,old}$$

$$y_{2md} = \left(\frac{\Delta t}{\tau_{1m}}\right) y_{1md,\,old} + \left(1 - \frac{\Delta t}{\tau_w}\right) y_{2md,\,old}$$

The subscripts 1 and 2 represent the first and second lags.

$$y_{2md,\,delayed} = y_{2md}(t - \theta_m)$$

$$y_{SPmd} = e - y_{m,d}$$

The second-order model requires a second-order filter on the reference trajectory.

$$r_{1d} = \left(\frac{\Delta t}{\tau_w}\right) y_{SPmd} + \left(1 - \frac{\Delta t}{\tau_w}\right) r_{1d,\,old}$$

$$r_{2d} = \left(\frac{\Delta t}{\tau_w}\right) r_{1d,old} + \left(1 - \frac{\Delta t}{\tau_w}\right) r_{2d,\,old}$$

The Act function is a second-order lead.

$$u_{1d} = \tau_{m,1} \frac{r_{2d} - r_{2d,\,old}}{\Delta t} + r_{2d}$$

$$u_{2d} = \tau_{m,1} \frac{u_{1d} - u_{1d,\,old}}{\Delta t} + u_{1d}$$

$$u_d = u_{2d} / K_m$$

$$u = u_d + u_0$$

(Because the model is in deviation variables, the bias must translate the MV.)

$$r_{1d,\,old} = r_{1d}$$

$$r_{2d,old} = r_{2d}$$

$$u_{1d,old} = u_{1d}$$

$$u_{2d,old} = u_{2d}$$

$$y_{1md,old} = y_{1md}$$

$$y_{2md,old} = y_{2md}$$

In the manual (MAN) mode, it is simplest to assume that the process is at a steady state and then initialize it. Alternately, you could run the model in the MAN mode so it is consistent with what has happened and ready with the right values when switched to AUTO.

$$y_{SP} = y_p$$

$$u_0 = u$$

$$r_{1d,old} = 0$$

$$r_{2d,old} = 0$$

$$u_{1d,old} = 0$$

$$u_{2d,old} = 0$$

$$y_{1md,old} = 0$$

$$y_{2md,old} = 0$$

9.3 IMC Control Simulation

Consider IMC control on a nonlinear third-order process with dead time. Figure 9-3 illustrates the process response to step changes in the CO.

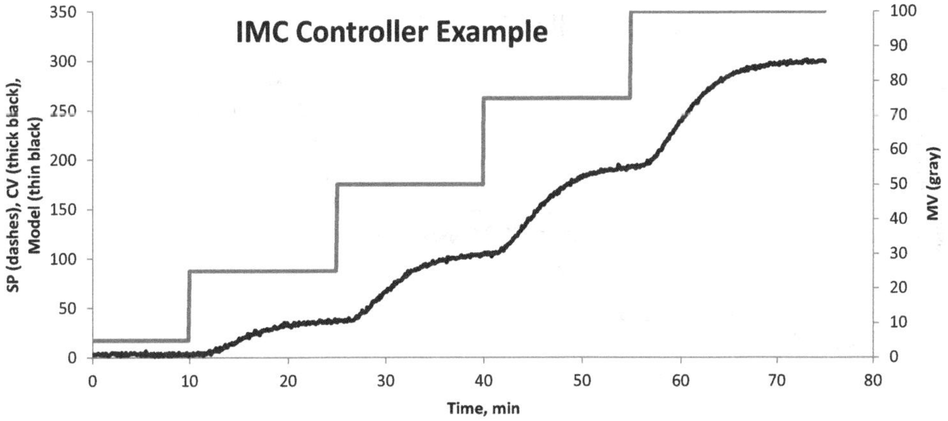

Figure 9-3. Illustration of process behavior.

The upper trace that makes steps is the CO (0% to 100%), and the CV response is the slightly noisy trace. In the MAN mode, the set point and modeled value track the process CV, so they are not visible in this figure. The process is nonlinear with a 5:1 gain ratio from the upper to the lower range. The process has a 5 min delay and is third-order with time constants of 1, 2, and 3 min each.

A reasonable first-order linear model (no dead time) has a gain of 3 and a time constant of 5 min. IMC control will use this simple model with a filter time constant of 4 min.

Figure 9-4 illustrates servo action. The controller is started in MAN, and the output is stepped to 20% at about 2 min. The process responds, and at 10 min, the controller is switched to AUTO. The set point is the dashed line, which is visible during the AUTO mode but matches the process value during MAN. The controller model (noiseless thin trace) also matches the process value at that time because it was initialized to do so. However, the process had not settled to a steady state, so as the controller lowers the MV (gray noisy trace) a bit to keep the CV (thick black noisy trace) at the set point, the model reveals a mismatch. There are two set-point changes.

Recognize these features in Figure 9-4:

1. The model mismatch goes from below the process to above it.
2. There is a bumpless transfer from MAN to AUTO.
3. In spite of the linear first-order model on this nonlinear third-order process, reasonable control is achieved.
4. There is zero steady-state offset but no integral in the controller.

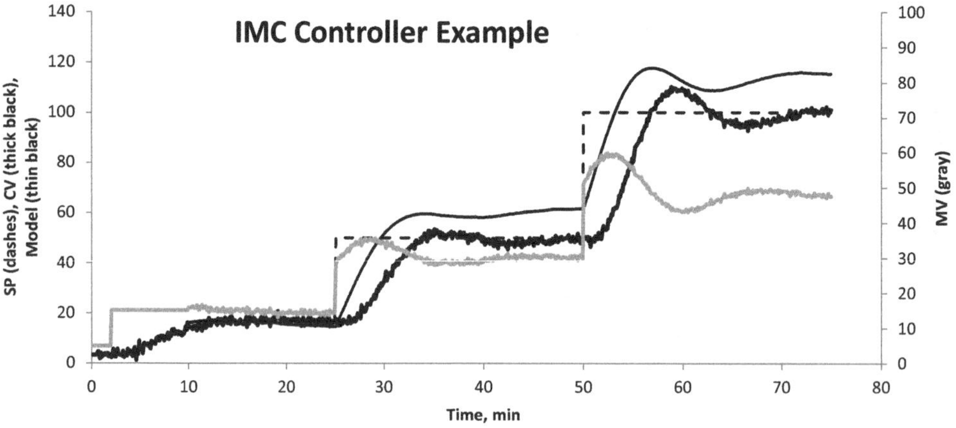

Figure 9-4. Illustrating servo action.

5. Anticipatory action is evidenced. The MV reverses direction long before the CV crosses the set point.

Figure 9-5 illustrates constraint recovery. The controller starts in the MAN mode with an output step and is switched to AUTO at 5 min. At 10 min, the set point changes to an excessive value, and even at 100% MV, the CV cannot reach the set point. At 30 min, the set point is lowered to a reasonable value, and the controller immediately responds. There is no delay in action (no integral to windup). Also, note that the aggressive response pushes the MV to the lower limit, but before the CV deviates below the set point, the controller anticipates that it will deviate and increases the MV to get the CV to the set point quickly.

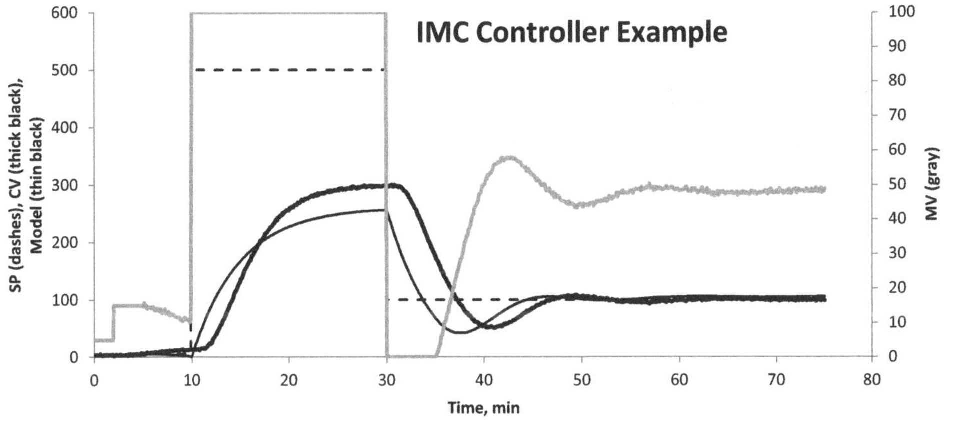

Figure 9-5. Illustrating constraint recovery in IMC.

This illustration is accessible on my website (*www.r3eda.com*) as an Excel VBA file. You can choose options, tests, process and controller models, noise and disturbance attributes, and controller tuning to explore. The software provides the integral of the squared error (ISE) and travel metrics. The code is open so you can see all the details.

9.4 Perspective

The introduction of IMC in 1986 led to much academic research attention, analysis, and extrapolation. However, there was little traction within the chemical process industry (CPI). IMC does not address the nonlinear or nonstationary issues. It can include the modeled impact of measured disturbances. Although, conceptually, it can extend to multiple-input, multiple-output (MIMO) processes with interaction, the formulation quickly becomes very complex. It does not address constraint management when the degree of freedom (DOF) is less than zero.

I chose to include IMC in this book for the following reasons:

- It shows that biasing the set point for the model by the pmm works.
- It reveals the three basic Model-Correct-Act functions.
- It shows that simple models in an MBC structure reduce to conventional control strategies.
- It shows that what we have become accustomed to, with three tuning factors in proportional-integral-derivative (PID), control is achieved with only one in MBC.
- It shows that control can be good, even with a poor model.

9.5 Essential Features

9.5.1 Bumpless Transfer

When the mode is switched from MAN to AUTO, the controller should not try to push the process to another place. In the MAN mode, the operator usually adjusts the CO manually to get the process to a desired operating state. Often, the operator holds it at that point long enough to see that it has settled as desired, nearly to a steady state. Usually, this means that in the MAN mode, the controller must initialize variables to be immediately correct for the AUTO mode. The following steps address this process:

1. To implement the set point following in the MAN mode, assign the set-point value to the present process value. See Example 9.2.
2. Initialize the model to be ready to go when switched to AUTO. If the process is relatively steady and the model is in deviation variables, initialize all reference values to the present value and initialize all model state variables to zero in the MAN mode. See Example 9.2.

However, if the MAN mode is intended as a temporary deviation, for instance, to bump a valve to observe responses, you might not want to reset the set point or reinitialize the model to steady-state values. In this case, the controller should have a switch for the operator to either choose set point following or to return to the prior set point option. Also, in this case, it would be best to continue to run the model in the MAN mode so the model states track what is happening for faster recovery when switched back to AUTO.

9.5.2 Constraint Recovery

This simple IMC does not have constraint avoidance on the CV or an auxiliary variable (AuxV). However, the CO does need to be limited to 0% to 100% of the full range (perhaps –6% to 106% to ensure that device calibration does not permit a full final element range). An excessive set-point value could push the output beyond the constrained MV value. Further, an override in the control system for safety, cavitation prevention, or many other reasons could change the actual MV value to the final element. A characteristic of conventional PID control is that the integral winds up when there is a constraint or override. The windup prevents the CO response from changing immediately when the constraint or override is removed. External reset feedback is a windup prevention strategy for PID control.

A model-based controller should not have a delay in responding when a constraint is removed. The MV value that goes to the model should be the actual MV implemented if an internal or external override changes the controller-calculated value.

9.5.3 Insensitivity to Model Mismatch

The controller model will not be true to the process (it is linear and reduced order, and the coefficient values will not exactly match the process). The mismatch between the model and the process should be inconsequential for control.

9.5.4 Execution Error Trapping

If the user inputs a negative value for a time constant, the control will move in the wrong direction, oscillate, and possibly go to an extreme limit. The controller must only accept feasible values. Gains, for instance, must have the appropriate signs and nonzero values.

If the user inputs a zero value for a time constant, the numerical method to solve the lags in Example 9.2, $\left(\frac{\Delta t}{\tau_w}\right)$, will encounter a divide by zero. More than just requiring nonzero and positive values, the $\left(\frac{\Delta t}{\tau_w}\right)$ value is the small Δt limit of the proper weighting value, $\left[1-e^{-\left(\frac{\Delta t}{\tau_w}\right)}\right]$, which is a valid approximation if $\Delta t < \approx 0.1\tau_w$. Not shown in Example 9.2, the number of samples for the dead time is $\theta/\Delta t$, which can only have integer values $n = ROUND\left(\frac{\theta}{\Delta t}\right)$. A numerical method heuristic is that the controller, Δt, should be at least 1/10 of any τ or θ value in the model or controller. A controller feature should include such a check and inform a user if a choice of Δt or any τ or θ violates the rule.

9.5.5 Noise Tempering

In IMC, the filter does both noise tempering and controller aggressiveness tuning. If greater noise tempering is desired, you could place an additional filter on the pmm.

Placing a filter on the measurement could create a lag difference between the model and process, which could amplify the pmm during transients. If you filter the measurement, either (1) base the model on the filtered process response or (2) filter the modeled output to calculate the pmm.

9.5.6 Both Servo and Regulatory Action

The single tuning factor in MBC should accelerate or temper the responses to a setpoint change (servo action) and an unmeasured disturbance (regulatory mode).

9.6 Takeaways

- If you implement IMC, convert the model (in deviation variables) to process variables.
- As Chapter 11 shows, when models are linear, the nonlinear control versions reduce to IMC. Because mathematical analysis of IMC has provided much support for the legitimacy and effectiveness of this approach, it also provides indirect support for the nonlinear control techniques.

Exercise

9.1 Example 9-1 showed that internal model control (IMC) reduces to conventional proportional-integral (PI) control with a first-order model. Example 9-2 partially completes a similar analysis. Finish it to determine the one line Laplace transform of the controller.

10
Generic Model Control with Steady-State Models

10.1 Introduction

The generic model control (GMC) control method was developed and documented in a paper[1] by Peter Lee and Gerry Sullivan in 1988, placing it in the era when computers were integrated with industrial process control. GMC has multiple industrial applications, one of which is discussed in the article "Generic Model Control — A Case Study"[2] by P. L. Lee, R. B. Newell, and G. R. Sullivan.

There are several versions of GMC. I choose to use the version that uses a steady-state (SS) model, which is relatively easy to obtain and ends up being classic proportional-integral (PI) control with output characterization. This version is probably implementable in your existing programmable logic controller or distributed control system.

1. P. L. Lee and G. R. Sullivan, "Generic Model Control," *Computers and Chemical Engineering* 12, no. 6 (1988): 573–580.
2. P. L. Lee, R. B. Newell, and G. R. Sullivan, "Generic Model Control — A Case Study," *Canadian Journal of Chemical Engineering* 67, no. 3 (June 1989): 478–484.

If the controller is multiple-input, single-output (MISO) and the major issue for control is nonlinearity, this is a good solution. However, it is not a first choice if the process is multiple-input, multiple-output (MIMO), has ill-behaved dynamics, and is subject to auxiliary variable constraints.

10.2 Generic Model Control (Steady-State Version) – Concept

This control strategy uses an SS model. The process response is y_{SS}, and the subscript indicates it is the SS (equilibrium or end-of-batch) response value. In Equation 10-1, the control action is u, and d represents measurable disturbances. Although the control has one controlled variable (CV) and one manipulated variable (MV), this is a MISO model because both u and d represent model inputs.

$$y_{SS} = f(u,d) \tag{10-1}$$

The control action is to identify the value of u that would make the SS model go somewhat beyond the biased set point (the set point for the model). This is illustrated in Figure 10-1 and mathematically represented by Equations 10-2 and 10-3. e is the traditional actuating error for control, the difference between the process set point and the process measurement. y'_{SP} is the biased set point, which is the set point (SP) for the model. Because the model will not match the process exactly, the set point for the model must be different from the set point for the process. If control seeks the SS control action to move the model to y'_{SP}, the process will asymptotically move to the set point. To accelerate the control action, temporarily seek the u value that would move the SS model beyond the biased set point to $y_{SS\ target}$, a temporary SS target.

$$y_{SS\,Targ} = y'_{SP} + K(y_{SP} - y_P) \tag{10-2}$$

$$e = (y_{SP} - y_P) \tag{10-3}$$

K in Equation 10-2 is the GMC controller gain and defines how far the temporary target should be beyond the biased set point. If $K = 0.2$, this means to aim for a temporary target that is 20% farther. For illustration convenience, in Figure 10-1, the beyond distance is about 80% of the actuating error, meaning $K = 0.8$, which would be a relatively

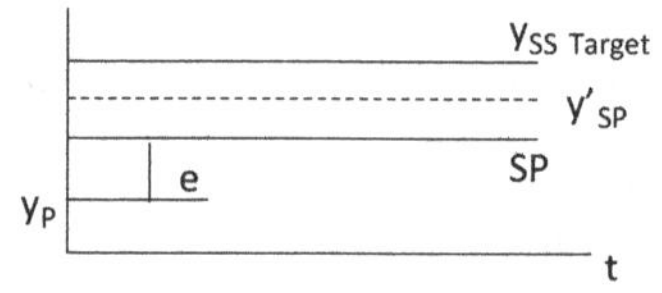

Figure 10-1. The GMC-SS concept.

aggressive controller. Values in the range of $0.1 < K < 0.3$ would be normal for the controller aggressiveness factor. As the process moves toward the set point and the actuating error diminishes, the beyond target relaxes to the biased set point.

After Equation 10-2 is used to determine the value of $y_{SS\,Targ}$, use Equation 10-4 to determine the control action.

$$u = f^{-1}(y_{SS\,Targ}, d) \tag{10-4}$$

This may be an iterative procedure if the control action cannot be explicitly solved from the inverse of Equation 10-1.

Because there is no dynamic model in GMC-SS, there is no past-to-now (yP2N) model to generate the process-model mismatch (pmm). GMC uses the actuating error, $e = (y_{SP} - y_P)$, as a measure of offset. However, this should only be valid at SS, not during a transient. Therefore, incrementally accumulate the pmm only at each sampling.

$$y'_{SP_i} = y'_{SP_{i-1}} + \alpha e_i \tag{10-5}$$

In Equation 10-5, the subscript i represents the sample number, the time counter, and α represents a small portion of the temporal error. Perhaps $\alpha = 0.1$.

Analysis shows that Equation 10-5 represents an integral action. First, represent $y'_{SP_{i-1}}$ by the prior values and incremental updates, then do the same for $y'_{SP_{i-2}}$.

$$y'_{SP_i} = y'_{SP_{i-2}} + \alpha e_i + \alpha e_{i-1} \tag{10-6}$$

$$y'_{SP_i} = y'_{SP_{i-3}} + \alpha e_i + \alpha e_{i-1} + \alpha e_{i-2} \tag{10-7}$$

Continuing the series back in time to the controller initiation and using N as the number of control actions in the time, t, that the controller has been in the automatic (AUTO) mode, $N = \frac{t}{\Delta t}$.

$$y'_{SP_i} = y_{SP} + \alpha \sum_{i=1}^{N=\frac{t}{\Delta t}} e_i \tag{10-8}$$

Multiplying and dividing the sum by Δt

$$y'_{SP_i} = y_{SP} + \frac{\alpha}{\Delta t} \sum_{i=1}^{N} e_i \Delta t \tag{10-9}$$

Using the rectangle rule of integration

$$y'_{SP} = y_{SP} + \frac{\alpha}{\Delta t}\int_0^t edt \tag{10-10}$$

And converting the integral scale factor to a time constant

$$\tau_i = \left(\frac{\alpha}{K\Delta t}\right)^{-1} \tag{10-11}$$

Then, Equation 10-2 becomes

$$y_{SS\,Targ} = y_{SP} + K\left(e + \frac{1}{\tau_i}\int edt\right) \tag{10-12}$$

The right-most term is the output of a PI control calculation. However, it does not represent the GMC-SS controller output. It represents the temporary set-point bias and the temporary beyond y'_{SP} combined as the beyond y_{SP} target. Using m to represent it in a two-step version of Equation 10-12, renaming the controller gain to the conventional K_c and indicating the required controller initial output bias, m_0, when switching from MAN to AUTO

$$e = (y_{SP} - y_P) \tag{10-13a}$$

$$m = m_0 + K_c\left(e + \frac{1}{\tau_i}\int edt\right) \tag{10-13b}$$

$$y_{SS\,target} = y_{SP} + m \tag{10-13c}$$

$$u = f^{-1}(y_{SS\,target}, d) \tag{10-13d}$$

For calculations in the MAN mode, see Section 10.5.

Reasonable values for K_c and τ_i are

$$0.1 \le K_c \le 0.3 \quad \textit{to target } 10\% \textit{ to } 30\% \textit{ beyond}$$

$$\tau_i \approx \tau_P \approx 1/4 \textit{ process open loop settling time}$$

This tunes intuitively like a PI controller, which is good for operators and technicians. A heuristic tuning approach is to remove the integral action, adjust K_c to set the

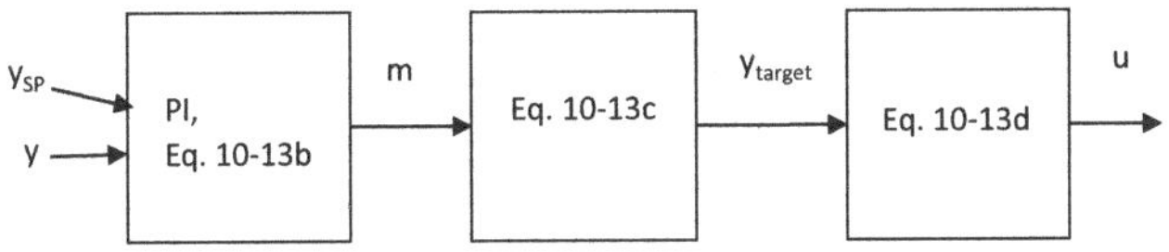

Figure 10-2. Controller structure concept – AUTO mode for GMC using an SS model.

aggressiveness in response to set-point changes, and then adjust τ_i to remove the SS offset in a desired manner.

The GMC-SS controller contains proportional action for aggressiveness to return to the set point and integral action for SS offset removal.

Equations 10-13a through 10-13d illustrate a classical control technique termed *output characterization.*

Figure 10-2 illustrates the sequence. The controller output is u, the MV and the signal to the valve. The variable m is the output of one part of the controller, the PI calculation.

This presentation considers that the controller output, m, has the same units and associated range as the $y_{SS\ Target}$ deviation from the y_{SP} would have. However, conventional controllers have a 0% to 100% input and a 0% to 100% output that does not permit negative values. Scaling is needed. Further, manual (MAN) model operation and overrides must be considered when $y_{SS\ Target}$ is infeasible. The next several sections address these issues.

10.3 Scaling for 0% to 100%

Variables m, m_0, e, y, and y_{SP} in Equation 10-13 all have the process variable units. The units for u might be those of the process engineer's view of the input to the process, such as a flow rate or a heat rate. The model was likely generated from a process perspective, and the units are not the classical controller 0% to 100% of the full-scale range. If the engineer is using a built-in conventional PI controller with both input and output in the 0% to 100% range, scaling is needed to convert the 0% to 100% m values to the CV values and range needed to add m to y_{SP} in Equation 10-13b. Then, convert the u value in engineering units to the 0% to 100% full scale.

First, consider scaling for variables that are in percentage of full scale, $m_{\%}$, $m_{0\%}$, $e_{\%}$, $y_{\%}$, and $y_{SP\%}$. Define the minimum and maximum CV values, then scale for the percentage of range.

$$y_{\%} = 0 + \frac{100-0}{y_{max} - y_{min}}(y - y_{\min}) = \frac{100}{y_{max} - y_{min}}(y - y_{\min}) \qquad (10\text{-}14a)$$

$$y_{SP\%} = \frac{100}{y_{max} - y_{min}}(y_{SP} - y_{\min}) \tag{10-14b}$$

Whether in the process units or the percentage of full scale, Equation 10-13a converts the actuating error on the CV to the target CV value. So, the value of K_c does not need to be scaled, and K_c retains its 0.1 to 0.3 nominal range. Also, the value of τ_i is unchanged.

After $m_{\%}$ is determined using Equation 10-13b, it must be unscaled for Equation 10-13c.

$$y = y_{min} + \frac{y_{max} - y_{min}}{100} m_{\%} \tag{10-14c}$$

Consider scaling to make the controller output, u, in the 0% to 100% range. The u in Equation 10-13d has the process units, such as flow rate or temperature. The conversion is simply

$$CO_{\%} = 100\frac{x_{desired} - x_{min}}{x_{max} - x_{min}} = 100\frac{u - x_{min}}{x_{range}} \tag{10-15}$$

10.4 Internal Override

The term *internal override* means that the IF-THEN conditional determining what to do if something exceeds a limit is within the controller. This contrasts with an external override (see Section 10.6) in which action external to the controller overrides or changes the controller influence on the process.

If $y_{SS\ target}$ from Equation 10-13c is excessive, the inverse of Equation 10-13d may encounter an execution error, and the calculation may need to be overridden by either setting $y_{SS\ target}$ at its maximum or minimum feasible value or setting u at its range extreme. This section of code might be written as

```
IF ytarget is feasible THEN
        u := f-1(ytarget,d)
ELSE
        u := specify the appropriate value, either 0% or 100%
END IF
```

If a PI controller (generating the m value) is not in charge for any reason (if something overrides its output, such as the code above), the integral can wind up. One simple anti-windup solution is to reset the integral to its extreme permissible value from the overridden value. If $y_{SS\ target} > y_{max\ feasible}$, then the maximum feasible value for m would be

$$m = y_{max\ feasible} - y_{SP} \tag{10-16a}$$

$$m_{\%} = 100\frac{y_{max\ feasible} - y_{SP}}{y_{range}} \tag{10-16b}$$

If $y_{SS\ target} < y_{min\ feasible}$, then the minimum feasible value for m would be

$$m = y_{min\ feasible} - y_{SP} \tag{10-16c}$$

$$m_{\%} = 100\frac{y_{min\ feasible} - y_{SP}}{y_{range}} \tag{10-16d}$$

Then, the value for the integral should be reset to

$$\int edt = \tau_i\left[\frac{m_{limit} - m_0}{K_c} - e\right] \tag{10-17}$$

An alternate technique to prevent windup is internal reset feedback. With this technique, the integral is replaced with a first-order filter on the controller output, m, and if that value is overridden by m_{limit}, the first-order filter input is m_{limit}. It is a relatively easy Laplace algebraic exercise to reveal that when the adjustable bias, $\hat{b}_{adj}$, is determined by filtering the controller output, as illustrated in Figure 10-3, and there is no override, the controller is equivalent to a PI controller where the integral of the error is used to determine the adjustable bias. When there is an override, the adjustable bias value asymptotically approaches the override value. Then, immediately upon leaving the limit, the adjustable bias relaxes. There is no integral windup, meaning no wind-down period is required to leave the limit. The hat over the symbol indicates a Laplace-transformed variable. The filter time constant, τ_i, is the same as the PI integral time constant.

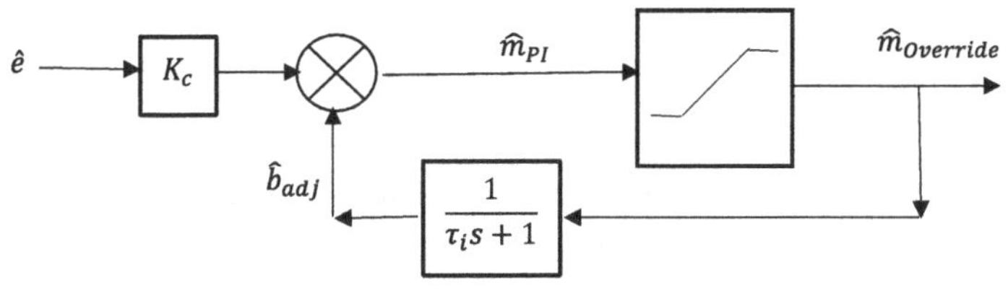

Figure 10-3. Internal reset feedback (irf) version of PI with an internal override.

The AUTO model PI-irf algorithm version is

$$
\begin{aligned}
&e := y_{SP} - y_{process} \\
&P := K_c \bullet e \\
&b := \lambda \bullet m + (1-\lambda) \bullet b \text{, where } \lambda = 1 - e^{-\Delta t/\tau_i} \\
&m := m_0 + P = b \\
&IF\, m > m_{max}\; THEN\, m := m_{max} \\
&IF\, m < m_{min}\; THEN\, m := m_{min}
\end{aligned}
$$

Here, the upper and lower limits on m would be determined as outlined in Equation 10-17.

10.5 MAN Mode Calculations

In the MAN mode, the operator would determine the value of u, the MV, the signal to the process in Equation 10-13d. This is not the value of m, the output of the internal PI algorithm of Equation 10-13b. This assumes, during the MAN mode, that the operator is setting the value of u so the process is at a desired value. The MAN mode must have several features: First, use set-point following to set $y_{SP} = y_{process}$. Second, presume that conditions are at an SS and that $e = 0$; then, for bumpless transfer from MAN to AUTO, initialize $\int edt = 0$, which makes $y_{SS\,target} = y_{SS}(u, d) = f(u, d)$, which assigns $m_0 = f(u, d) - y_{process}$.

The resulting MAN mode code for GMC with an SS model would look like this:

```
IF mode = "MAN" THEN          'Controller is in MAN'
        ymodel := f(u,d)
        ySP := yp
        integral := 0
        m0 := ymodel-yp
END IF
```

10.6 External Override

There may be an override within the controller to an excessive controller-calculated output. Section 10.4 shows how to implement it and prevent integral windup. There also could be external overrides. For instance, the controller might calculate that the output to the process should be 83%, but there might be an external override limiting the signal to the final element to a value of 71%. The override might be due to a safety-related aspect, priority of an auxiliary variable, or other reason. If there is an override and the controller is not in charge, the integral will wind up.

To prevent windup to a limit, use the reset feedback version of the PI controller, but instead of using the integral maximum or minimum feasible value for the controller output as the feedback signal, use the actual override u value to determine the equivalent m value. This is external reset feedback (erf).[3] The model can convert the u_{actual} value to the $y_{SS\ target}$ value. Using process variable units

$$y_{SS\ target} = f(u_{actual}, d) \tag{10-18}$$

In scaled variables

$$m_{erf\%} = \frac{100}{y_{range}} f\left(u_{\%} \frac{y_{range}}{100}, d\right) \tag{10-19}$$

10.7 GMC Demonstration

Figure 10-4 shows the process and the SS model response to step changes in the controller output in MAN mode.

The upper steps are the controller output (right vertical axis, 0% to 100%), and the lower steps are the SS controller model (left vertical axis). The lower trace is the second-order plus dead time nonlinear gain and nonlinear time constant process. The process gain has a 30:1 ratio over the MV range, and the SS controller model has about a 5:1 gain

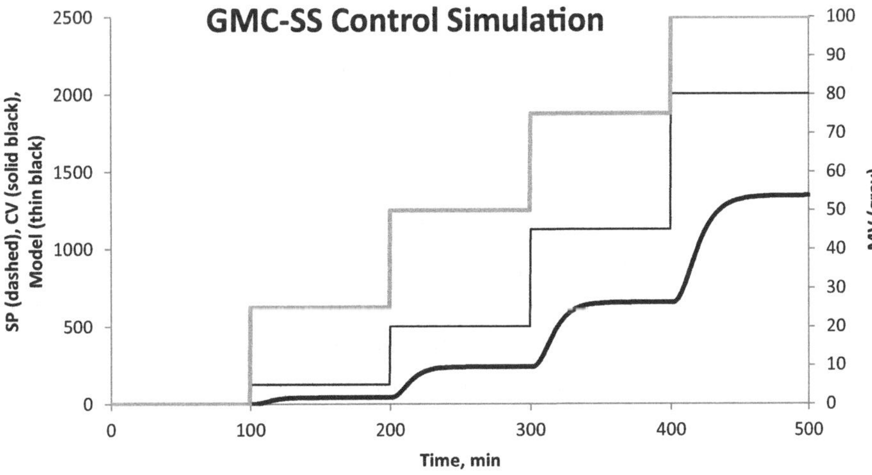

Figure 10-4. Process and model comparison.

3. J. Ou, G. Narayanswamy and R. R. Rhinehart, "External Reset Feedback for Generic Model Control," *ISA Transactions* 37, no. 3 (1998): 189–199.

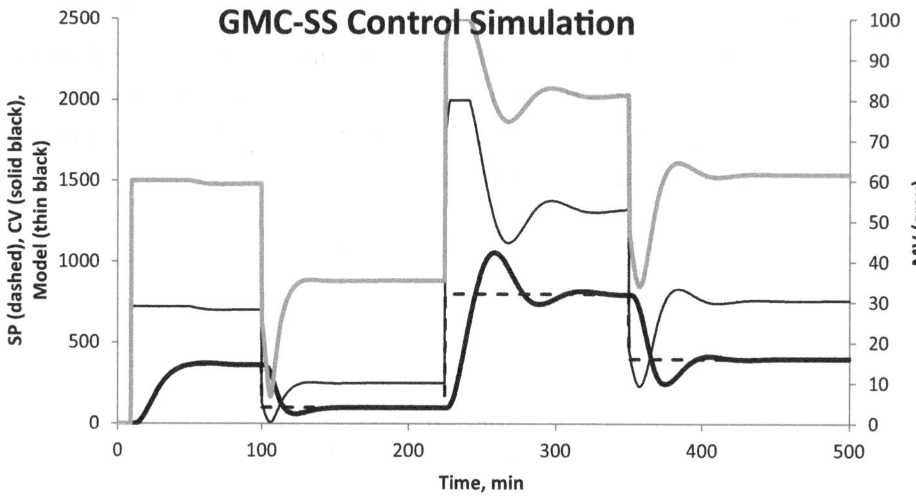

Figure 10-5. Set-point changes.

ratio. Although the controller model is nonlinear, it is not a good fit. The process settling time is about 30 min in the low range and about 70 min in the upper range. In MAN mode, the set point follows the CV so it is not visible on this figure.

Figure 10-5 reveals bumpless transfer and set-point tracking. The controller gain is 1.2, and the integral time is 10 min. The controller starts in MAN mode. At 20 min, the output is set to 60% and the process moves toward a CV (the thick black line) value of about 300. At 50 min, the controller is set to the AUTO mode and initializes its set point to that CV value. Because the process had not quite settled and was still rising, the controller had to begin to reduce the MV to keep the process at the set point. Then, starting at 100 min, there is a sequence of three set-point changes. Note that the dynamic response is similar for all three despite the process nonlinearity and model-process mismatch.

Figure 10-6 shows the same controller start-up and sequence of set-point changes, but in this case, an external override limits the MV to the process to 70%. Notice that in Figure 10-5, the MV had to be in the 80% to 100% range for the period centered on about 300 min. In Figure 10-6, the MV is limited to 70%, and the CV cannot reach the set point. At a time of about 350 min, the set point is changed, and the controller responds immediately without windup. The PI controller in this simulation uses the filtered erf signal as the adjustable bias instead of the integral of the error.

This example is accessible on my website (*www.r3eda.com*) as an Excel VBA file. You can choose options, tests, process and controller models, noise and disturbance attributes, and controller tuning to explore. The software provides the integral of the squared error (ISE) and travel metrics (see Chapter 6). The code is open so you can see all the details.

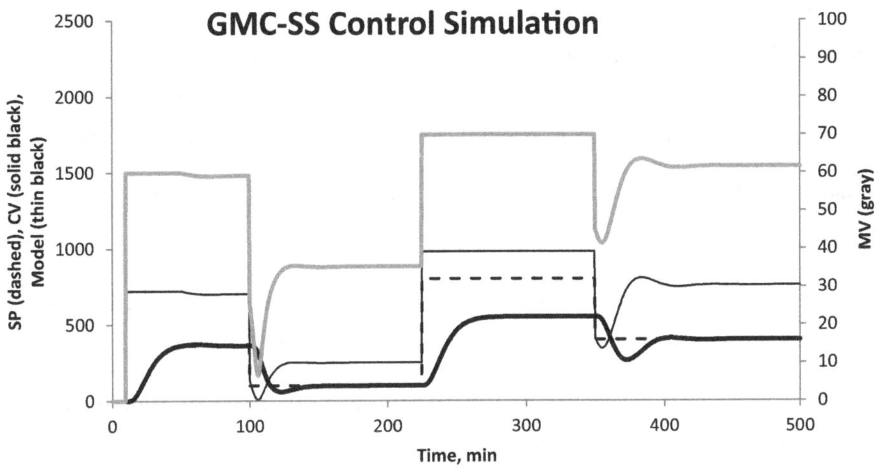

Figure 10-6. Constraint and recovery without windup.

10.8 Observations

GMC-SS has some desirable and undesirable aspects.

- **Desirable**
  - The inverse SS model contains process nonlinear characterization.
  - The inverse SS model does not require the yP2N model calculation.
  - The PI-style tunes like operators expect and understand.
  - The values for K_c and τ_i are bounded, and nominal values are intuitively consistent.
  - The model does not contain e-anticipated, lead, or other helpful dynamics. It might be missing something, but it is beneficial to keep it relatively simple.
  - This is an SS model, which is conventional, well-understood by engineers, and probably available.
  - The inverse does not have to be a one line equation as implied by Equation 10-13d. It could be a subroutine.
- **Undesirable**
  - The model has two tuning coefficients, not just one like simple model-based control.
  - The dynamics of the controller do not change with the process.
  - The inverse SS model may require an iterative solution.

A simple MBC was presented in Section 1.5. The advantages of GMC with an SS model over the simple MBC are as follows:

- It uses an SS, not a dynamic model, making it easier to specify the model.
- There is no dynamic model to update at each sampling, making it computationally simpler.
- It is implemented as an output characterization to a standard PI controller, making it easy to implement within many commercial products.
- It uses a standard PI control algorithm, which already incorporates desired MAN and AUTO mode features.
- It is comfortable to implement. There is little concern over details in the code that may seem to work well as you start it but have a "gotcha" moment at 2:00 a.m. on a Sunday.
- It is familiar to tune because it is a standard PI algorithm. Choose a gain that provides the desired aggressiveness (K_c should be a value of about 1.1 or 1.2), then an integral time that removes the offset at a desired rate (which should be about the value of the process time constant).

However, GMC-SS has some disadvantages relative to simple MBC:

- The integral method of removing offset causes the same windup and initialization problems in GMC that it has in conventional PI control. Its role is to correct the process-model mismatch. However, when there is no process-model mismatch and no justification for shifting the bias for the target, the integral continues to work during (1) set-point changes while the controller is moving the process to the new set point, (2) dead time periods while the controller is waiting for the process to respond, (3) transients for high-order processes where the process response is slow, and (4) times in which constraints or overrides prevent moving to the set point.
- Even though the static gain of the GMC controller is nonlinear, the implied dynamics of the GMC model are linear and stationary. If the process dynamics change with flow rates, levels, or other conditions, the τ_i value in the controller may need retuning.
- The model does not adapt and provide secondary, useful information about the process health (although one could incrementally adapt the model).

10.9 Takeaway

- GMC-SS may be the easiest entry to using nonlinear models in control. If the controller is MISO and the major issue for control is nonlinearity, this is a good solution (simple and effective).

Exercise

10.1 Implement generic model control (GMC) on your test case. Alternatively, visit *www.r3eda.com* to access my simulator with GMC and explore the following:

- Tuning
- Set-point values that place the manipulated variable (MV) on limits
- The transfer between manual mode and automatic mode
- External override with and without the external reset feedback option
- The impact of noise
- The impact of process-model mismatch

11 Simple Model-Based Control – MISO, Unconstrained

11.1 Introduction

This chapter discusses the case of a single controlled variable (CV) and a single manipulated variable (MV), but with measured disturbances also affecting the CV—the multiple-input, single-output (MISO) case. The model may have several internal states, but only one is to be controlled. This approach was presented by R. R. Rhinehart and colleagues in various publications. Two of the earliest are "Process Control Through Nonlinear Modeling"[1] by Rhinehart and J. B. Riggs and "Process-Model Based Control of a Pilot-Scale Fluidized Bed Coal Gasification Reactor"[2] by Rhinehart, R. M. Felder, and J. K. Ferrell.

1. R. R. Rhinehart and J. B. Riggs, "Process Control Through Nonlinear Modeling," *Control* 3, no. 7 (July 1990): 86–90.
2. R. R. Rhinehart, R. M. Felder, and J. K. Ferrell, "Process-Model Based Control of a Pilot-Scale Fluidized Bed Coal Gasification Reactor," *Advances in Instrumentation and Control* 43, Part 1 (October 1988): 341–349.

Substantially restating Section 1.5 but including alternate options, there are three functions in model-based control (MBC): Predict, Correct, and Action (referred to in this text as "Act"). The structure can be represented by the operations in Figure 11-1.

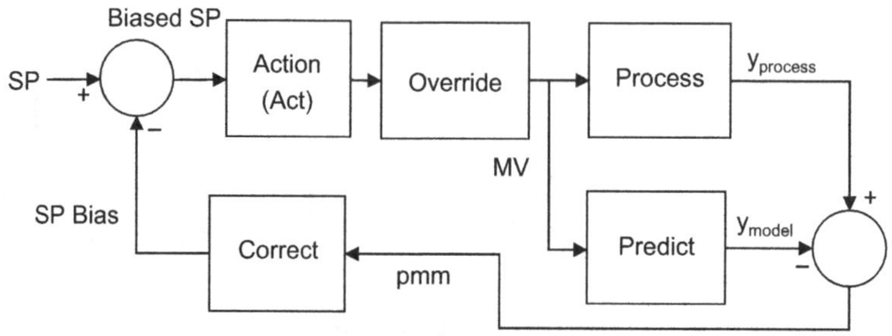

Figure 11-1. MBC representation.

11.2 Predict

The first MBC function uses the model to predict or mimic what the process output is expected to do. This is the lower right box in Figure 11-1. This updates the past model value with the immediate past disturbance and control actions to predict the updated (current or new) modeled output. With the conventional practice of 30 control samplings in an open-loop transient (or 10 samplings within a time constant), the control interval, Δt, is usually small enough to permit Euler's explicit finite difference method to solve the model numerically.

The dynamic model could be stated as a system of algebraic and differential relations.

$$\underline{g}(\underline{y},u,\underline{d},\underline{p})=\underline{0} \tag{11-1}$$

$$\frac{d\underline{y}}{dt}=\underline{f}(\underline{y},u,\underline{d},\underline{p}) \tag{11-2}$$

This could represent a first-principles dynamic model or a first-principles steady-state (SS) model with first-order dynamics for each state (Wiener or Hammerstein types). u is the MV. From a process viewpoint, this might be the flow rate to the process; from the controller viewpoint, this would be the 0% to 100% signal sent to the process. If you developed your model thinking that u represents flow rate, add another algebraic relation to convert the controller output percentage version to the process input. $\underline{y}$ represents the vector of state variables, with one y being the controlled variable. $\underline{d}$ represents the vector of measured disturbances, and $\underline{p}$ represents the vector of model coefficients (parameters).

Desirably, many of the states respond much faster than the key one. Presuming that the fast states are pseudo-steady states adds some algebraic relations to Equation 11-1 but reduces Equation 11-2 to one differential equation.

Also, desirably, Equations 11-1 and 11-2 can be formulated to provide explicit solutions, rather than the implicit formulation indicated above, resulting in

$$\underline{y} = \underline{g}(u,\underline{d},\underline{p}) \tag{11-3}$$

$$\frac{dy}{dt} = f(\underline{y},u,\underline{d},\underline{p}) \tag{11-4}$$

Adhering to the keep it simple and safe (KISS) principle, seek to reduce the model to one differential equation in the state variable to be controlled. If you cannot, you can still use the simple MBC techniques presented in this chapter.

Using the forward finite difference approximation to the differential, Euler's explicit method, with the subscript i representing the time counter, the model equations become

$$\underline{y}_{i-1} = \underline{g}(u_{i-1},\underline{d}_{i-1},\underline{p}_{i-1}) \tag{11-5}$$

$$y_i = y_{i-1} + \Delta t \bullet f(\underline{y}_{i-1},u_{i-1},\underline{d}_{i-1},\underline{p}_{i-1}) \tag{11-6}$$

Equations 11-5 and 11.6 represent the past-to-now model (yP2N), which uses the prior model and process input values to estimate what the current modeled value is expected to be. Use the prior modeled y value, not the process measurement. Use the implemented u value, not the desired MV value from the controller.

Although the time subscript is illustrated, there is no need to remember all past i values; only the most recent values are needed to update the model. In computer assignment statements, for a four-state model with one state being controlled, the equation set becomes

$$y_1 := g_1(u,\underline{d},\underline{p}) \tag{11-7a}$$

$$y_2 := g_2(u,\underline{d},\underline{p}) \tag{11-7b}$$

$$y_3 := g_3(u,\underline{d},\underline{p}) \tag{11-7c}$$

$$y_{CV} := y_{CV} + \Delta t \bullet f(\underline{y},u,\underline{d},\underline{p}) \tag{11-8}$$

The set of assignment statements is not necessarily one line of code and might be expressed in a subroutine or function.

Example 11.1

Use Model Examples 1 and 2 from Chapter 2 to generate code for a yP2N function.

The signal from the controller to the valve is u in percent. The valve stem response to u is

$$\tau \frac{dx}{dt} + x = 0.01u$$

The flow rate response to x is

$$F_h = Cv \bullet f(x) \sqrt{\frac{\Delta P}{G\xi + \alpha \bullet Cv^2 \bullet f^2(x)}}$$

If F_h is being used to control the temperature, the response is

$$\frac{V}{(F_c + F_h)} \frac{dT}{dt} + T = \frac{F_c T_c + F_h T_h}{(F_c + F_h)}$$

Assuming that the valve time constant is much smaller than the process time constant, we can use the pseudo-steady approximation for the valve position.

$$x = 0.01u$$

Using Euler's explicit method, the numerical version of the dynamic model (calculating the yP2N value from the prior conditions) is

$$T_{next\,\Delta t} = T + \Delta t \frac{(F_c + F_h)}{V} \left[\frac{F_c T_c + F_h T_h}{(F_c + F_h)} - T \right]$$

Then, the function code for the yP2N model would be

$$x := 0.01u$$

$$F_h := Cv \bullet f(x) \sqrt{\frac{\Delta P}{G\xi + \alpha \bullet Cv^2 \bullet f^2(x)}}$$

$$T := T + \Delta t \frac{(F_c + F_h)}{V} \left[\frac{F_c T_c + F_h T_h}{(F_c + F_h)} - T \right]$$

Example 11.2

Repeat Example 11.1, but do not presume the pseudo-steady state.

The signal from the controller to the valve is *u* in percent. The valve stem response to *u* is

$$\tau \frac{dx}{dt} + x = 0.01u$$

Then, the forward finite difference approximation to the valve model is

$$x_{next\ \Delta t} = x + \Delta t(0.01u - x)/\tau$$

The yP2N function code would be

$$x_{new} := x + \Delta t(0.01u - x)/\tau$$

$$F_h := Cv \bullet f(x)\sqrt{\frac{\Delta P}{G\xi + \alpha \bullet Cv^2 \bullet f^2(x)}}$$

$$T_{new} = T + \Delta t \frac{(F_c + F_h)}{V}\left[\frac{F_c T_c + F_h T_h}{(F_c + F_h)} - T\right]$$

$$x := x_{new}$$

$$T := T_{new}$$

Note: In the explicit numerical procedure, the values used in the right-hand side of the equation represent the past values. The first line in the code updates the stem position for the future time, but the past stem position (not the new one) is required to determine the flow rate. Therefore, the old and new values must be kept separate. This is inconsequential in a two-dynamic-state model and even less so for small Δt, but in a model of many internal dynamic states, keeping past and future values separate is recommended.

11.3 Correct

The second function in MBC is a correction for process-model mismatch. Process-model mismatch (pmm) is the difference between the yP2N modeled prediction and the process measurement, as represented by the circle difference operation on the lower right side of Figure 11-1. The modeled value will not exactly match the process value for a variety of reasons, including model error or unmeasured disturbances.

If the model does not match the process, using the process set point and the inverse of the model to determine the MV value will make the model match the set point, but then the process will not match the set point. Consider these SS relations for the model and process.

$$y_m = f_m(u) \tag{11-9}$$

$$y_p = f_p(u) \tag{11-10}$$

If u is determined from the inverse of the model to make the model match the set point

$$u = f_m^{-1}(y_{SP}) \tag{11-11}$$

Then, the SS process value will be

$$y_p = f_p(f_m^{-1}(y_{SP})) \tag{11-12}$$

And $y_p = y_{SP}$ only if the model and process functions are identical at an SS.

11.3.1 Bias the Model Set Point with the pmm

The simplest approach is to bias the set point for the model by the pmm.

$$pmm = y_p - y_m \tag{11-13}$$

The biased set point for the model is

$$y'_{SP} = y_{SP} - pmm = y_{SP} - y_p + y_m \tag{11-14}$$

Then, when $y_m = y'_{SP}$,

$$y_p = y_m + pmm = y'_{SP} + pmm = y_{SP} - pmm + pmm = y_{SP} \tag{11-15}$$

The model may also predict an auxiliary variable (AuxV) value, z. There will also be a pmm for that variable. In this case, bias the constraint for the modeled AuxV by the pmm of the AuxV.

$$pmm_z = z_p - z \tag{11-16}$$

$$z'_{limit} = z_{limit} - pmm_z \tag{11-17}$$

The pmm bias for the model target is a simple method to correct for the model mismatch in the process. However, there are diverse alternate methods.

11.3.2 Bias Set Point or Constraint Limit by the Filtered Value pmm

A first-order filter is commonly used to temper noise. Using z to represent any signal and z_f as the filtered value

$$\tau \frac{dz_f}{dt} + z_f = z \tag{11-18}$$

The *pmm* could be filtered, which will temper both noise and any dynamic mismatch between the model and the process.

$$z = pmm, \;\; z_f = pmm_f \tag{11-19}$$

Using a forward finite difference for the derivative, the numerical method calculation of the filtered pmm is

$$pmm_{f,i} = \lambda\, pmm_i + (1-\lambda) pmm_{f,(i-1)} \tag{11-20}$$

where $\lambda = 1 - e^{-\Delta t/\tau}$

This equation has many alternate names, such as *exponentially weighted moving average* (EWMA) and *autoregressive moving average of orders 1 and 1* (ARMA(1,1)).

Then, the biased value for a set point or constraint value would be determined by

$$y'_{SP} = y_{SP} - pmm_{y,f} \tag{11-21}$$

$$z'_{Constraint} = z_{Constraint} - pmm_{z,f} \tag{11-22}$$

There are many variations on filtering. The first-order approach is shown here. It could be a moving average or any of many statistically based approaches; my preference is a filter based on a statistical process control cumulative sum (see Chapter 17).

Primarily, a filter is used to temper noise. If noise on the process variable (PV) were the only concern, you could filter the PV and calculate the pmm from the deviation between the yP2N model and the filtered CV, $pmm = y_{p,f} - y_m$. However, this puts a lag on the measurement and will cause a dynamic mismatch if the model does not include that additional lag. During transients, the additional mismatch will make the pmm larger. If the model represents the process, then filter the model and calculate pmm from the two lagged values, $pmm = y_{p,f} - y_{m,f}$. Because this requires two operations, which must have the same time constant or filter factor, it would be simpler to filter the pmm and let the y_p and y_m values represent what is happening, not the lagged version.

As a further benefit, filtering the pmm will temper the dynamic mismatch between the model and the process. If, for instance, the process is high-order, it starts responding along an S-shaped curve. If the model is first-order plus dead time (FOPDT), it does not respond until after the dead time (see Figure 7-1). Then, when the model responds, it moves faster than the process. Eventually, the model and the process should settle at about the same time, but in the early response to input changes, there could be a significant mismatch in one direction and then the other. This could misdirect the pmm adjustment for the biased model set point, causing oscillations in the MV that create a dynamic mismatch. Filtering the pmm tempers this effect.

If tuning (adjusting the Act objective) does not seem to make the controller less aggressive, it may be because of a dynamic mismatch of the model to the process. First, consider revising the model delay and/or time constants. If that is not practicable, then consider filtering the pmm. However, filtering the pmm causes a lag in correcting the biased set point for the model, which will increase settling time.

11.3.3 Adjust Model Prediction by the pmm

This approach is characteristic of model predictive control using linear models. Rather than biasing the set point for the model, the model is translated (shifted) by the residual (the pmm).

$$\underline{Y_m N2F'} = \underline{Y_m N2F} + \underline{1}^* pmm_f \tag{11-23}$$

$$\underline{Z_m N2F'} = \underline{Z_m N2F} + \underline{1}^* pmm_{z,f} \tag{11-24}$$

This is impractical if the model is nonlinear, especially if it has multiple states.

11.3.4 Forecast the pmm

The prior two methods presume that the current pmm value will last far into the future and that a step-and-hold impact caused the pmm. However, the deviation may be progressively increasing or decreasing in time. If you think you can predict the pmm trend from past data and extrapolate it into the future, then do so. If the past disturbance is increasing progressively, then so will the pmm. If the linear model and nonlinear process begin to differ, that will likely continue as the operating region changes.

However, in my experience, getting a reasonable trend in the residual from past data is speculative, and using the pmm value now to forecast that it stays the same into the future is fully adequate and much simpler.

11.3.5 Incrementally Accumulate pmm from the Actuating Error

Integrate the actuating error (this is the generic model control [GMC] approach, and useful if there is not a yP2N model).

$$e = y_{SP} - y_{process} \tag{11-25}$$

$$pmm = \frac{\propto}{\Delta t} \int e dt \tag{11-26}$$

Then, bias either y_{SP} or y_{model} with the pmm. The problem is integral windup on a constraint or override.

11.3.6 Filter the pmm

If the process measurement is noisy, the pmm will reflect that noise, and so will the biased set point for the model. This will also make the controller action noisy. So, filter the pmm. I recommend a first-order filter because it is simple, effective, and well-understood.

Filtering adds a lag. Filtering the PV to temper the noise but not the controller model-predicted value adds a lag to the apparent process but not the model. It increases the mismatch during transients. So, filter the pmm, not just the PV.

A dynamic mismatch between the model and the process may differ from the SS mismatch. During a transient, such as a set-point change, the pmm will go from its SS value through the dynamic mismatch patterns and back to the SS value. During a transient, the set point for the model will be incorrect, degrading controller performance. This might make it appear that there is excessive derivative action. To temper this, you can filter the pmm to use the lag to temper the dynamic mismatch.

11.3.7 Incrementally Change a Model Parameter at Each Sampling Based on the pmm

This is not so much for control as it is for process monitoring, complementary use of the model for forecasting constraints, process analysis, and optimization. An incremental model adjustment online will improve the control action. However, as long as the model is reasonable, and there are about 30 control actions within a settling time, perfectly matching the model to the process is not necessary for good control (see Chapter 9).

11.4 Act – An Elementary Choice – Simple MBC

The third MBC function is calculating the control action to determine the controller MV value. This is represented as the box labeled "Act" in Figure 11-1. This function starts with a user-defined, desired performance objective for the controller to achieve. A simple desire is to calculate a u value that would push the model toward the biased set point at a rate proportional to the deviation from the biased set point. (This strategy is effective for first-order models, or nearly so, with inconsequential delay).

Mathematically, this simple desire is

$$\left.\frac{dy_m}{dt}\right|_{desired} = \frac{y_{SPbias} - y_m}{\tau_{want}} = K_c(y_{SPbias} - y_m) \qquad (11\text{-}27)$$

The single tuning factor, τ_{want}, is the desired first-order trajectory time constant for the model to return to the biased set point. Alternately, K_c, the reciprocal of τ_{want}, can be used so a larger value of K_c leads to a more aggressive controller action. For some, this is more convenient, but it masks the understanding that the controller objective is based on the time constant of a desired trajectory.

Notes:

1. The biased set point, y_{SPbias}, is for the model, and y_m is the modeled value.
2. The numerator in the middle of Equation 11-27 is the traditional actuating error.

$$y_{SPbias} - y_m = y_{SP} - pmm - y_m = y_{SP} - y_p + y_m - y_m = y_{SP} - y_p = e \qquad (11\text{-}28)$$

The right-hand side of Equation 11-27 is $K_c e$, which is the normal model for proportional action. However, rather than stating either in a conventional equivalent and masking the concept of the desired model trajectory, I will preserve the Equation 11-27 structure.

Note: $\frac{y_{SPbias} - y_m}{\tau_{want}}$ is the user-desired rate of change of the modeled output, y_m. If there were no pmm, this would also be the consequential rate of change in the process, y_p. However, the model does not match the process behavior exactly.

Insert the desired rate of change to the model and solve the model inverse to determine the controller output, u.

Usually, the model is used to calculate the rate of change of the modeled response, dy_m/dt, given process influences and state. However, in control action, the solution must determine the value of the MV, u, that would cause the desired rate of change. This is the inverse of the normal modeling solution procedure, often termed *solving the inverse.*

If you are lucky, the inverse will be mathematically tractable. You will be able to explicitly solve for u. However, it might not be possible to rearrange the equation to do so, and a numerical method might be needed. Root-finding might be a first thought, and it works. However, optimization also works and has the advantage of being able to cope with constraints, which are topics for subsequent chapters. This book focuses on optimization as the approach to solving nonlinear implicit equations.

Whether the model function is expressed as one equation or as a computer subroutine or function, the mathematical model represents the rate of change.

$$\frac{dy}{dt} = f(y,u,d,p) \tag{11-29}$$

Matching the modeled rate of change of Equation 11-29 to the desired rate of change of Equation 11-27

$$\frac{y_{SPbias} - y_m}{\tau_{want}} = f(y,u,d,p) \tag{11-30}$$

If you cannot explicitly solve for u, then rearrange Equation 11-30 for root-finding, then optimization.

$$g(y_{SPbias}, y, \tau_{want}, u, d, p) = \frac{y_{SPbias} - y_m}{\tau_{want}} - f(y,u,d,p) = 0 \tag{11-31}$$

Then, square it.

$$g^2(y_{SPbias}, y, \tau_{want}, u, d, p) = 0 \tag{11-32}$$

The value of u that makes $g^2(\ldots) = 0$ is the same value that makes $g(\ldots) = 0$. It is also the same value that makes the model rate of change equal to the desired rate. The optimization statement is

$$\begin{gathered} \underset{\{u\}}{min}\; J = g^2(y_{SPbias}, y, \tau_{want}, u, d, p) \\ \text{S.T. } 0 \le u \le 100 \end{gathered} \tag{11-33}$$

There may be additional constraints.

There are many effective optimization procedures. Two will be presented in Chapter 20. You may have convenient access to many others.

Example 11-3

Show the Correct and Act functions for Example 11-1.

The Correct function is to bias the process set point by the pmm to become the set point for the model.

$$pmm = T_p - T_{mP2N}$$

$$T_{mSP} = T'_{SP} = T_{pSP} - pmm$$

where T_p is the process measurement, T_{mP2N} is the model prediction of what the process would be, T_{pSP} is the true set point for the process, and $T_{mSP} = T'_{SP}$ is the set point for the model.

The Act objective is to move the model toward T'_{SP} at a rate defined by the deviation.

$$\left.\frac{dT_{N2F}}{dt}\right|_{desired} = \frac{T'_{SP} - T_{mP2N}}{\tau_w}$$

Note: $T'_{SP} - T_{mP2N} = e$, the traditional actuating error in PID control.

Inserting the desired rate into the now-to-future (N2F) model in Example 11-1 and initializing T with T_{mP2N},

$$\frac{V}{(F_c + F_h)}\left(\frac{T'_{SP} - T_{mP2N}}{\tau_w}\right) + T_{mP2N} = \frac{F_c T_c + F_h T_h}{(F_c + F_h)}$$

Solving for the desired hot water flow rate needed to effect the desired N2F model rate of change,

$$F_h = \frac{V\left(\frac{T'_{SP} - T_{mP2N}}{\tau_w}\right) - F_c(T_c - T_{mP2N})}{T_h - T_{mP2N}}$$

Then, inverting the valve equation to determine the associated $f(x)$ value,

$$f(x) = \sqrt{\frac{F_h^2 G\xi}{Cv^2(\Delta P - \alpha F_h^2)}}$$

If the valve's inherent characteristic is linear, then $f(x) = x$. If the characteristic is ideally equal percent, then $f(x) = R^{x-1}$. Solve for the desired valve position, x, from $f(x)$. Then, from the desired valve position, determine the controller output.

$$u = 100x$$

11.4.1 A Variation of Controller Objectives

The controller Act function could have a variety of objectives that specify how to determine the MV value. Considering simplicity and effectiveness, I would choose the first option in the list below. However, the other options in the list also work.

- Option 1: Desire the model to move toward the biased set point at a rate proportional to the deviation, as illustrated in Figure 11-2.
- Option 2: Desire the model to follow a reference trajectory toward the biased set point. Figure 11-3 illustrates a first-order trajectory. However, if the model is second-order with two consequential time constants, it may not be feasible, or rational, to make it track a first-order response. So, ask the model to follow a second-order trajectory. This is the internal model control (IMC) approach.
- Option 3: Modify either option 1 or option 2 for ill-behaved dynamics. For instance, either have the model follow the set point after the delay or after the inverse has passed or, as with IMC, use the invertible model rather than the full model.
- Option 4, coincidence point: Identify the step-and-hold control action that makes the model hit the biased set point at a desired future time. Choose a time beyond ill-behaved dynamics—perhaps 85% of the open-loop settling time. This is predictive functional control, illustrated in Figure 11-4, which is the topic of Chapter 13.

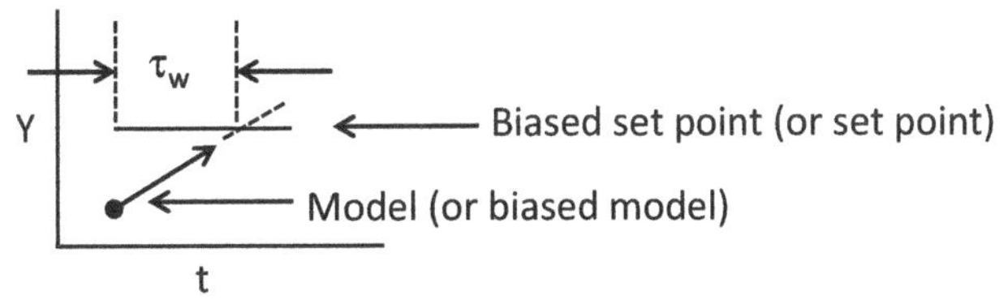

Figure 11-2. Move the model toward the biased set point at a rate proportional to the deviation.

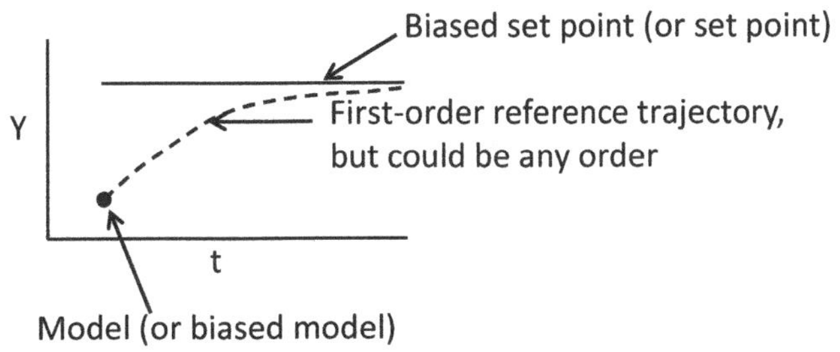

Figure 11-3. Move the model toward a biased set point along a reference trajectory.

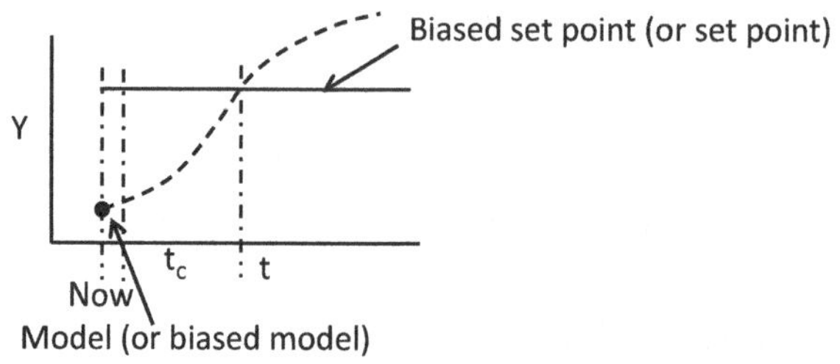

Figure 11-4. Move the model to the coincidence point.

- Option 5: Identify the sequence of steps in U that best shapes the future model response to follow the reference trajectory. See Figure 11-5. This is horizon predictive control (HPC), or model predictive control (MPC), with CV damping, which is the topic of Chapter 14. Determining a reference trajectory

$$\min_{\{\underline{U}\}} J = \sum_{now}^{H} (r_i - Y_i)^2 \tag{11-34}$$

- Option 6: Identify the sequence of U steps to minimize the sum of squared deviations from the set point and temper control by a penalty for a large ΔU. See Figure 11-6. This is MPC or HPC with MV damping. Desire that the CV be at the set point but penalize large MV moves.

$$\min_{\{\underline{U}\}} J = \sum_{now}^{H} (SP_i - Y_i)^2 + f \sum_{now}^{H} (\Delta U_i)^2 \tag{11-35}$$

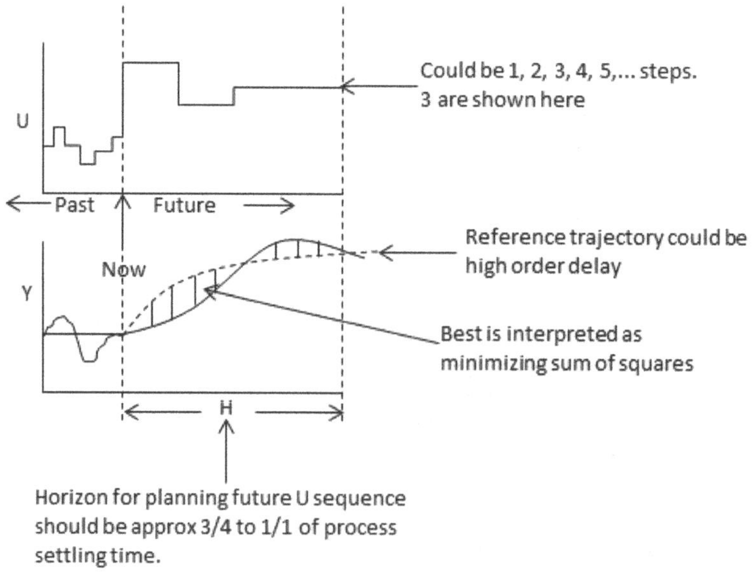

Figure 11-5. Move the model toward the biased set point somewhat along a reference trajectory.

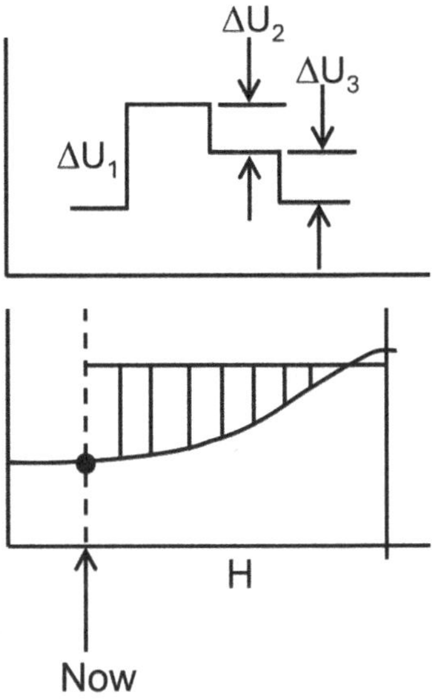

Figure 11-6. Move the model to the biased set point with a penalty for large MV moves.

- Option 7: If an SS model is used for control, target for a position beyond the set point. This is an element in the GMC (with an SS model) approach, illustrated in Figure 11-7.

$$Y_{target} = Y_m + K(Y'_{SP} - Y_m) \tag{11-36}$$

A value of 1.2 for K means an attempt to move 20% farther. With the SS model represented as

$$Y_{SS} = f_{SS}(U) \tag{11-37}$$

the control action is the SS model inverse to reach Y_{target}.

$$U = f^{-1}(Y_{target}) \tag{11-38}$$

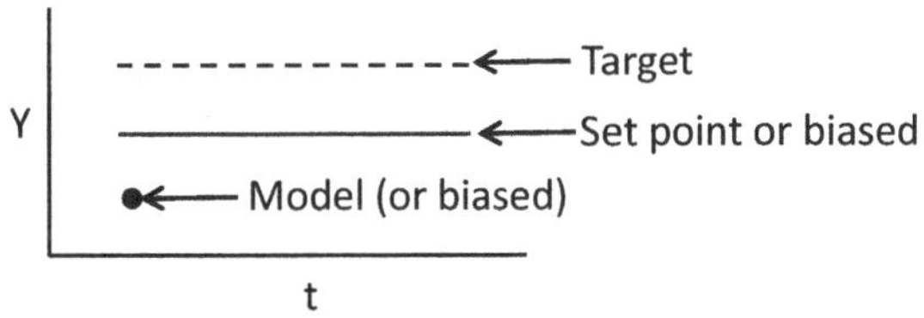

Figure 11-7. Move the SS model toward a value beyond the biased set point.

This does not account for the dynamics. However, for most chemical processes, the dynamic response is simple (high-order or low-order plus dead time, stable), but the problem is the nonlinear Y_{SS} response to U_{SS}.

For most chemical processes, the SS models are well-known, conventional, central to process design and optimization, and available.

As control approaches the set point, the target for the model becomes the biased set point, as illustrated in Figure 11-8.

- Option 8: Go to a constraint. The maximum throughput is often on some operational constraint, as illustrated in Figure 11-9. Frequently, production schedules require maximizing throughput.

  Constraints can be modeled as linear. "Can be" does not mean that they are, but often linear modeling is adequate and permissible, leading to the computational convenience of linear programming (LP), which can be used to efficiently identify a boundary (a conjunction of linear constraints) that minimizes a linear objective function.

These techniques have advantages and disadvantages. A common advantage of options 1 through 6 is that there is a single tuning parameter per CV (or MV).

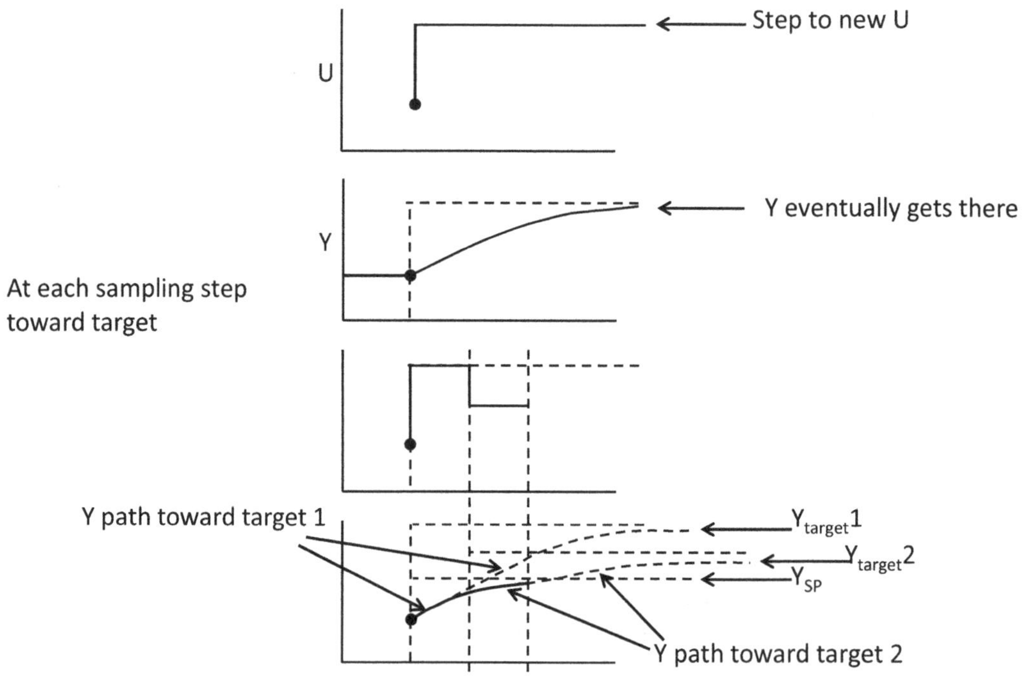

Figure 11-8. Illustrating the continually changing target as control approaches the set point.

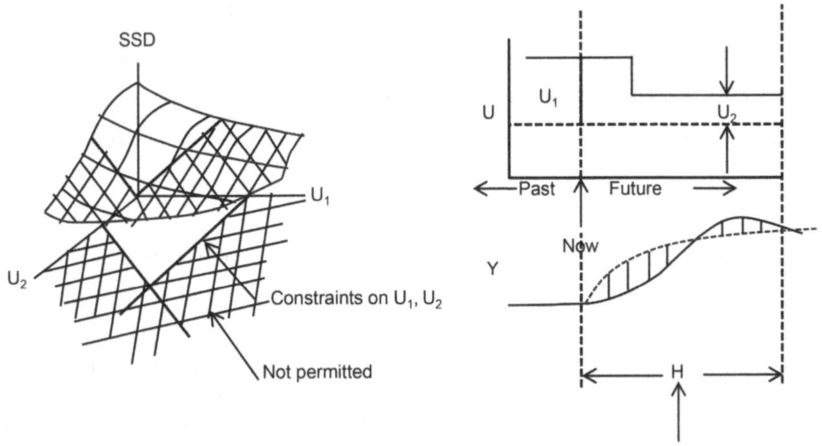

Figure 11-9. Control to a constraint.

A common advantage of options 4, 5, and 6 is that regardless of ill-behaved dynamics, they work.

A common advantage of 1, 2, 3, and 7 is that they are simple to implement yet fully adequate for unconstrained MISO loops. However, if no explicit inverse exists, they require an iterative search for the MV value.

I prefer option 5 over option 6 (CV damping over MV damping). Although dynamic matrix control (DMC, a premier early MPC) used MV damping, and the academic presentation of MPC uses MV damping in multivariable control, I think it is difficult to relate the move suppression factor on each MV to a collective impact on the CV. I prefer CV damping, where you define the desired response rate for each CV.

11.4.2 Infeasibilities

The controller may encounter a square root of a negative number or a divide by zero or other operations that could lead to an execution error. Usually, the override is obvious, but it must be included in the executable code.

11.5 Comments

This set of procedures for the Predict, Correct, and Act functions represent the controller strategy termed *simple MBC.*

Returning to Figure 11-1, each of the three functions (Predict, Correct, Act) could be specified with various options or executed with various techniques. The multitude of choices has led to a variety of controller algorithms and associated acronyms.

If the model is linear and first-order, the three-stage method described here is equivalent to an internal reset feedback version of a standard proportional-integral (PI) controller with the integral time equal to the process time constant. It is comforting that idealizations lead to a familiar solution, confirming the correctness of the control steps.

However, if the model is linear and first-order, use the conventional PI control technique, which will be comfortingly familiar to many of your stakeholders.

Note: For single-input, single-output (SISO) applications, the controller output, *u*, could be the signal to the final control element (e.g., a valve, variable speed drive, or electrical heater). This is often termed *direct digital control*. However, in more complicated applications, it is desirable to have the model-based controller determine set points (e.g., flow rates and temperatures) for secondary loops to manipulate the final control element. The secondary loops are often PI, but if they have nonlinear and ill-behaved dynamics, the secondary loops could also be SISO model-based.

11.6 Multiple Model Use – A Caution

The model for the process is used in two places in the controller: the Predict function and the Act function.

The Predict function updates the past modeled value with the actual implemented MV to predict what the CV is doing now. This is a one-time-step update using one-step past values. I use the name *past-to-now* (P2N) to refer to the model. Initialize the one-time-step update with the prior P2N modeled value.

The Act function uses the controller model to forecast what might happen in the future, associated with a trial solution for the MV value. Guess at an MV and see what the modeled CV does. Re-guess at an MV value until the future modeled CV matches the desired rate of change. I call this the *now-to-future* (N2F) model. Initialize the N2F prediction with the current P2N modeled value. After each MV guess, the N2F model must be re-initialized with the current P2N value.

In control simulation and testing, a process model is also used as a surrogate for the process response. This third model should be different from the controller models. We can never describe Nature exactly; it seems more complicated and less ideal than our concepts. Process responses are also affected by internal noise (turbulence and incomplete mixing) and unmeasurable and continually changing disturbances (environmental effects). It would be cheating to demonstrate a controller using the same model in the process simulator as in the controller. Ensure the surrogate process model is a few steps less ideal than the controller model.

Note: The "process model" in the controller is not the true model of the process. The controller model cannot be the true process or the model used in the simulator. Although the controller model is intended to be a model of the process, and although it is classically called a "process model," it is not.

11.7 Model Sufficiency/Perfection

The process model in the controller must adequately represent the process, but it does not need to be an exact or perfect representation. We use FOPDT models effectively in control. These idealizations mimic a high-order response with a reasonable fit. Empirical determination of the gain, dead time, and time constant for the FOPDT "process model" are all somewhat in error, subject to the experimental vagaries. Finite impulse response models seem more perfect, but these too are linear and fit to experimental data. However, they are all adequate for control.

Alternatively, one could use rigorous models attempting to explain every process nuance perfectly. However, balancing sufficiency with perfection, first-principles models are fully adequate. Contrasting modeling perfection, consider that humans can catch a fly ball or steer a car through a curve without rigorous models, using just intuitive continual correction. Likewise, PI controllers based on linear, stationary, simplistic (FOPDT), inexact models are also adequate for control.

11.8 MAN Mode Calculations

Controllers operate in the manual (MAN) or the automatic (AUTO) mode. The AUTO mode operations were described earlier in this chapter. In the MAN mode, the controller must initialize variables needed in the AUTO mode. These include the yP2N value. In the MAN mode, run the yP2N model at each time step.

11.9 Internal or External Override

If there is an override of the calculated *u* value, either internally by limiting *u* between extremes of 0% or 100% or externally by another select/limit/safety/override/operator action, use the actual implemented *u* value for the yP2N model, not the prior desired value from the controller.

11.10 Illustration of Simple NMBC

Figure 11-10 shows the car speed response (Section 2.2.9) to MV changes with the controller in MAN mode. The steps (gray line) are the MV values (0% to 100%) in time. Of the two first-order-ish traces, the car speed (thick black line) (km/h) is generally the highest trace, and the modeled value (thin line) is generally the lower trace. The CV

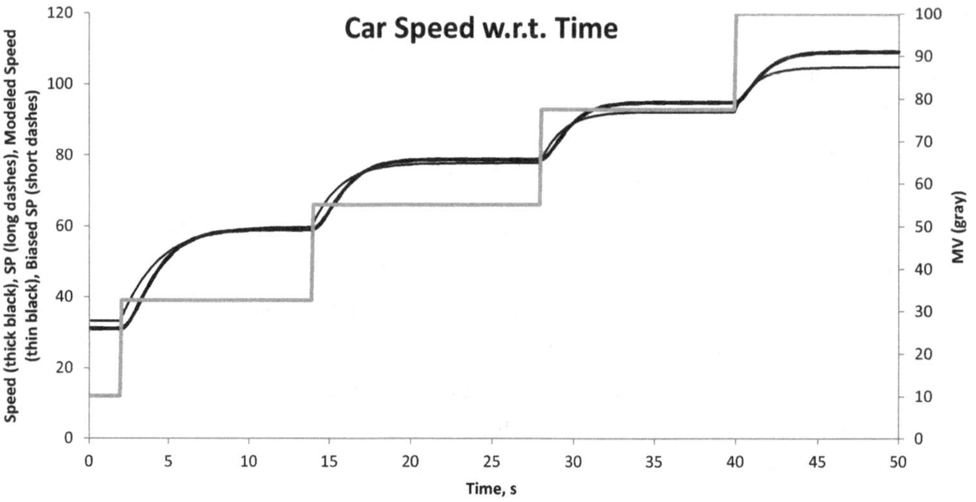

Figure 11-10. Model and process response to MV steps.

has about a 4:1 gain ratio from low to high speed values, and the pmm switches sign from the low speed to the higher speed values. At low speeds, the CV settling time is about 10 s, and that of the model is about 15 s. At high speeds, the CV settling time is about 5 s, whereas the model takes about 7 s. The process and model are both nonlinear and nonstationary, and they are not the same. Because the controller is in MAN mode with a set point following feature, the SP and biased SP values are the same as the CV and modeled CV.

Figure 11-11 shows the transfer from MAN to AUTO and set-point changes. Initially, the controller is in MAN mode with the MV (solid gray line) at 10%. At a time of 5 s, the MV steps to 25% and the car speed and model prediction begin to rise. At 10 s, the controller is switched to AUTO mode, and because of the previous set point following, the set point (line with long dashes) begins a horizontal dashed trace at the CV value just before the mode changes at about 49 km/h. The MV remains near the 25% value, but because the process speed was still rising, the controller MV begins to drop to keep the speed at the set point. This is bumpless transfer.

Set point changes are made at times of 20 s and 35 s. The line with long dashes shows the set point. In each case, the MV (gray line) hits a limit of either 100% or 0% but comes off the limit without windup before the CV overshoots the set point. The biased set point for the model is the line with short dashes, and the modeled value is the thin black curve that chases it.

Figure 11-12 shows the simple MBC response to a constraint. Until the time of 15 s, the events are identical to those in Figure 11-11. However, at 15 s, the speed set point was

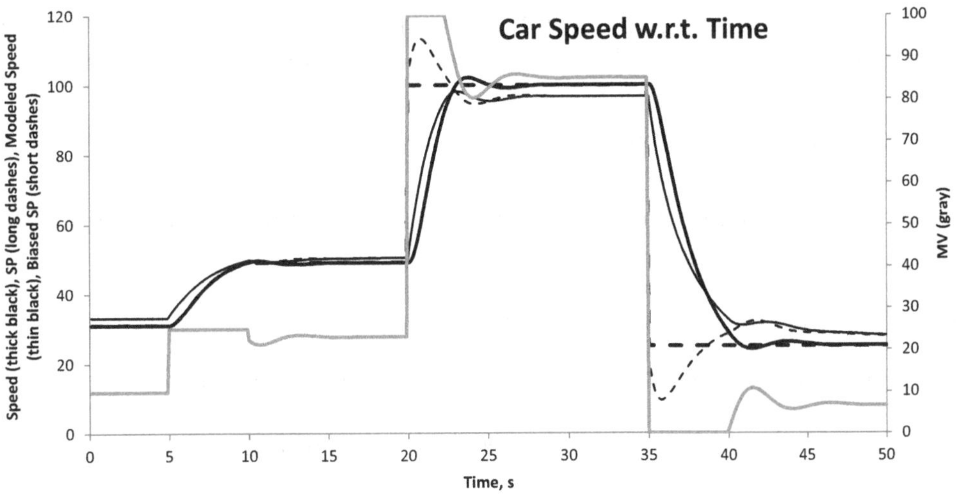

Figure 11-11. MAN-to-AUTO and set-point changes.

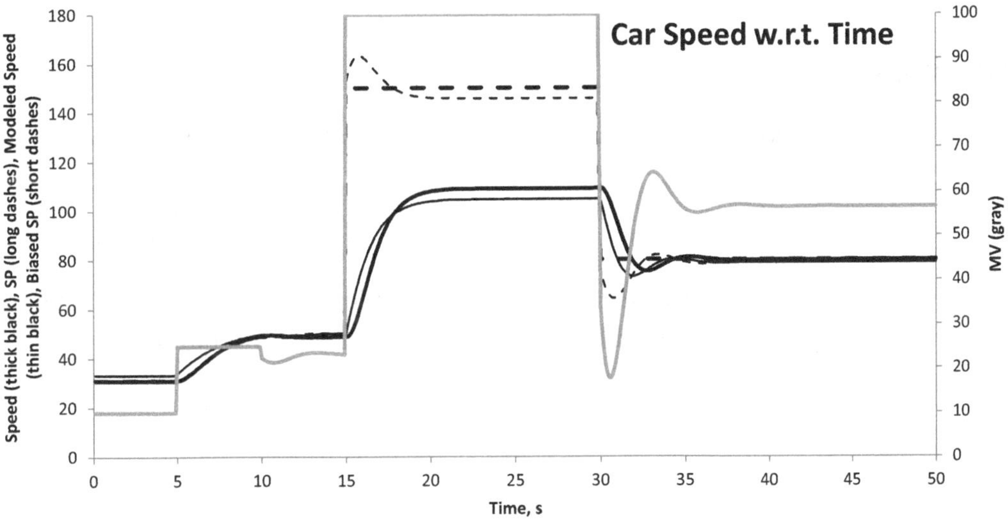

Figure 11-12. Constraint handling.

set at the unattainable 150 km/h value. The MV (gray line) went to 100% and held that value, but the car speed rose to only about 110 km/h. The modeled car speed during this period was about 105 km/h, again revealing the process-model mismatch. At 30 s, the set point (line with long dashes) was returned to a feasible value, and the controller responded immediately.

This example is accessible on my website (*www.r3eda.com*) as an Excel VBA file. You can choose options, tests, process and controller models, noise and disturbance attributes, and controller tuning to explore. The software provides the integral of the squared error (ISE) and travel metrics. The code is open so you can see all the details.

The example shows the following:

- Control is good and similar throughout the entire nonlinear operating range in spite of several sources of pmm.
- Only one tuning coefficient is needed.
- Initialization in the MAN mode permits bumpless transfer.
- Recovery is immediate when the constraint is removed.

11.11 Takeaway

- You can choose from among many options for each controller function. Keep it as simple as the application permits.

Exercises

11.1 Develop the Predict, Correct, and Act equations for any of the six models in Section 2.2.

11.2 Develop the Predict, Correct, and Act equations for any model of your choice.

11.3 Access my simulator on *www.r3eda.com* and explore tuning, servo performance, regulatory performance, bumpless transfer, and constraint handling.

12 Simple Process-Model-Based Control–Unconstrained MISO

12.1 Introduction

Like Chapter 11, this chapter discusses the case of a single controlled variable (CV) and a single manipulated variable (MV), but with measured disturbances also affecting the CV—the multiple-input, single-output (MISO) case. The model may have several internal states, but only one is to be controlled. Multivariable and constraint avoidance topics are presented in Chapters 23 and 24.

The key difference between the simple model-based control (MBC) in Chapter 11 and the process-model-based control (PMBC) in this chapter is that coefficient value(s) in the PMBC model are adjusted incrementally to keep the model truer to the process. As long as the model is a good, or even just a fair, representation of the process, the model set-point bias by the process-model mismatch (pmm) is fully adequate for control. Although model adjustment has a bit of improvement in automatic control, the main benefits are associated with active, real-time process monitoring, for example, in forecasting constraints, predictive maintenance, process analysis, and such supervisory aspects.

R. R. Rhinehart and J. B. Riggs published an article on online incremental model adjustment titled "Two Simple Methods for On-Line Incremental Model Parameterization" in 1991.[1] This chapter follows the method presented in Rhinehart's book *Nonlinear Regression Modeling for Engineering Applications: Modeling, Model Validation, and Enabling Design of Experiments*,[2] which has been demonstrated in many studies, including "A Comparison Between Two Nonlinear Process Model Based Controllers"[3] by Riggs and Rhinehart and "Cascaded Process Model Based Control: Packed Absorption Column Application,"[4] by A. Govindarajan and colleagues.

Again, there are three functions in MBC: Predict, Correct, and Act. The structure can be represented by the operations in Figure 12-1. This is the familiar figure, but now dashed lines are added from the Correct function to both the Predict and Act functions. The new lines represent model coefficient values that are adjusted in the Correct function. Exiting the left of the Correct function are still the pmm values used to bias the set point for the model.

This chapter focuses on the Correct function in the control structure.

12.2 Correct

The second function in MBC is a correction for pmm, which is the difference between the modeled prediction and the process measurement, as represented by the circle difference operation on the lower right side of Figure 12-1. The modeled value will

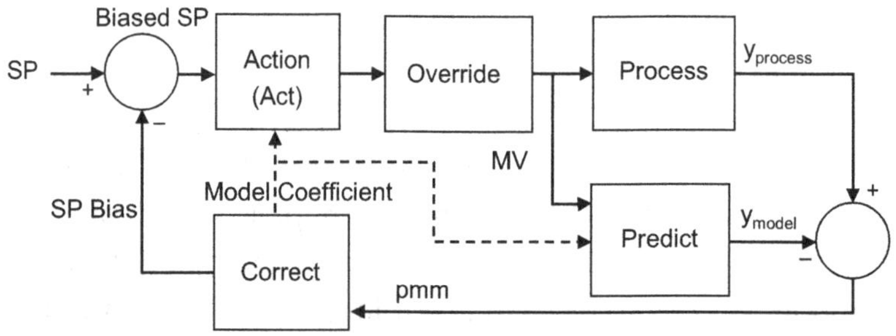

Figure 12-1. MBC representation.

1. R. R. Rhinehart and J. B. Riggs, "Two Simple Methods for On-Line Incremental Model Parameterization," *Computers & Chemical Engineering* 15, no. 3 (1991): 181–189.

2. R. R. Rhinehart, *Nonlinear Regression Modeling for Engineering Applications: Modeling, Model Validation, and Enabling Design of Experiments* (New York: John Wiley & Sons, 2016).

3. J. B. Riggs and R. R. Rhinehart, "A Comparison Between Two Nonlinear Process Model Based Controllers," *Computers & Chemical Engineering* 14, no. 10 (October 1990): 1075–1081.

4. A. Govindarajan, S. K. Jayaraman, V. Sethuraman, P. R. Raul, and R. R. Rhinehart, "Cascaded Process Model Based Control: Packed Absorption Column Application," *ISA Transactions* 53, no. 2 (2014): 391–401.

not exactly match the process value for a variety of reasons, including model error or unmeasured disturbances.

The main approach for control, for both MBC and PMBC, as developed in Chapter 11, remains to bias the set point for the model with the pmm. This provides a simple and immediate correction to remove offset. Filtering the pmm to temper the effects of noise and dynamic mismatch may still be desirable.

The unique feature of PMBC is that it incrementally changes a model parameter at each sampling based on pmm. This seeks to match the modeled CV to the process.

A good way to adjust model coefficient values is to fit the model to historical data. Usually, this would be a least-squares regression, requiring nonlinear optimization. If historical data includes transients, then as well as adjusting coefficients that predict steady-state (SS) values, the adjustment can include coefficients that relate to the dynamic (transient) behavior. This is also termed a *batch model adjustment*. Collect historical data and fit the model to the batch of data. Do not overlook this approach in either MBC or PMBC when initializing models or for periodic model calibration.

In contrast, the PMBC method incrementally adjusts model coefficient values in real time as data is gathered. This makes the model true to the process (or more so than without coefficient adjustment) as the model is being used.

This is not so much for control as it is for process monitoring, complementary use of the model for forecasting constraints, process analysis, and optimization. Although incremental model adjustment online will improve the control action, do not consider this to be the primary motivation. As long as the model is reasonable, and there are about 30 control actions within a settling time, a perfect match of model to process is not necessary for good control, as illustrated in prior chapters.

The adjustment differs in concept and purpose from that used in adaptive control. In adaptive control, the model is generic and typically linear, and model coefficient values are adapted to make the model reasonably match the process. In PMBC, the model is a first-principles process model developed to match the process, but it is imperfect. Here, the adjustment in a model coefficient value is intended to make the first-principles model have greater fidelity to the process and to be useful for process monitoring.

12.2.1 Choosing the Adjustable Coefficient in Phenomenological Models

The choice of which model coefficient to adjust should be compatible with the following five principles:

1. The model coefficient must represent a process feature of uncertain value. (If the value could be known, measured, or otherwise calculated, incremental adjustment would not be needed.)
2. The process feature must change over time. (If it did not change, adaptation would be unnecessary.)
3. The model coefficient value must significantly impact the modeled input-output relation. (If it has an inconsequential impact, there is no justification to adjust it.)
4. The model coefficient should have an SS impact if the data used for model adjustment come from SS periods. (If the coefficient is not related to SS, such as when the coefficient is a time constant, then when the process is at or near an SS, the model coefficient value will be irrationally adjusted in response to process noise. This is important because continuous processes mostly operate at steady conditions.)
5. Each coefficient to be adjusted must have an independent response variable.

If one or more of the five principles are not true, there is no sense in adapting that model coefficient online.

12.2.2 A Simple Incremental Approach

There are many ways to update model coefficient values and adapt models in real time, which have led to a variety of adaptive control methods. A simple and effective approach to adjusting a phenomenological model is to desire that the model coefficient be adjusted so the pmm approaches zero in a first-order manner when the process is near steady conditions. The mathematical statement of that desire is

$$\tau_{pmm}\frac{dpmm_{SS}}{dt}+pmm_{SS}=0 \tag{12-1}$$

where

$$pmm=y-\tilde{y} \tag{12-2}$$

with y being the process measurement and $\tilde{y}$ being the modeled value from the past-to-now (yP2N) model.

The transient process model is represented as

$$\frac{d\tilde{y}}{dt} = f(\tilde{y}, u, d, p) \tag{12-3}$$

where $\tilde{y}$ is the modeled response, u is the controller influence on the process, d represents measurable disturbances, and p represents the coefficient that will be incrementally adjusted.

At an SS

$$0 = f(\tilde{y}_{SS}, u, d, p) \tag{12-4}$$

which can be rearranged to solve for $\tilde{y}_{SS}$.

$$\tilde{y}_{SS} = g(\tilde{y}_{SS}, u, d, p) \tag{12-5}$$

Equation 12-5 might be an explicit relation for $\tilde{y}_{SS}$, or possibly, it will be an implicit solution as indicated.

If the coefficient value p changes, then $\tilde{y}_{SS}$ will change. If the process is near an SS condition, the d and u values do not change. Then, using the calculus chain rule, the change in $\tilde{y}_{SS}$ w.r.t. time is

$$\frac{d\tilde{y}_{SS}}{dt} = \frac{\partial g}{\partial p}\frac{dp}{dt} + \frac{\partial g}{\partial \tilde{y}_{SS}}\frac{d\tilde{y}_{SS}}{dt} \tag{12-6}$$

Rearranged as

$$\frac{d\tilde{y}_{SS}}{dt} = \left(\frac{\partial g}{\partial p}\frac{dp}{dt}\right) / \left(1 - \frac{\partial g}{\partial \tilde{y}_{SS}}\right) \tag{12-7}$$

Substituting Equation 12-7 into 12-1 and rearranging

$$\left(\frac{\partial g}{\partial p}\frac{dp}{dt}\right) / \left(1 - \frac{\partial g}{\partial \tilde{y}_{SS}}\right) = \frac{pmm}{\tau_{pmm}} \tag{12-8}$$

Rearranging for incremental model coefficient adjustment using a simple numerical solution (Euler's explicit finite difference) for $\dfrac{dp}{dt}$

$$p_i = p_{i-1} + \left[\frac{\Delta t}{\tau_{pmm}}\left(1 - \frac{\partial g}{\partial \tilde{y}_{SS}}\right)\right]\frac{pmm_i}{\partial g/\partial p} \tag{12-9}$$

Equation 12-9 is recognizable as Newton's method of incremental, recursive coefficient updating (root-finding), with the bracketed term as a tempering factor. The user's choice of the value of τ_{pmm} prevents excessive changes in the model coefficient value and tempers measurement noise effects. The sampling interval, Δt, is in the same time units as the time constant and provides a normalizing factor for the frequency at which the model is updated. If the model of Equation 12-5 is explicit in $\tilde{y}_{SS}$, then the $\dfrac{\partial g}{\partial \tilde{y}_{SS}}$ term in Equation 12-9 has a value of zero.

The use of partial derivatives in Equation 12-9 should not be off-putting (although it will likely challenge an engineer's analytical ability). Use a numerical approximation. Substitute $[g(\tilde{y}_{SS}, u, d, p+\delta p) - g(\tilde{y}_{SS}, u, d, p)]/\delta p$ for $\partial g/\partial p$ where δp represents a small perturbation in the p coefficient value—small enough to remain locally linear but large enough not to cause digital truncation error.

The $g(\tilde{y}_{SS}, u, d, p)$ relation does not have to be a single equation line. It might represent several equations in a multistate process or require a root-finding procedure. If so, use a computer programming function or subroutine to determine the value of $g(\tilde{y}_{SS}, u, d, p)$ and determine sensitivities numerically.

In computer assignment statements, the subscripts in Equation 12-9 are unnecessary because past values are used to assign the new value.

The value of τ_{pmm} should be large compared to model dynamics (so model adjustment during set-point changes when there is dynamic pmm does not interact with control calculations), but the value of τ_{pmm} should be small compared to the time period over which the process attribute changes (for rapid tracking of the process).

Example 12-1: Mixing in a Continuous Flow-Through Tank

For example, consider a mixing tank model (Model 3, Section 2.2.6) in which two streams of concentrations, c_1 and c_2, enter a tank at a stationary liquid level. Perhaps F_1 and c_1 represent wild flow properties, and F_2 is the controlled flow. The net inflow concentration is

$$c_{in} = \frac{c_1 F_1 + c_2 F_2}{F_1 + F_2} \tag{Ex12-1-1}$$

Note that c_{in}, the net inflow concentration, is a nonlinear response to F_1 or F_2. Usually, the mixing dynamics in the tank define the outflow concentration as

$$\frac{V}{F}\frac{dc}{dt}+c=c_{in} \tag{Ex12-1-2}$$

Where V is the tank volume and F represents the total flow rate ($= F_1 + F_2$), the ratio V/F has the units of a time constant.

This represents a static nonlinearity of Equation Ex12-1-1 followed by linear dynamics of Equation Ex12-1-2, which is often termed a *Hammerstein model*. The two equations can be combined to represent one nonlinear dynamic model.

$$\tau\frac{dc}{dt}+c=\frac{c_1F_1+c_2F_2}{F_1+F_2} \tag{Ex12-1-3}$$

While developed here as separate equations, then combined, the same result, Equation Ex12-1-3, could be derived from a transient material component balance on the tank.

Consider that c_1, the wild flow concentration, is not measurable, it changes in time, and it is chosen as the model coefficient to be incrementally updated. Then, applying Equation 12-9, the adjustment rule becomes.

$$c1 := c1 + \Delta t * pmm * (F1+F2)/(tauwpmm * F1) \tag{Ex12-1-4}$$

The reader should recognize that if F1 is zero, the calculation cannot be executed. Should F1 be zero, there is no information that can be used to adjust c1 by the process output. Accordingly, there is no justification for attempting to use the method to update the model coefficient, c1. The code should be

```
If F1 > threshold THEN
        c1 := c1 + Δt * pmm *(F1 + F2)/(tauwpmm * F1)
END IF
```

12.2.3 An Alternate Approach

The model may be a dynamic model represented as

$$\frac{d\tilde{y}}{dt}=f(\tilde{y},u,d,p) \tag{12-10}$$

At an SS, the derivative is zero, leaving

$$f(\tilde{y},u,d,p)\big|_{SS}=0 \tag{12-11}$$

This might be a functional relation that can be solved explicitly for the SS modeled y value, leading to an Equation 12-5 representation. If so, follow the procedure in Section 12.2.2. However, if Equation 12-11 does not admit an explicit relation, it can be solved by an iterative root-finding algorithm such as Newton's method. If the guess for $\tilde{y}_{SS}$, $\tilde{y}_g$, is close to the true value, then one iteration will lead to a very close representation of the $\tilde{y}_{SS}$ value.

$$\tilde{y}_{SS} \cong \tilde{y}_g - \frac{f(\tilde{y}_g, u, d, p)}{\left.\frac{\partial f}{\partial \tilde{y}}\right|_g} \tag{12-12}$$

The "approximately equal" symbol acknowledges that Newton's method does not jump to the true value but recursively approaches it.

Again, using Equation 12-1 to express the desire that the model coefficient be adjusted so the pmm approaches zero in a first-order manner when the process is near steady conditions, and using Equations 12-12 and 12-2, the chain rule of differentiation, and the finite difference approximation to the time derivative

$$p_i = p_{i-1} + \left[\frac{\Delta t}{\tau_{pmm}} \frac{\partial f}{\partial \tilde{y}_g}\right] \frac{pmm_i}{\left.\left(\partial f/\partial p - f\left(\frac{\partial^2 f}{\partial p \partial y}\right)/(\partial f/\partial y\right)\right|_g} \tag{12-13}$$

A good value for $\tilde{y}_g$ would be either the prior $\tilde{y}_{SS}$ or the current process SS value. Here, using the equal sign accepts that Newton's and the finite difference approximations are adequately true.

12.3 Simple PMBC

The Predict and Act functions remain the same as those in the simple MBC in Chapter 11.

This set of procedures for the Predict, Correct, and Act functions represents the PMBC controller strategy.

This illustration considers the car speed control in Section 11.9. Figure 12-2 illustrates the car speed when the controller is in manual (MAN) mode with a constant output and two disturbances impacting the speed. The upper horizontal line is the MV with a constant value of 80%. With that input, the modeled speed (the lower horizontal line)

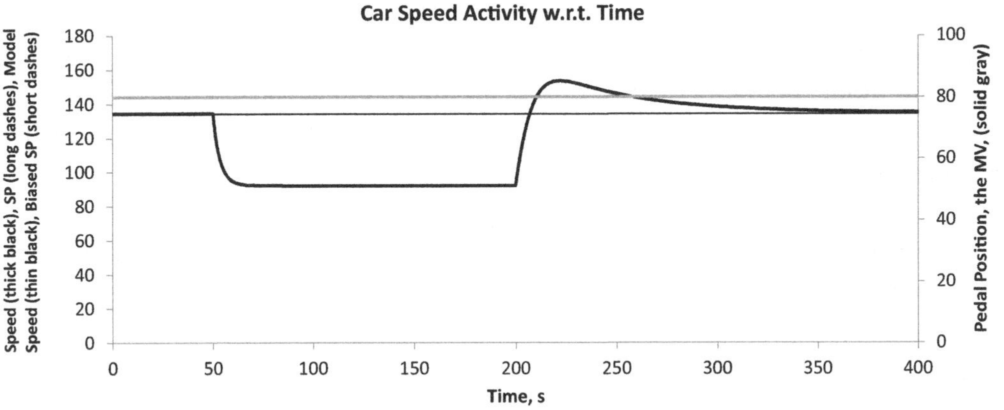

Figure 12-2. MAN with disturbance and no model adjustment.

is also constant, with a value of about 130 km/h. Disturbances happen at times of 50 s and 200 s, which represent an immediate slope up to a hold and then a slope down approaching zero. The car speed (the process) starts with the middle horizontal line, drops when the first event happens, increases as the slope returns to zero, rises with the second upset, and approaches the initial speed as the disturbance slope returns to zero. Set points and biased set-point traces are not visible because they follow the CV and modeled speed in MAN.

Using the first method to adjust the drag coefficient in the model with a desired time constant of 10 s, Figure 12-3 illustrates the same disturbance events. However, the model does not remain constant; it seeks to approach the process incrementally.

Not illustrated here, with a smaller time constant to match the model to the process, the adaptation is faster, and with a larger time constant, the adaptation is slower.

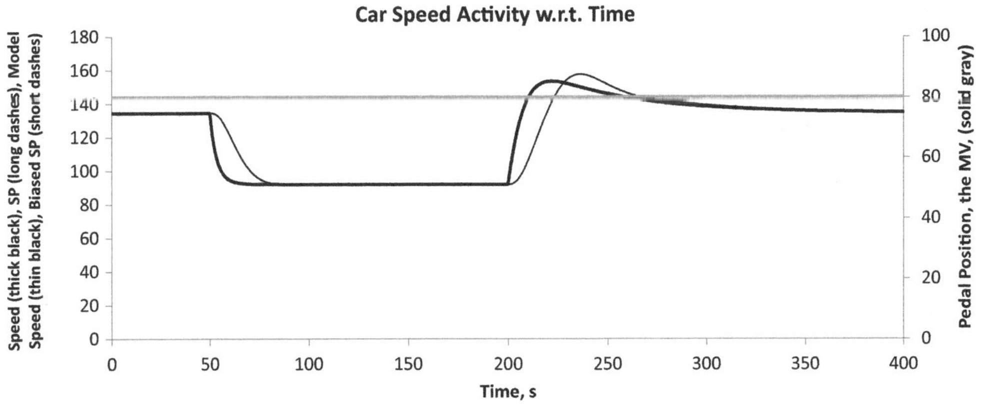

Figure 12-3. MAN with a disturbance with incremental model adjustment.

In Figure 12-4, the controller is switched to automatic (AUTO) mode at a time of 10 s and given a speed set point of 95 km/h. The upsets happen at times of 100 s and 250 s. There is no model adjustment. The upsets are fairly large, causing rapid 7 to 10 mph deviations. Note that the biased set point (dashed) leads the model, which returns the process to its set point. Control works well without model adaptation.

Figure 12-5 shows the model moving to match the process after each set point or disturbance change.

Figure 12-6 shows how the modeled drag coefficient changes to make the model match the process.

Figure 12-7 shows how the maximum speed is predicted. The maximum speed would happen if the MV was set and held at 100%. The thick line that makes sharp changes is the truth about the process. The thin line that lags is the model-predicted, constrained

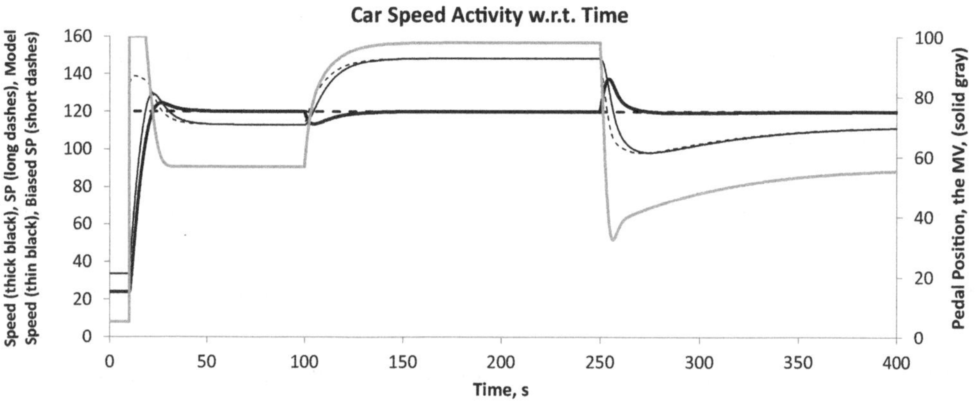

Figure 12-4. MAN-to-AUTO with disturbance and no model adjustment.

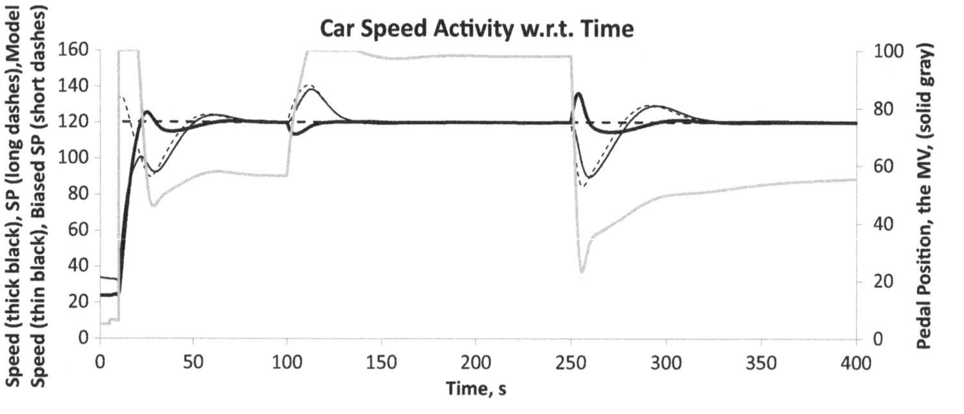

Figure 12-5. MAN-to-AUTO with disturbances and incremental model adjustments.

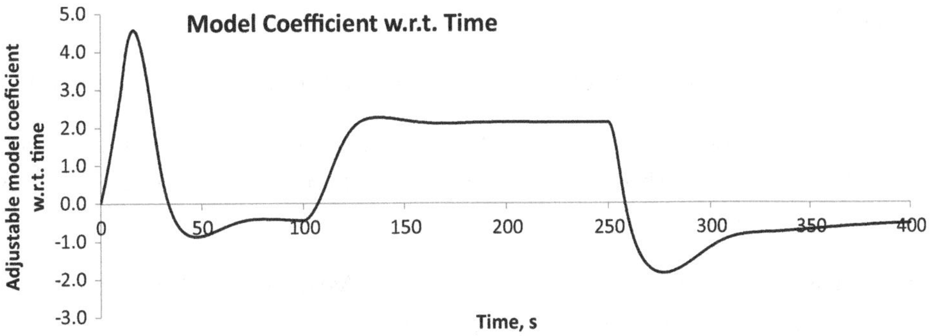

Figure 12-6. Incremental adjustment of the model drag coefficient.

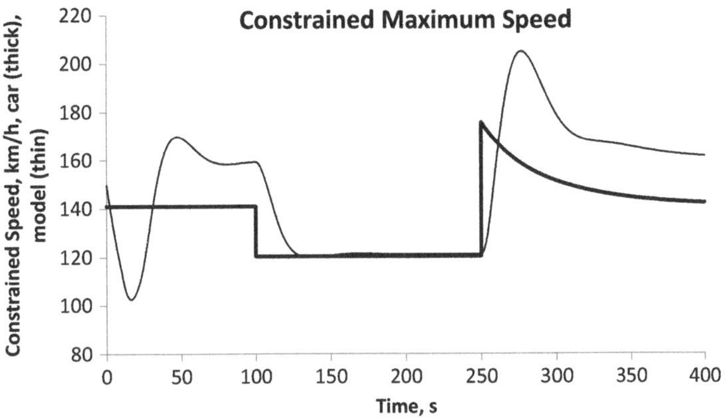

Figure 12-7. Constraint prediction from incremental model adjustment.

speed value. The car speed is held at the set point of 95 km/h. During the first time period (0 s to 150 s), the model predicts that about 130 km/h will be the maximum speed, whereas the actual speed is about 126. This deviation is due to the pmm that is not eliminated by the local matching to the speed of 95 km/h. It is roughly a 4 km/h error on a 35 km/h projection. The model was purposely designed not to match the process.

In Figure 12-7, the car was kept at a speed of 95 km/h, which is not close to the actual constrained limit of about 130 km/h, so pmm causes the offset in Figure 12-7. However, as the MV increases toward its limit of 100% and the process approaches its constrained speed limit, the model prediction will match the process (not shown).

12.4 Takeaway

- The control benefit of incremental model adjustment is inconsequential. The important benefit is real-time, online monitoring of the process characteristics for forecasting maintenance needs, unit switching, and constraint operation.

Exercises

12.1 Use the procedure in Section 12.2.2 to derive Equation Ex12-1-4 from Ex12-1-3.

12.2 Apply the procedure in Section 12.2.2 or 12.2.3 to your preferred model. How will you evaluate the derivatives if the model is not a one line equation but a function or subroutine?

12.3 Access the process-model-based control (PMBC) simulator at *www.r3eda.com* and explore the impact of the τ_{pmm} value.

13 Simple Predictive Functional Control – Unconstrained MISO

13.1 Introduction

Predictive functional control (PFC) was introduced by J. Richalet, A. Rault, J. L. Testud, and J. Papon in "Model Predictive Heuristic Control: Applications to Industrial Processes," published in 1978.[1,2] The benefit of PFC over the simple model-based control (MBC) is that it can cope with high-order, dead-time-dominated, or inverse-acting modeled dynamics. Unlike the simple MBC and process-model-based control (PMBC) strategies, PFC does not presume that a single time constant dominates the model dynamics.

In PFC, the Act function determines a manipulated variable (MV) value that, if implemented in a single step-and-hold, would make the modeled y value hit the biased set point after a user-defined time. The future point is called a *coincidence point*. In PFC, the Correct function simply transfers the process-model mismatch (*pmm*) as the set-point

1. J. Richalet, A. Rault, J. L. Testud, and J. Papon, "Model Predictive Heuristic Control: Applications to Industrial Processes," *Automatica 14*, no. 5 (1978): 413–428.
2. J. Richalet and D. O'Donovan, *Predictive Functional Control: Principles and Industrial Applications* (London: Springer, 2009). doi: 10.1007/978-1-84882-493-5.

bias for the model; however, one could also use the Correct function to adjust the model incrementally, as shown in Chapter 12. The Predict function remains the one-time-step update of the model, the past-to-now modeled value (yP2N) function.

The generic MBC structure is shown in Figure 13-1.

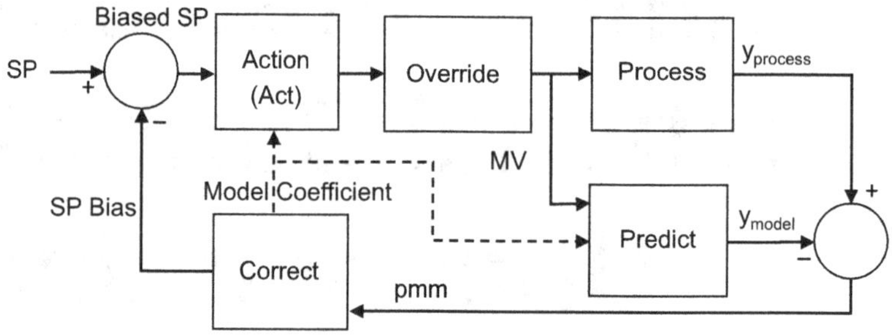

Figure 13-1. The basic MBC structure.

13.2 The Act Function in PFC

The PFC coincidence point concept is to determine a step-and-hold MV value that gets the model to move to the coincidence point (the biased set point at a future time). This is illustrated in Figure 13-2. The left graph (13-2a) shows the MV in time, and the right graph (13-2b) shows the modeled value in time. The time-axis point labeled "Now" is the present time. To the left is what happened in the past, which is indicated by the yP2N model incremental updates in the Predict function. To the right is what might happen in the future, as forecasted by the now-to-future modeled value (yN2F) model within the Act function.

Notes:

1. In Figure 13-2a, the past control actions (step-and-hold for each Δt control interval) are indicated on the MV graph. The control action is based on one step-and-hold long into the future.
2. In Figure 13-2b, the past modeled controlled variable (CV) values resulting from the control action are indicated by straight lines (connect the dots) between incremental predictions. The past model values were calculated in the yP2N model. This is not the process CV, it is the modeled CV. The CV graph shows the biased set point for the model.

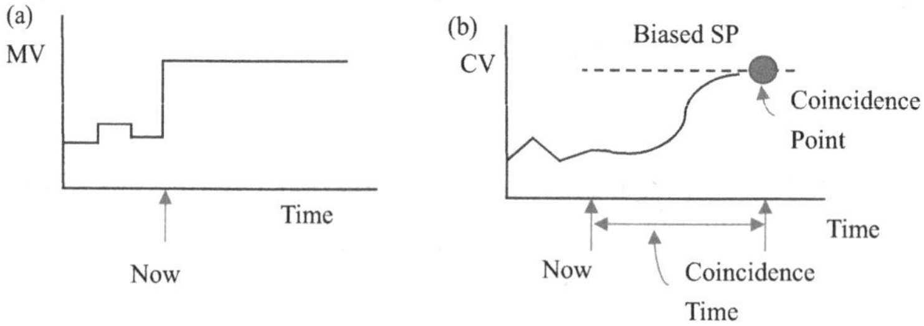

Figure 13-2. Coincidence point concept: 13-2a is the MV action, and 13-2b is the modeled response.

Note: "Now" is indicated on the graphs, and the future MV is a single step-and-hold for the coincidence time. The forecast CV is the yN2F model calculated within the Act function, not the model Predict function. Note that there is no discontinuity at the transition between past and future because the yN2F model is initialized by the yP2N state value(s).

Although the step-and-hold trial solution seeks to get the model to the biased set point, the controller will not hold that value for that coincidence time duration. The Act calculations are recalculated at each small Δt, as the past MV and modeled CV actions indicate. Updating on the small Δt interval enables the controller to respond to set-point or disturbance changes rapidly.

The model does not have to express well-behaved dynamics (a presumption in generic model control with an SS model and the simple MBC and PMBC strategies), nor does it need to be initialized at an SS.

The coincidence time should be about 85% of the open-loop settling time. A shorter coincidence time leads to more aggressive control action. A longer coincidence time makes a controller less aggressive. The coincidence point must be substantially beyond the dead time or the inverse action duration of the model.

Maybe you will be lucky enough to have an explicit inverse of the dynamic model to be able to determine the u value. More likely, you will need an iterative trial-and-error approach. If a guess at the MV amplitude step is too small or too large, adjust the step until the yN2F value is close enough to the biased set point. Any number of root-finding algorithms will work. Interval halving is robust and easily implementable for a multiple-input, single-output (MISO) case. Alternately, one could square the deviation at the coincidence time and use an optimization algorithm to seek the MV value that minimizes $\left[yN2F(t+ct)-y'_{SP}\right]^2$. In either case, it is essential that the iterative procedure is robust. I recommend using optimization to prepare for constraint handling in either a multiple-input, multiple-output (MIMO) or an auxiliary variable situation.

After each MV trial solution, before seeing the impact on the forecast modeled value, the yN2F model must be re-initialized with the most recent results of the yP2N model values.

Also, for an iterative procedure, there must be convergence criteria (or one criterion) to indicate when the iterative search is close enough to the optimum MV value. Realize that (1) the coincidence point is far into the future, and there will be about 20 control actions between "now" and "then." So, if the convergence criterion stops when the model prediction is still 5% from the coincidence point, there will be many future actions to trim that current action. Also, (2) the model does not match the process perfectly, so even if the iteration makes the model exactly match the biased set point, the process will not exactly match the set point. However, with many control actions between "now" and "then," the offset will be corrected. Finally, (3) if the iterative search makes small changes in the MV guess (the trial solution) on the order of 0.1%, these changes will not have an observable impact on the process. In fact, such changes may be at the discretization limit of digital devices (resolution) or smaller than lost motion effects in the final control element, and they may have zero impact on the implementable MV. All in all, "loose" convergence thresholds on either the MV trial solution or the CV value are acceptable and are desirable to minimize the computational burden on the search algorithm.

After converging on a u value, implement it. If there is an override, acquire that implemented u value. After the sampling interval (control interval, scan time, Δt), feed that implemented u value to the yP2N model to update the predicted y value.

Be sure to initialize the yP2N model with the prior yP2N output, not the coincidence point values from the yN2F model. Then, calculate the pmm, bias the set point, and run Act to calculate the next u value.

The Act function is an iterative calculation of the model inverse. It identifies the value of u to make the model do what you want. Be sure that on each iteration in the search, the yN2F model is re-initialized with the most recently calculated yP2N values.

Calculating the future MV value is either an optimization or a root-finding application.

The stages in the Act function are expressed as a flowchart in Figure 13-3. The dashed lines indicate the transfer of information; the solid lines indicate the transfer of control.

The optimizer is initialized, then the first trial solution is initialized. During regulatory periods (with no set-point changes), the best initial trial solution will probably be the prior MV value. Depending on the algorithm, this initialization step might involve generating new initial trial solutions (players) or determining the local sensitivity of the objective function to the MV values.

The optimizer then chooses a trial solution, u, a future MV value, which could be constrained by limits (e.g., 0% to 100%) or rates of change (e.g., $|u_i - u_{i-1}| < 5\%$). This trial future MV is transferred to the yN2F model, which projects what the future modeled CV will be from now to the coincidence time. The model may have several internal states, which must be initialized with the states, $\underline{y}_m$, from the yP2N model.

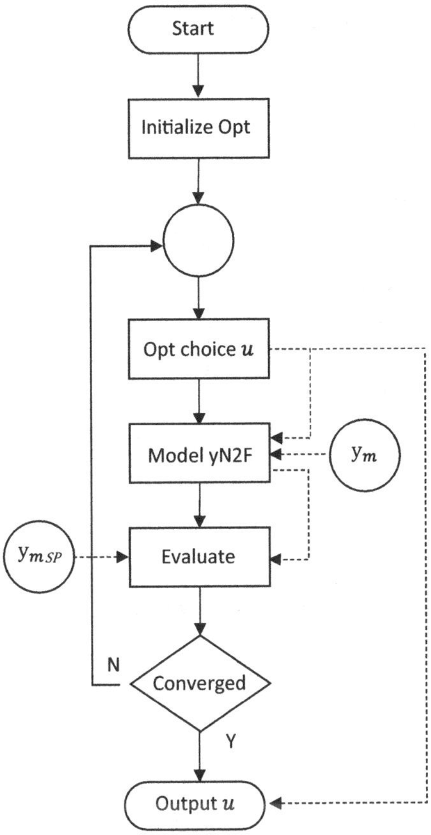

Figure 13-3. Detail of the Act function in PFC.

The Evaluate function determines the objective function value, the deviation between the forecast CV and the biased set point. In optimization, the objective function value would be either an absolute value or the squared deviation.

If the optimization has converged (if the change in the MV value from the prior iteration is small enough), then output the MV value. What is small enough? I think this is easily related to the magnitude of an MV change that has no visual impact on the CV. This might be a 1% change. An alternate convergence criterion could be based on the objective function, the deviation from the biased set point, which would be inconsequential.

If the optimization has not converged, return to the optimizer for the next trial solution. Be sure that the yN2F model is re-initialized with the present yP2N states.

The P2N and N2F models must be the same (same number of states, same functionalities, same coefficient values, same measured disturbances, etc.). If a coefficient is adjusted online, the P2N and N2F models must use the new values.

13.3 Illustration of a Simple PFC

In this illustration, the process is nonlinear, fourth-order, with a delay. The model is linear, second-order, with a delay.

Figure 13-4 illustrates the model and process with the controller in manual (MAN) mode making equal steps in the MV. In the 75 to 175 s range, the lower trace is the process response. Note that the process is nonlinear with about a 4:1 gain ratio. The modeled value (generally in the middle) shows a significant mismatch in the process. First, the bias changes from about +10% of the full range at initial times to –15% in the 250 s period. Second, the model responds more slowly than the process at early times, about the same in the 125 to 150 s period, and faster at the higher values. Third, the modeled delay is smaller than the process delay. Fourth, the controller model is linear with gain, time constant, and delay constant over the entire range.

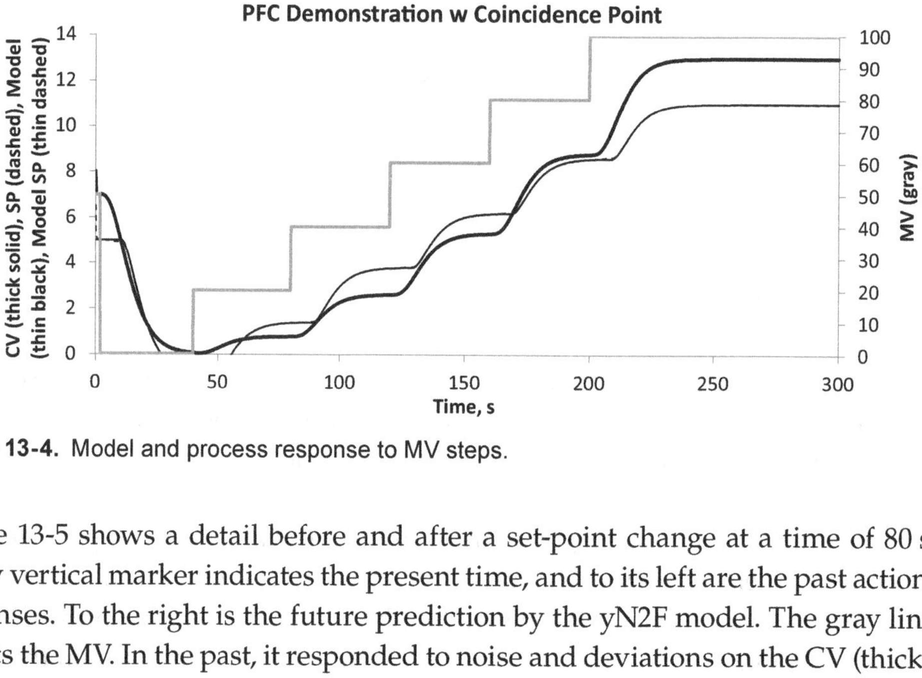

Figure 13-4. Model and process response to MV steps.

Figure 13-5 shows a detail before and after a set-point change at a time of 80 s. The nearly vertical marker indicates the present time, and to its left are the past actions and responses. To the right is the future prediction by the yN2F model. The gray line represents the MV. In the past, it responded to noise and deviations on the CV (thick black line), aiming to keep the CV at the set point (black dashed line). At a time of 80 s, the set point made a step from about 4 to 3. At that time, the PFC controller immediately sent the MV to about 37%, but because of model and process delays, about five control actions later, neither responded.

The coincidence point is indicated by the dot at about 100 s on this snapshot, 20 s into the future. The dashed line to the right of the now marker (the horizontal line at 81 s) is the biased set point for the model, and the changing trace is what the yN2F model

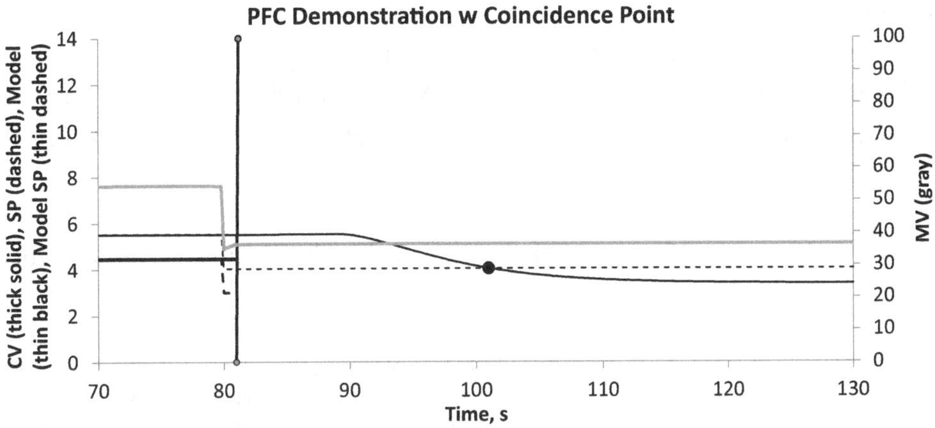

Figure 13-5. Detail of one MV calculation.

function is projecting. The optimizer found the *u* value to make the model match the biased set point.

Note: The simulation shows time extending beyond the coincidence point. This is to reveal that the coincidence point is not at a steady state. The control action that pushes the model to the biased set point at the coincidence time will be a bit excessive and push the model beyond the biased set point if it is left in effect.

Figure 13-6 shows the MAN-to-AUTO transfer and response to set-point changes with a temporary constraint. Up to a time of 20 s, the controller is in MAN. However, note that the model is running and responding to the MV; the middle smooth trace represents its values. When the control mode is switched to automatic (AUTO) mode, the CV value becomes the set point and the dashed line is revealed. At this time, the dashed line for the biased set point is also revealed.

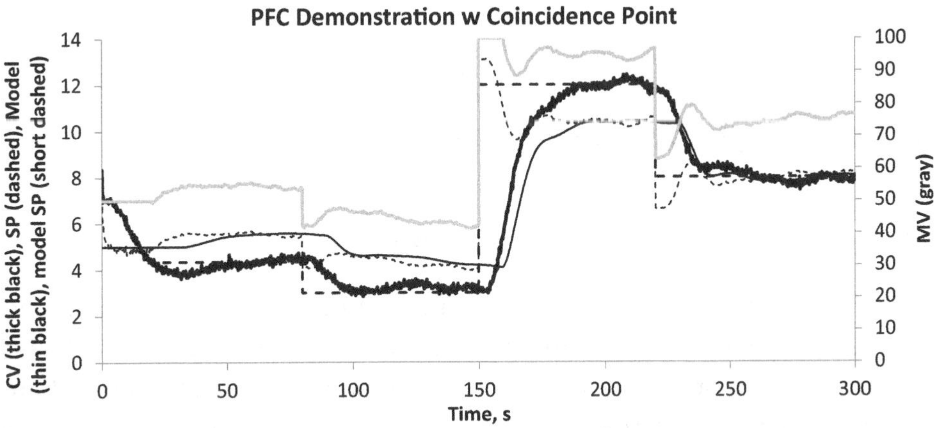

Figure 13-6. Bumpless transfer and set-point changes with noise and disturbances.

Throughout the graph, the CV trace is noisy. As a result, the noise is reflected in the MV value (gray line, generally the upper trace). The set point changes at times of 80, 150, and 210 s (the dashed horizontal lines). Also, throughout the graph, the disturbance makes random deviations with a persistence of about 20 s. Control keeps the CV at the set point.

The set-point change at 150 s caused the MV to hit the limit and remain there for about 15 s.

Note: The MV began moving off the constraint before reaching the set point at about 175 s. This reveals some predictive action (requiring derivative action in proportional-integral-derivative [PID] control), and there is no windup (which would result from integral action in PID control). Yet, there is only one tuning coefficient: the user's choice of the coincidence time.

The pmm is filtered with a filter factor of 0.2, which tempers the noise and the dynamic mismatch between the model and the process.

The coincidence time is the single tuning factor. In Figure 13-6, it is 40 s into the future. In Figure 13-7, it is 15 s, leading to a more aggressive controller.

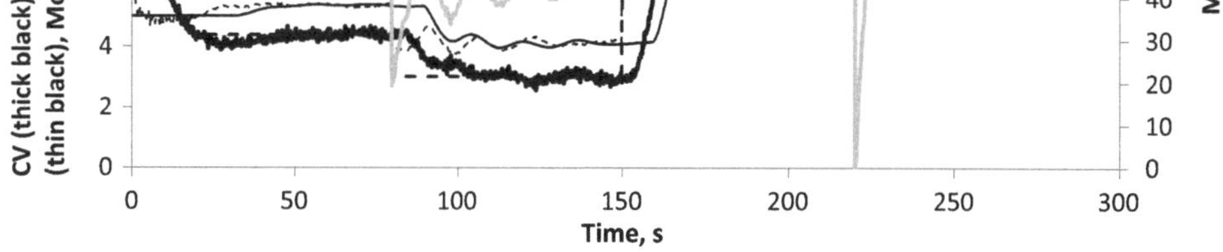

Figure 13-7. Same as Figure 13-6 but more aggressive.

13.4 A Comment on Predicting the Future

Any predictive control that uses a model to predict the future also must forecast the pmm value and any measured disturbance values. The pmm value is caused by several factors (unmeasured disturbances and process-to-model mismatch), and it changes in time as disturbances and process behavior change. Measured disturbances change in time. We know the past values of the pmm and measured disturbances, but we should have the future values when using the model to predict what will happen in the future and to have the appropriately biased model set point as the target.

One could use any empirical or phenomenological model fit to past pmm values to predict the future pmm. It could be an autoregressive moving average model or a first-order model. However, it seems safer to consider that the disturbance makes a step-and-hold, holding at its current value into the future. This is the way we treat residuals in linear model predictive control. The model must be carefully developed to not respond to noise and to retain rational forecasting limits. Attempting to model the future trend in disturbances is predicated on there being no new events in the control horizon, which would make the future process behavior inconsistent with past behaviors.

Although there is no basis for this presumption, we presume that the current set point will be the set point at the coincidence time. Similarly, we presume that the present values of the pmm and the measured disturbances will not change in the near future.

13.5 Takeaway

- PFC is a relatively simple, nonlinear, model-based approach that can handle ill-behaved dynamics. In this chapter, it was demonstrated in a single-input, single-output application with simple constraints. It can be extended to MIMO applications.

Exercises

13.1 Obtain the predictive functional control (PFC) simulator from *www.r3eda.com* and explore the impact of the coincidence time, noise, disturbances, and constraints in regulatory and servo control.

13.2 Discuss the relative merits and disadvantages of PFC with respect to generic model control (GMC) with a steady-state model, simple nonlinear model-based control, and process-model-based control (PMBC).

13.3 Apply PFC to a simulator of your choice.

14 Simple Horizon Predictive Control – MISO with Constraints

14.1 Introduction

The benefit of horizon predictive control (HPC) over simple model-based control (MBC) is that it can cope with high-order, dead-time-dominated, or inverse-acting modeled dynamics. Unlike the simple MBC and process-model-based control (PMBC) strategies, HPC does not presume that a single time constant dominates the model dynamics. Predictive functional control (PFC) (discussed in Chapter 13) solves this issue by targeting the model to hit the biased set point at a future coincidence point using a single step-and-hold manipulated variable (MV) action. In contrast, HPC solves the issue of ill-behaved dynamics by identifying a postulated sequence of MV actions over the future that seeks to have the model best fit a reference trajectory path toward the biased set point. The advantage of HPC over PFC is response path shaping. HPC can propose a future sequence of MV actions that pushes hard to make the process respond rapidly now and back off as the process approaches the set point. HPC can also shape the postulated MV actions to avoid constraints.

HPC is also called *model predictive control* (MPC) or *advanced process control* (APC), and early procedure names were *dynamic matrix control* (DMC) and *identify and command*

(IdCom). HPC was independently conceived by Cutler and Ramaker while working at Shell in the United States and by Richalet and colleagues while working for Adersa in France.[1,2]

In HPC, the Act function determines an MV value sequence that, if implemented, would make the modeled y value follow a desired path to the biased set point. The desired path is called the *reference trajectory,* and it is usually a first-order return to the biased set point. The now-to-future modeled value (yN2F) model does not need to follow the reference trajectory exactly. In fact, if the reference trajectory is first-order and the model has dead time, it cannot be made to follow exactly. In HPC, the MV sequence is optimized to make the model match the reference trajectory in a least-squares manner.

Like PFC, the control action in HPC is not followed over the entire planned future. Set-point changes and disturbances could happen. So, the first MV action is implemented and held for the control interval, Δt, and all the control calculations are redone at the end of the small time interval.

For example, when some people walk their dog, they make the dog "heel" to exactly follow their path. Others walk their dog with a loose leash, generally keeping the dog moving along their path but permitting the dog to move about that path in a manner that matches the dog's nature. Similarly, HPC defines a path for the controller model to move from its current location to the biased set point. But if the model has a delay and the desired path (termed the reference trajectory) is first order, no control action can make the model exactly follow that path. HPC does not require the CV to exactly follow the reference trajectory, but permits it to generally follow the desired path.

The diagram of the controller functions in Figure 14-1 is the same as that for MBC, PMBC, and PFC.

In HPC, the Correct function simply transfers the process-model mismatch (pmm) as the set-point bias for the model. However, one could also use the Correct function to incrementally adjust the model and filter the Correct function to temper noise and the dynamic mismatch between the process and the model. The Predict function remains the one-time-step update of the model, the past-to-now modeled value (yP2N)

---

1. C. R. Cutler and B. L. Ramaker, "Dynamic Matrix Control: A Computer Control Algorithm," *IEEE Transactions on Automatic Control* 17 (1979).
2. J. Richalet, A. Rault, J. L. Testud, and J. Papon, "Model Predictive Heuristic Control: Applications to Industrial Processes," *Automatica* 14, no. 5 (1978): 413–428.

function. However, in HPC lingo, the difference between the model and the process is termed the *residual*, not pmm.

The generic MBC structure is illustrated in Figure 14-1.

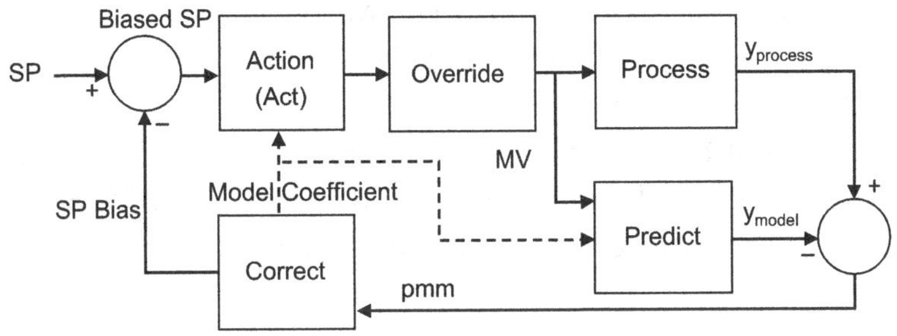

Figure 14-1. MBC structure.

14.2 Background

HPC means that the control action makes the process generally follow a desired path over a future time horizon. The horizon is roughly 80% of the open-loop settling time.

The academic term for HPC has become MPC, whereas the process industry term has become APC. Initial HPC products used linear, empirically obtained, finite impulse response models, which require extensive step-testing of the process. The optimization procedure to determine the sequence of MV actions that best fit the model prediction to the reference trajectory was linear programming (keeping control action on a constraint) or successive quadratic (requiring gradients and special treatments for constraints). There are many other product-specific features.

This chapter describes a structure to use first-principles process models in APC, with the benefits of preserving the engineer's process knowledge and requiring much less invasive process testing. Also, the structure shows the use of a direct search multiplayer optimizer, Leapfrogging, which can identify interior solutions, handle constraints, and cope with the characteristics of nonlinear models. This chapter also presents a simple simulation of a nonlinear process to demonstrate nonlinear APC features.

As mentioned earlier, MPC has many alternate names and acronyms. The common attribute is that MPC uses a dynamic process model and optimization to plan a sequence of future control actions to make the model match a desired future path while avoiding constraints. The first control action (MV) in the plan is implemented. The entire plan is not executed, only the first step. After one control interval, the future

MV sequence plan is recalculated to accommodate changes in set points, disturbances, and the pmm. Then, the first step in the new plan is executed, not the entire plan.

The models in MPC can reveal disturbances and interactions, as well as dead time and other ill-behaved dynamics. The optimization in MPC compensates for the model features, has override action, and can choose the best MVs to manage the process when there are extra options. Although such features could be implemented in classic advanced regulatory control (ARC) with feedforward, decouplers, ratio, and override, once an implementation exceeds about two inputs and two outputs, the complexity and number of tuning and model coefficients become complicated to manage. MPC provides a unified structure, with only one tuning coefficient for each controlled variable (CV).

A rule of thumb is that MPC (or any control strategy with equivalent features) will halve the CV deviations associated with conventional regulatory control. This enables assigning set points closer to specification limits, which translates to many economic benefits, such as higher throughput, less waste, and lower energy and raw material use. Although it is relatively expensive to license an MPC product and the process of step-testing to develop empirical models takes time and moderately upsets the process, industry has accepted MPC as one of the best control approaches in continuous chemical processing.

Since about 1980, industries have taken many approaches to nonlinear process control. It seems that the industrial acceptance of digital control devices and distributed control systems (DCSs) of the 1970s revealed a vision to many investigators that MBC could be practicable. The following are just a few examples, listed in order of increasing complexity:

- Generic model control (GMC) provides a structure to use steady-state (SS) models in an output characterization of a proportional-integral (PI) controller. If the process is single-input, single-output (SISO) and the problem is nonlinear gains, this is an excellent solution.
- PMBC is a single step-ahead controller (it does not forecast a future plan or handle ill-behaved dynamics) that can handle multiple-input, multiple-output (MIMO) processes with constraints, disturbances, and nonlinear and nonstationary features. It was developed to adapt the model to changing process attributes such as catalyst reactivity, tray efficiency, heat exchanger fouling, and drag coefficients. If delay, high-order response, or inverse action is strong, PMBC is not a good solution.

- PFC was developed to handle nonlinearity and ill-behaved dynamics (delays, high-order, and inverse action). After implementation, it uses a process model to calculate a step-and-hold MV action that would make the model match its set point at some future time (the coincidence point) well past the influence of the ill-behaved dynamics.
- MPC can also use first-principles models as opposed to a vector representation of linear finite impulse responses, which is the topic of this chapter.

These are all practical, industrially proven techniques. Complying with the keep it simple and safe (KISS) principle that guides applications, the first three strategies should be considered prior to MPC using first-principles models for multivariable, constraint-handling control.

14.3 The Act Function in HPC

This method chooses a future sequence of MV actions to shape the CV future response to match a desired reference trajectory and to avoid constraints now and into the future. This is also subject to constraints (MV, CV, auxiliary variable [AuxV], future, and rate of change).

Like PFC, the MV action is forecast for a time horizon, but it is not a one-step-and-hold action. Like PFC, it is recalculated at each control interval, and only the first in the sequence of forecast actions is implemented. Like when planning a trip, you might have a path. However, you may encounter a detour and have to recalculate a new path. You cannot stick to the prior plan.

The HPC method calculates a future MV sequence (perhaps 3 or 30 for each MV)—a plan for the future MV actions that best shape the desired CV response, avoiding constraints over some forecast time horizon. The time horizon should be about 85% of the open-loop settling time.

I have been dividing the control horizon into three time segments, such as now-to-1/4, 1/4-to-1/2, and 1/2-to-end, and calculating three MV values, one for each period that shapes the CV response to best fit a first-order reference trajectory back to the set point. I have also explored alternate spacing patterns. I have been happy with the effectiveness and low computational burden of just three future control actions. However, there seems to be a one-upmanship contest among vendors to increase the number of future planned MV moves.

In HPC, the Act function forecasts several future MV moves (this book uses three for each MV) to

- shape the future modeled CV response to best fit a reference trajectory (here, first-order, from current model past-to-now (P2N) value to the biased set point for the model); and
- steer variables clear of constraints, limits, or balance violations when a constraint is unavoidable.

Calculating the future MV sequence is an optimization application.

The stages in the Act function are expressed as a flowchart in Figure 14-2. The dashed lines indicate the transfer of information, and the solid lines indicate the transfer of control. The block labeled "Determine" uses the biased set point and the modeled CV

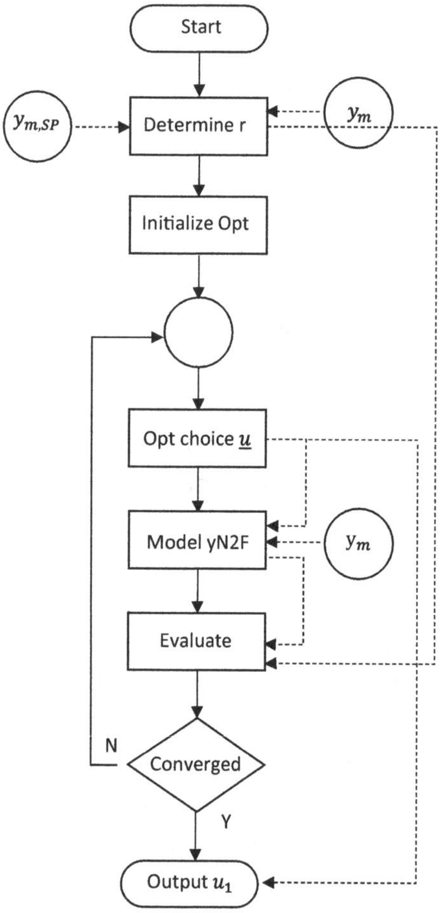

Figure 14-2. Detail of the Act function in HPC.

to determine the desired future modeled CV values, called the reference trajectory. This is usually a first-order path from y_m to $y_{m,SP}$. The reference trajectory is transferred to the Evaluate operation down the chain.

The optimizer is initialized. The initial trial solution will probably be the prior MV sequence. Depending on the algorithm, this step might involve generating new trial solutions (players) or determining the local sensitivity of the objective function to the MV values.

The optimizer then chooses a trial solution, $\underline{u}$, a set of future MV values, which could be constrained by limits (e.g., 0% to 100%) or rates of change (e.g., $|u_i - u_i - 1| < 5\%$). This trial set of future MVs is transferred to the yN2F model, which projects what the future modeled CV will be from now to the time horizon. The model may have several internal states, which must be initialized with the states, $\underline{y}_m$, from the yP2N model.

The Evaluate function then determines the objective function value. Conventionally, this is the sum of squares deviations between the forecast CV and the reference trajectory. However, a person could include additional measures such as the cost of the utilities.

If the optimization has converged, if the changes in MV values from the prior iteration are small enough, then output the first MV value. There is no need to output the sequence. Only the first is implemented. What is small enough to claim convergence? I think this is easily related to the magnitude of an MV change that has no visual impact on the CV. This might be a 1% change. An alternate convergence criterion could be based on the objective function, the sum of squares deviation from the reference trajectory. However, it is easier to relate to and quantify inconsequential MV changes.

If the optimization has not converged, return to the optimizer for the next trial solution. Be sure that the yN2F model is re-initialized with the present yP2N states.

14.3.1 HPC Objective 1 – Following the Reference Trajectory

Figure 14-3 illustrates the Act function. The left vertical axis represents a process variable (PV), the right vertical axis represents the MV value (0% to 100%), and the horizontal axis represents time.

The time of 20, marked by the vertical line, is the current time. Past events are on the left of the time marker, and a possible future plan containing three MV steps (the gray line) extends beyond a time of 35 s on the right. The plan has not happened and might not happen. Like a trip plan, which is revised when you hit a detour or decide to take a side visit, the original might never happen.

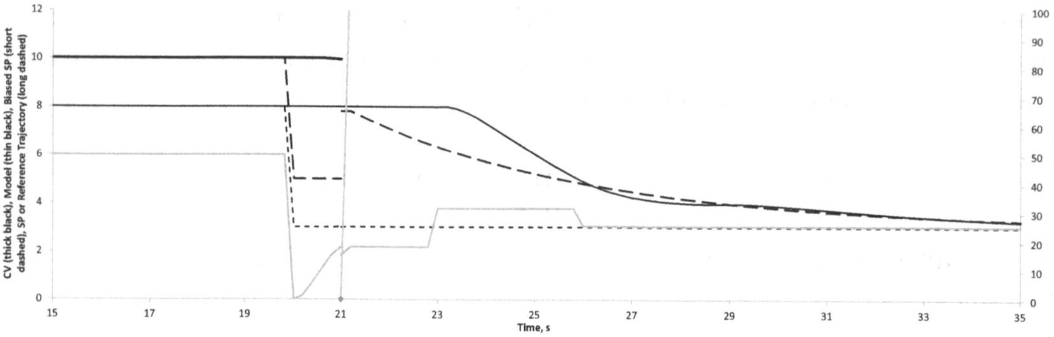

Figure 14-3. Forecast MV moves to make the model best match a desired path.

Beginning on the left side of the marker, the process started at an initial steady state of about 10 (the thick black line). This is the CV line, which is nearly at the set point of 10 (the long-dash line). The gray line is the MV. The upper lines were the model predictions (the model is the thin black line, and the short-dash line is the biased set point). The difference between the modeled and actual CV values is the pmm, which is about 1.5 CV units. At a time of 20 s, the set point changed to a value of 5, which is when the MV dropped to nearly zero and made the set point and biased set point dashed lines visible. Neither the CV or the modeled value changed because both have a deadtime of about 3 s.

To the right of the time marker are the possible future MV outcomes. The short-dash line is the biased set point for the model, the current target. The long-dash line that makes a first-order transition from the current modeled CV to the biased set point is the reference trajectory, the desired first-order path for the controller MV plan to move the modeled CV value. This was chosen to be a first-order path. The MV plan (gray line) is the three steps, the possible MV sequence. The uppermost thin black line is the model-projected CV value response to the three possible MV steps. It starts at the present modeled value and shows a further delay of about 2 s, then a higher-order lag, and then a subsequent path that substantially follows the reference trajectory. To get to the modeled CV path, the controller makes a large MV change to move the modeled CV quickly; if held there, the modeled CV would fall substantially lower than desired. Continuing the plan, the controller raises the MV to prevent an undershoot from the reference trajectory, which would cause the modeled CV to remain above the reference trajectory longer into the future. The third step in the plan is to lower the MV a bit to prevent an overshoot of the trajectory and to keep the modeled CV nearly on the remaining trajectory.

Notes:

1. The future forecast time is termed the control horizon, referring to the horizon in HPC. Although it could, the horizon does not extend to the settling time, to an SS for either the

modeled or process CV or for the reference trajectory, but typically to 70% to 90% of the settling time. Accordingly, the reference trajectory does not reach the biased set point, as Figure 14-3 illustrates. This does not matter because only the first MV step will be implemented. The future plan is recalculated at the next sampling with information about new disturbance values, new pmm evidence, and new set points.

2. The control interval is not the length of the first MV step-and-hold. This simulation has about 30 control intervals in the future plan. The first MV step-and-hold in the current plan lasts for about 15% of the horizon (about five control intervals), the second step-and-hold for about 25% (about seven control intervals), and the last for about 60% (about 24 intervals).
3. This three-step forecast MV pattern is not essential. More steps would give a more precise matching of the model to the reference trajectory, but in my experience, three is fully adequate and minimizes the optimization burden. One could use alternate spacing patterns for the MV transitions. Because the first step will be implemented and it shapes the early CV response, and because subsequent MV steps trim the plan and are farther out into the future, I like a plan that has more MV moves earlier. This plan roughly spaces MV moves according to the golden ratio, but other options work just as well. You might be more comfortable with 1/4, 1/4, and 1/2.
4. The plan shown is a near-final plan. The three MV values are the optimization decision variables (DVs). Optimization keeps fine-tuning the DV values, seeking the best set to minimize the sum of squared deviations (SSDs) between the reference trajectory and the model forecast trend until the convergence criterion is met.
5. In optimization, convergence is normally on the objective function (the SSD). However, there is no sense in continuing the optimization search when the iteration-to-iteration change in the MV is less than what might be visually perceptible. In an application, my choice for convergence is when the incremental changes in the MV values are less than 1%. However, for simulations where smaller changes are perceptible on a plot, I usually choose 0.1%.

The optimization objective is not to force the model to match the reference trajectory exactly but to minimize the forecast CV deviations from r, the reference trajectory. If the model dynamics are not simply first-order, then it would be impossible to make the model match the reference exactly. The delay in the model, which is different from the delay in the process, also makes a perfect match impossible. Here, the objective function is a least-squares deviation from the reference trajectory over the control horizon.

$$\underset{\{u_1, u_2, u_3\}}{Min} \quad J = \sum_{i=1}^{H} (r_{1i} - \tilde{Y}_{1i})^2 \tag{14-1}$$

In Equation 14-1, i is a counter for future intervals and H is the number of intervals in the control horizon.

If the reference trajectory is first-order, the time constant, τ, is the controller tuning factor and the reference trajectory is initialized at the yP2N value.

$$\tau \frac{dr}{dt} + r = SP', \qquad t_{t=now} \le t \le H, \qquad r_{t=now} = \tilde{Y}_{P2N,\,now} \tag{14-2}$$

If Δt, the control interval, is small relative to τ, the time constant for the reference trajectory—as it should be, then using Euler's explicit method to numerically solve the first-order differential equation to determine *r*(*t*)

$$r_{t+\Delta t} = \frac{\Delta t}{\tau} SP' + \left(1 - \frac{\Delta t}{\tau}\right) r_t \tag{14-3}$$

Equivalently, discretize the analytical solution to a first-order differential equation and use

$$r_{t+\Delta t} = \lambda \cdot SP' + (1-\lambda) r_t \tag{14-4}$$

where $\lambda = 1 - e^{-\frac{\Delta t}{\tau}}$.

Note: For some MPC algorithms, the objective is to make the model match the biased set point in a least-squares sense throughout the entire horizon, creating an aggressive control action. For these, the control action is tempered by a move suppression factor, a penalty for large MV moves. This is called *MV damping.* What I have described with a reference trajectory target is termed *CV damping.* Because tuning with MV damping does not explicitly relate to any of the CV dynamics, I prefer CV damping. CV damping is based on choosing τ for the CV reference trajectory, which creates a single tuning factor that directly relates to that CV response speed.

14.3.2 HPC Objective 2 – Avoiding Constraints

Another objective is to plan the future MV sequence to avoid constraints. The constraint might be on a CV or any other PV, often termed an auxiliary variable (AuxV). The constraint might represent a safety or specification limit, a desired resting value, or a rate of change. There are diverse ways to penalize a constraint violation or proximity of the PV to the constraint. The "soft" penalty method for a deviation penalizes a violation proportional to the square of the violation.

$$Penalty_1 = \begin{Bmatrix} 0, & \text{if } AuxV_1 \text{ is within the limit} \\ (PV_1 - L_1)^2, & \text{if } AuxV_1 \text{ exceeds the limit} \end{Bmatrix} \tag{14-5}$$

Here, *PV* represents any process variable forecast by the yN2F model, and *L* is the limit biased by that variable pmm.

However, because constraints could happen at any of several future times, sum all the penalties. With u_1, u_2, and u_3, the three future possible MV values, the optimization statement is

$$\underset{\{u_1, u_2, u_3\}}{Min} \sum_{i=1}^{H} Penalty_{1,i} \tag{14-6}$$

If several variables are subject to constraints, calculate the penalties for each.

Constraints on the MV, such as rate-of-change limits or values beyond the 0% to 100% range of feasible values, could be handled as either soft constraints in Equation 14-6 or as hard constraints, limiting the MV choices in the optimization. I prefer to treat MV constraints as hard, limiting the feasible values for the optimizer.

CV or AuxV constraints should be soft. A disturbance, noise, or such near a constraint could move the AuxV beyond the constraint. As it has already happened, no future control action can undo that. In such a case, all attempts to identify the MV sequence to eliminate the constraint violation will fail. To prevent such infeasible corrections, treat CV and AuxV constraints as soft.

Alternately, some prefer to use a barrier function that (1) begins to assign a penalty in the proximity of the constraint before the limit violation and (2) increases the penalty toward infinity at the constraint.

14.3.3 Combining All Optimization Considerations

The two soft penalty considerations (CV and AuxV) represent undesirable aspects. The controller should determine the control action that minimizes all of them. Now that the considerations are quantified, a standard way of combining several desirables is to sum all the measures of badness and seek the MV sequences that minimize the sum of all undesirables.

$$\underset{\{\underline{u}\}}{Min}\ J = \frac{\sum_{i=1}^{H}(r_i - \tilde{y}_i)^2}{EC_1^2} + \frac{\sum_{i=1}^{H} Penalty_{1,i}}{EC_2^2} + \frac{\sum_{i=1}^{H} Penalty_{2,i}}{EC_3^2} + \cdots \tag{14-7}$$

Note: The optimization DVs, the possible sequence of future $\underline{u}$ values, are underscored, representing vectors. The MV has several possible future values (three MV values for the one MV have been illustrated here).

Because each term in Equation 14-7 will have unique units and some aspects will be more meaningful than others, weighting factors are required to unify units and balance relative importance. This representation uses equal concern (EC) factors. They scale each quantity by a factor with the same units as the numerator and represent equal concerns for violating the desired value.

Note: If there are no constraint violations, the penalty terms in Equation 14-7 are all zero. If the model CV can exactly follow the reference trajectory (or if the model is at an SS at the biased set point), the first term could also be zero.

The EC weighting factors could be used for tuning aggressiveness, but that is not their role. The tuning factor is the time constant for the reference trajectory.

14.3.4 Setting the EC Factor Values

Here is a procedure for setting EC factor values.

1. Choose the most critical term. For example, let's assume that $\sum_{i=1}^{H}(r_i - \tilde{y}_i)^2$ is of greatest concern and should be minimized.
2. Choose a moderate $\tilde{y}$ deviation from perfection. It might be the deviation from the reference trajectory along the future time or the deviation from the biased set point at the SS. This value is an EC factor. EC_1 is equal to the $\tilde{y}$ deviation considered to be moderate. For example, when driving a car, perhaps a 3 mph deviation from the speed limit is a moderate concern; on a 1 to 10 scale, it might rank as a 4. As another example, maintaining an indoor temperature at a deviation of 2°F from a desired value might also have a concern rank of 4 out of 10.
3. Feel the concern this deviation creates. Use both qualitative and quantitative aspects. Rate the concern level on a 1 to 10 basis.
4. Assign EC values to each term in Equation 14-7 to represent each deviation from perfection with the same level of concern. Consider the following example.
   - $EC_1 = 0.5\ wt\%$ *product deviation from SP* has a 1 to 10 concern rating of 3.
   - To have the same concern rating of 3, assume the deviation from a target flow rate is 15 lb/min. Then $EC_2 = 15\ lb/min$.
   - Also, to have the same concern rating of 3, assume the deviation from a process temperature limit is 10 *F*. Then $EC_3 = 10\ F$.

The EC values are situation-dependent, meaning they change with the situation. The situation includes expenses to make a product, the impact of competition on product quality, customer demand on product throughput, the personal agenda of a new plant manager, and other factors.

Note: Traditional optimization uses Lagrange-type multipliers to weight terms in a sum. Here, those values are the reciprocal of the EC denominator values. I think the weighting method using EC factors is intuitive and makes it easier to assign values that provide relative importance.

14.3.5 Optimization to Determine the Possible Future MV Values

Unlike simpler model-based controllers, the inverse (to determine the MV value from the model) is not a simple equation. This is an optimization procedure. The yN2F model should be in a function or subroutine, and the optimizer should send it trial solutions (the possible MV sequence). The yN2F function will return what the modeled

CV and AuxV will do in the future time horizon, and then an evaluation function can calculate the objective function value. The optimizer can then determine whether the trial solution is improving things and "guess" at a new MV sequence. When the guesses are close enough to the optimum, claim convergence and send the first MV value to the process.

The optimization statement is

$$\underset{\{\underline{u}\}}{Min} \; J = \frac{\sum_{i=1}^{H}(r_i - \tilde{y}_i)^2}{EC_1^2} + \sum_{k=1}^{M} \frac{\sum_{i=1}^{H} Penalty_{k,i}}{EC_k^2} \tag{14-8}$$

S.T: *hard constraints on the MVs (Value or Rate of Change)*

where M = the number of soft constraints, and H = control horizon – the number of Δt calculations in the future time projection. This should go to about 85% of open-loop settling time.

Note: Because the EC factors have the same units as the CV or constrained variable and the numerators in Equation 14-8 are squared, the EC factors in the denominators are also squared.

Classic MPC products use linear programming or successive quadratic optimization. Several constraint-handling, gradient-based techniques could also be used. My preference is direct search, multiplayer optimization methods. Multiplayer methods initially have many randomized initial trial solutions (test points—called *individuals, particles,* or *players*). Multiplayer methods are "global," and direct search methods are not confounded by surface aberrations (nonidealities, discontinuities, numerical striations, etc.) and can handle hard constraints and nonlinear responses. Leapfrogging is my choice, considering its effectiveness, speed, simplicity, and robustness. Visit *www.r3eda.com* for a description and examples, or refer to *Engineering Optimization: Applications, Methods, and Analysis* by R. R. Rhinehart.[3]

14.3.6 pmm and the Biased Set Point

Part of the Correct function in Figure 14-1 is to pass along the pmm to bias the set point for the model. If the process measurement is noisy, the biased set point will be noisy, making the controller output noisy. So, filter the pmm before passing it along to bias

3. R. R. Rhinehart, *Engineering Optimization: Applications, Methods, and Analysis* (Hoboken, NJ: John Wiley & Sons, 2018).

the model set point. I use a first-order filter because it is conventional, easy to understand, and easy to adjust to reduce the noise amplitude.

The process-to-model mismatch will reflect the dynamic mismatch as well as the static (SS) mismatch. The control concept uses pmm to bias the set point as an SS correction. If there is a substantial dynamic mismatch, the pmm will erroneously "correct" the model set point, redirecting the MV output, which will subsequently need correcting and cause another dynamic response and mis-correction. One solution could be to recalibrate the model to match the process better. A much more convenient solution is to filter the pmm so it slowly changes during transients. Although this might cause a lag in eliminating SS offset, in my experience, such lag is not noticeable.

Therefore, in the Correct function, use a first-order filter to temper the pmm and the effects of noise or model mismatch on the biased set point.

14.3.7 Direct Digital Control or Not?

Direct digital control (DDC) sends signals from the controller directly to the final control element (valves, heaters, etc.). If you have a single-loop controller (either SISO or multiple-input, single-output [MISO]) and the MBC calculation is within the programmable logic controller (PLC) or DCS, this may be permissible.

However, model-based controllers are typically implemented external to the PLC or DCS where computational power is available. Although the model-based controller decisions regarding the signal to the final control element could be transferred to the DCS or PLC to be relayed to the process, experience has shown that it is best to transfer set points for the process primitive variables (flow rates, temperatures, levels, etc.) and let the proportional-integral-derivative algorithms in the DCS determine the signals to the final elements. This is a cascade type of implementation. The primitive or lower-level loops typically respond faster than the primary variables that must be managed by constraint-handling HPC. So, this transfer of set points does not realistically slow control response. The advantages are that the DCS or PLC has many auxiliary functions related to signal processing and validation that should be used, and it is designed to be robust to power and communication failures. If the computer where MPC is implemented gets confounded or fails, the DCS or PLC will continue the process operation with the last set of set-point values. The process will not shut down, even if the MPC computer does.

14.4 Illustration of Nonlinear HPC

This section discusses nonlinear, horizon predictive, constraint-avoiding MBC for the car speed example. The process, representing the real car, is a simulation. To make the illustration legitimate, the simulation equations for the process are not the same as those for the controller model. The simulation is more complex; it has alternate functionalities for effects and coefficient values that are different from the controller model. One cannot perfectly model the reality of a process. Simulations of control should reflect that reality.

For control, this is a SISO process with the controller output to the final control element. However, the process also has an AuxV with an upper limit constraint that is intended to represent cabin noise.

Figure 14-4 represents a deterministic (noise- and disturbance-free) trend in the simulated process and controller model, subject to ideal step changes in the MV.

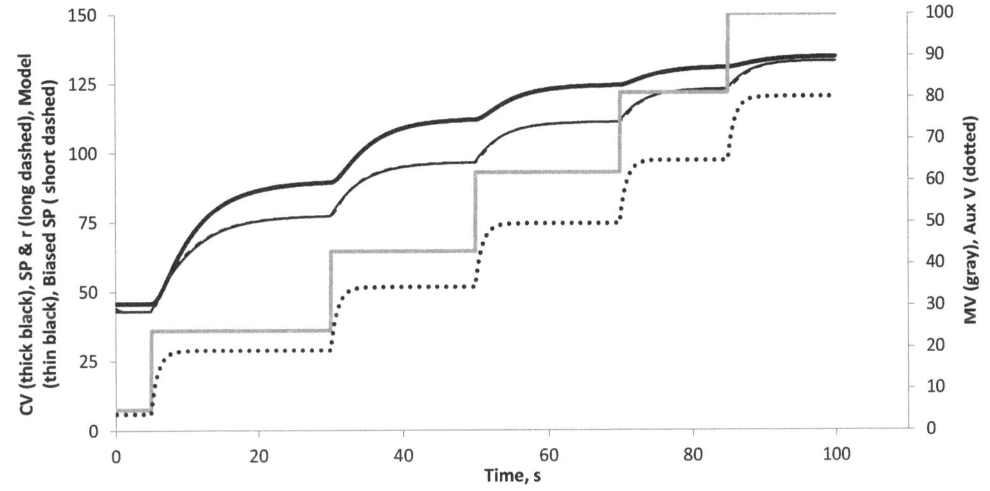

Figure 14-4. Trends in MAN mode.

The car accelerator pedal position, the MV, is the gray line that makes step changes, with values displayed on the right vertical axis. Each step is an increase of 20%. The upper, thick black curve is both the car speed and the set point, which are attached to the left-hand vertical axis. In manual (MAN) mode, the set point follows the CV for bumpless transfer to automatic (AUTO) mode. Note that the first MV step increases the CV by about 42 km/h, but the last equivalent MV step increases it by only about 6 km/h. The gain change ratio is about 7:1, which is evidence of nonlinearity. Also note that the settling time is about 15 s at the higher speed, whereas at the first MV step at the lower speed, the process is still rising after 25 s. This nearly 2:1 change in settling time is the second piece of evidence that the process is nonlinear.

The modeled speed value represented by the thin black line is nonlinear with dynamics that change with operating speed. However, the model features do not match the process exactly. The model is usually lower than the process, and the pmm progressively increases with speed and then decreases. However, because the dynamics differ, the model temporarily rises above the process at the first MV step.

The lowest curve, represented by the dots, is the AuxV. In this case, it is a simple representation of noise magnitude associated with the engine and car movement.

Note: At about 100 s, the MV is 100%, the corresponding feasible speed is a bit over 125 km/h, and the consequential AuxV value is about 80. These will be constrained as the example continues.

Figure 14-5 illustrates several features in AUTO mode. Initially, at time = 0, the controller is in MAN mode; it is placed in AUTO at 2 s. Because of set-point tracking and yP2N model calculation in MAN mode, there is no bump in the transfer to AUTO. Set-point changes occur at about 5 s, 30 s, 50 s, and 70 s, indicated by the long-dash line that makes steps. The thick black curve represents the process CV (car speed), and despite the 7:1 gain nonlinearity, the 2:1 settling time ratio, and the changing pmm, the overshoot magnitude and rise time in the first and fourth set-point changes are very similar.

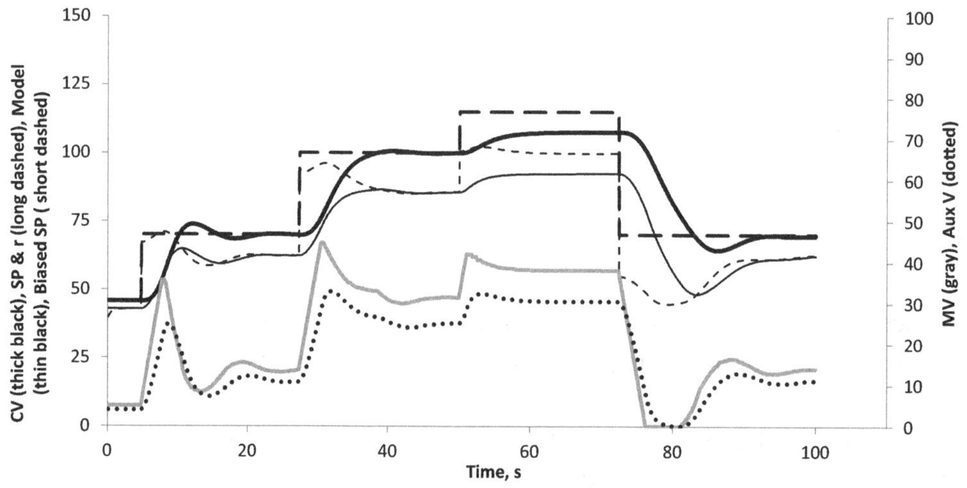

Figure 14-5. Events in AUTO mode, example 1.

Three categories of constraints are also illustrated in Figure 14-5. The third and fourth set-point changes represent constraint conditions. The AuxV (large dots) has a soft constraint of 30 (right vertical axis), so even though Figure 14-3 shows that a speed set point of 110 km/h is feasible at MV = 85%, at that MV value, the AuxV has a value of about 70, which would exceed the constraint. In Figure 14-5, the EC factors make constraint violations more important than meeting the speed set point, and the optimizer limits the MV (gray) to about 40%, preventing the speed from reaching its set point. When the set point is dropped to a speed of 80 km/h (left vertical axis), there is no delay in the MV action (there is no windup in the controller). That was a soft constraint. A second constraint happens at the final set-point change, where the controller wants to place the MV below zero. The MV stays at the constrained value limit until responding (again without windup). That is a hard constraint. The third constraint is a rate-of-change constraint on the controller, another hard constraint, which is evidenced by the linear ramp change in the MV immediately after each set-point change.

Figure 14-6 illustrates the same events as in Figure 14-5, except that the rate-of-change constraint is removed, and the EC factor for the AuxV violation is decreased to place greater importance on the AuxV, which means lesser importance on the CV to set-point response. Without the rate-of-change constraints, note the spikes in the MV action just after the set-point changes. Also, notice the greater SS deviation of the CV to the set point and associated smaller deviation of the AuxV from the limit of 30 during the 50 s to 70 s period.

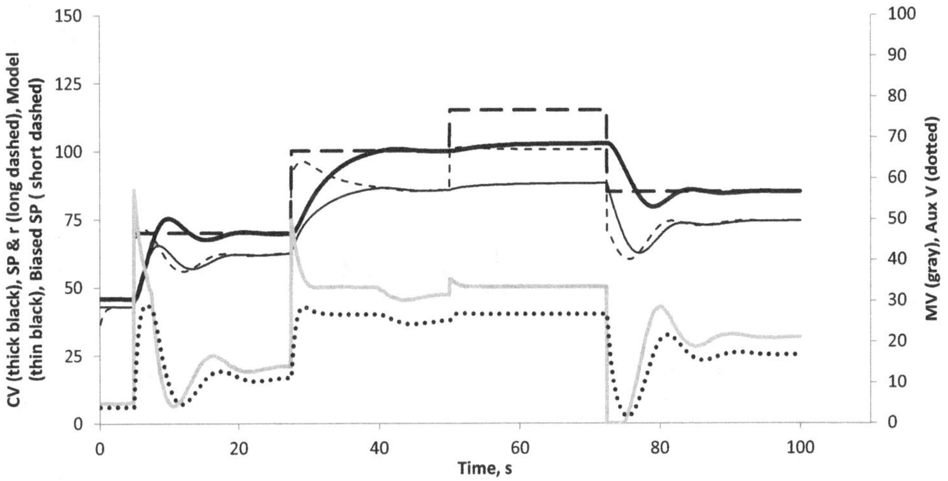

Figure 14-6. Events in AUTO mode, example 2.

In HPC, there is only one tuning factor for each CV: the reference trajectory time constant. In Figure 14-7, the time constant is significantly larger, making the controller less aggressive. Compare Figure 14-7 to Figure 14-6, with the same control choices (no rate-of-change constraint and duplicate EC factors) except for aggressiveness. Notice the smaller jumps in the MV action and the nearly overdamped response of the CV to the set point.

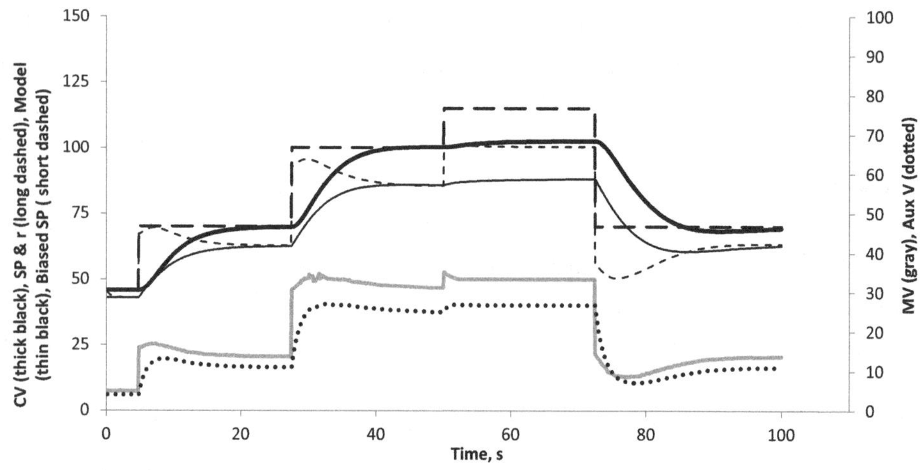

Figure 14-7. Events in AUTO mode, example 3.

Finally, in Figure 14-8, measurement noise and random disturbance patterns are included in the process. Compare this to Figure 14-6, which has the same tuning and EC values. The noise is evident in the perturbations on the CV value throughout. The disturbance is evident in the trend in the MV during the 40 s to 60 s period and the up-and-down action in the MV during the 20 s to 50 s period, even though the CV remains close to the set point. The disturbance is unmeasured, not included in the controller model, but it does affect the process.

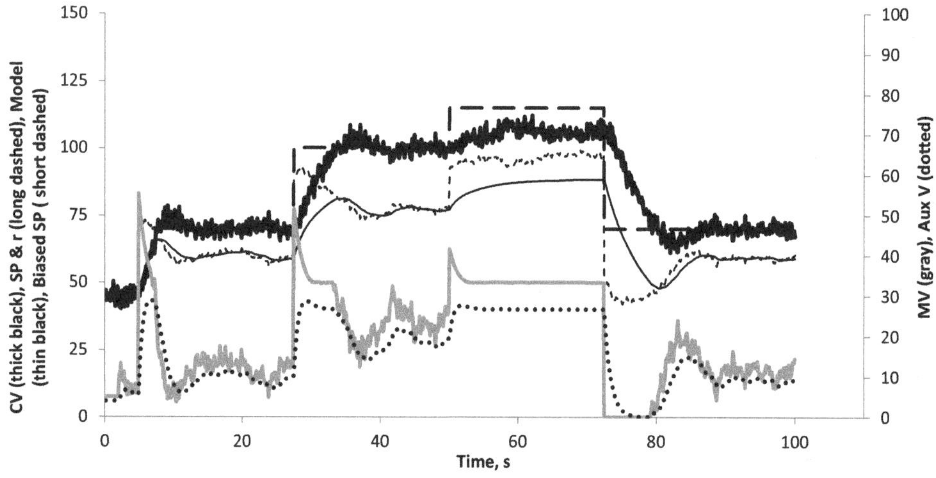

Figure 14-8. Events in AUTO mode, example 4.

The pmm is filtered to temper the impact of measurement noise on the control action. The noisy CV is not filtered, which has a benefit of being able to display the noise pattern and the real state of the process to the control room operators. If the process CV were filtered, creating a significant lag, the modeled CV should be equivalently filtered.

14.5 End Notes

- Any predictive control that uses a model to predict the future must also forecast the pmm value and any measured disturbance values. The pmm value is due to several factors (unmeasured disturbances, noise, and process-to-model mismatch), and it changes in time as disturbances and process behavior change. We know the past values of the pmm and measured disturbances, but we should have the future values when using the model to predict what will happen in the future and to have the appropriately biased model set point as the target.
- One could use any empirical or phenomenological model fit to past data to predict the future. It could be an autoregressive moving average model or a first-order model. However, it seems safer to consider that the disturbance makes a step-and-hold, holding at its current value into the future. This is the way we treat residuals in linear MPC. A model must be carefully developed to not respond to noise and retain rational forecasting limits. Attempting to model the future trend in disturbances is predicated on there being no new events in the control horizon. If there were, they would be currently unknown and could not be modeled.
- Although there is no basis to presume that the current set point will not change in the future, we presume that the current set point will be the set point in the future. Similarly, we presume that the present values of the pmm and the measured disturbances will not change in the near future.
- The model does not have to be perfect, great, very good, or even good. As this simulation reveals, a fair model works.
- Implement advanced control in stages. Start with the model shadowing the process to affirm model validity. Test bumpless transfer. Then, observe the controller that is suggesting actions. If you are confident it is making good decisions, let the controller take action while supervised. Finally, implement autonomous control, displaying goodness of control messages.
- Implement HPC showing the operator the MV predictions during the future horizon. Otherwise, the decoupling (with multivariable processes) and predictive constraint avoidance consideration might lead to unexpected decisions and operator discomfort.
- MVs may be removed for temporary maintenance. These should be removed from optimization. Include on-off switches so the controller understands the temporary condition.

- Be sure that the measurements are valid. Faulty inputs lead to bad decisions. This is no different for any level of complexity in control. You may need sensor redundancy or data reconciliation for critical measurements.
- You may need to change the model when the process changes (perhaps piping or units are reconfigured).
- Provide a trap for any possible infeasible calculations and a plan B action. Cascading MPC outputs as set points to lower-level PLC or DCS controllers is one strategy to keep the process running even if the supervisor MPC gets stuck.
- Keep it simple and safe (KISS). If gain-scheduled PI works, use it; if classic advanced regulatory control solves the problem, use it; if generic model control with a steady-state model solves the nonlinearity, use it.
- Consider incremental model parameter adjustment online to make the model adapt to the process. By observing the adapting model coefficient, one can identify a clue for predictive maintenance, and the model will be better at forecasting the constrained limits of process operation.
- The justifications for first-principles models in control include the following:
    - One-model consistency for many uses
    - Ability to preserve process mechanistic knowledge
    - Process analysis, monitoring, and constraint forecasting
    - Ability to handle nonlinearity, nonstationarity, and disturbances that invalidate linear models
    - Ease of tuning
    - Ease of and reduced process testing for model calibration (adjustment)
    - Wider valid modeling range than with linear models
- How can one economically justify using a nonlinear model? APC (linear multivariable control) experience shows a halving (or better) of process variance, which permits operating closer to constraints and specifications. This permits higher throughput, reduced waste, higher quality, and lower operating costs, providing economic benefits. In my assessment, any control scheme (e.g., ARC, fuzzy, APC, nonlinear MBC) with equivalent disturbance-rejecting, feedforward, and decoupling features has the same benefits as any controller with those same features. The nonlinear model-based controllers will have those benefits. They will also work over a wider operating range, require fewer recalibrations, and be less expensive to recalibrate than controllers with empirical models.

14.6 Takeaways

- Because it combines solutions to all control-related issues, nonlinear HPC could be considered the best strategy. However, it is also the most difficult to understand and most complex to program.

- Some advise using the simplest strategy that addresses the needs of a particular application. However, others advise using HPC for all applications, which unifies all implementations into one strategy and prepares control engineers on simple loops with the skills needed for more complicated loops. Using the simplest solution for each application matches the KISS principle I was raised with. On the other hand, standardizing on one control approach makes training and maintenance simpler.

Exercise

14.1 Access the nonlinear HPC simulator on *www.r3eda.com* and explore the features of bumpless transfer, equal concern values, events, optimizer convergence criterion, noise, constraint handling, and so on.

15 More on MISO Models

15.1 Introduction

The simple controllers explained so far in this book are single-loop controllers with one controlled variable (CV) and one manipulated variable (MV). However, several measured disturbances or loads that are process inputs could also influence the CV. In this case, the appropriate model could be multiple-input, single-output (MISO).

In control texts, the model is typically represented on one line of a mathematical representation, such as second-order plus dead time. However, phenomenological process models usually have several lines, one for each stage. For example, in controlling reactor temperature by jacket cooling water flow, the stages modeled could be (1) the flow rate response of a valve to the MV, (2) the jacket fluid temperature, (3) the heat transfer coefficient dependence on reaction content viscosity and jacket fluid velocity, (4) the heat removal rate, (5) the reaction content temperature, and, finally, (6) the new rate of reaction and its associated heat generation rate. Further complicating this, the heat transfer coefficient may also have separate equations relating to laminar or turbulent flow conditions, and if a reactor contents are not well mixed, the reactor could be modeled as having several zones. This series of steps would probably be solved in a function or subroutine, not as a single-line calculation.

Obtaining a complete first-principles model may not be possible. For instance, we commonly use dimensionless group models to give us the transfer and loss coefficients in heat transfer, mass transfer, and fluid flow. These were best fit to empirical data; they are not phenomenological. In reaction rates, we use kinetic models. Although they may be predicated on a first-principles reaction mechanism, the coefficient values in these models were best fit to empirical data.

There are many modeling methods. You will likely need to mix and match models and methods to your preference. Here are some considerations.

15.2 Empirical Trims to Steady-State Phenomenological Models

We commonly use dimensionless group models to predict transfer and fluid drag coefficients. Although the dimensionless group terms can be defined by converting a primitive partial differential equation into a dimensionless form, dimensionless groups are formulated more frequently in a heuristic manner, perhaps following a Buckingham Pi procedure after an intuitive listing of all the factors and properties that might influence the behavior. The common mathematical form for a dimensionless group model is a product of exponentiated dimensionless groups.

$$N_1 = aN_2^b N_3^c \tag{15-1}$$

which could be expressed as a sum of logarithms

$$Log(N_1) = a' + b \cdot Log(N_2) + c \cdot Log(N_3) \tag{15-2}$$

This form is not the form of an analytical solution to the differential equation, nor is it the mechanistic form related to a phenomenological concept. From experience, it is a convenient form that provides reasonably good models to match a limited range of experimental data. This approach has a long tradition in process modeling.

Truncated Taylor series models of function $z = f(x)$ have theoretical origins and proofs that a power series model can fit any function within a chosen error with enough terms. The truncated series form can also be rearranged to obtain a simple power series of polynomial model. For example, here is a three-term truncated Taylor series expansion about the base value $x = x_0$.

$$z(x) = z(x_0) + \frac{1}{1!}\frac{dz}{dx}\bigg|_{x0}(x - x_0) + \frac{1}{2!}\frac{d^2z}{dx^2}\bigg|_{x0}(x - x_0)^2 \tag{15-3}$$

Considering that x_0 is a fixed value and $z(x_0)$ and the derivatives evaluated at x_0 are constants, Equation 15-3 can be rearranged to the conventional quadratic power series model.

$$z(x) = a + bx + cx^2 \tag{15-4}$$

where, for example, $b = \left.\frac{dz}{dx}\right|_{x0} - x_0 \left.\frac{d^2z}{dx^2}\right|_{x0}$, a rearrangement of the constants in Equation 15-3.

Higher-order, truncated Taylor series models similarly reduce to higher-order power series models, providing a somewhat credible basis for using a power series model. But, again, the coefficients will be fit to match data, and the model may only be locally valid.

Any number of mathematical forms, including neural network models or human language models (fuzzy, for instance), could be used to fill in the gaps in phenomenological models.

15.3 Empirical Conversion of Steady-State to Transient Models

You may have an SS model and want to convert it to a transient model by presuming some form of linear dynamic response. This would be done using a Wiener or Hammerstein approach.

For instance, if the SS model is $y_{SS} = g(u,d,p)$ and it is presumed that y responds over time to u in a first-order manner, the blend of presumed linear dynamics with the SS model is

$$\tau \frac{dy}{dt} + y = g(u,d,p) \tag{15-5}$$

You could use transient process data to determine the τ value that makes this model best fit the data.

15.4 Empirical Trims to Transient Phenomenological Models

A transient model might be represented by a first-order differential equation.

$$\tau \frac{dy}{dt} + y = Ku \tag{15-6}$$

Using a forward difference for the derivative

$$\tau \frac{y_{i+1} - y_i}{\Delta t} + y_i = Ku_i \tag{15-7}$$

With $\frac{\Delta t}{\tau} \ll 1$, the numerical solution is

$$y_{i+1} = \left(1 - \frac{\Delta t}{\tau}\right) y_i + \frac{\Delta t}{\tau} K u_i \tag{15-8}$$

With constant values for all coefficients, this becomes

$$y_{i+1} = a y_i + b u_i \tag{15-9}$$

which is termed an *autoregressive* (the new y value depends on the prior y value) moving average (the new y value depends on the time-associated u value) model. It is an autoregressive moving average (ARMA) model with one past y and one past u of orders 1 and 1, ARMA(1,1).

The model for the prior y value could be inserted in Equation 15-9.

$$y_{i+1} = a^2 y_{i-1} + b u_i + a b u_{i-1} \tag{15-10}$$

which is now an ARMA model of orders 1 and 2 and a delay of 1. Successive substitutions could convert it to a higher-order moving average (MA) with a longer delay.

A fourth-order ordinary differential equation can be numerically converted to an ARMA(4,1) model.

$$y_{i+1} = a_0 y_i + a_1 y_{i-1} + a_2 y_{i-2} + a_3 y_{i-3} + b_0 u_i \tag{15-11}$$

Here, a_k represents the combined coefficient values.

Again, substitutions could convert this into a higher-order MA model with a delay.

$$y_{i+1} = \sum_{j=0}^{n} a_j y_{i-j} + \sum_{k=0}^{m} b_j u_{i-j} \tag{15-12}$$

If you do not know the order of your process, you do know that whatever it is, the phenomenological model can be converted to an ARMA(n,m) form. So, for empirical modeling, you choose orders n and m and then adjust the $n + m$ coefficient values to best fit the model to experimentally obtained data. Conveniently, the structure of Equation 15-12 makes this a linear regression.

Experience has shown that the most parsimonious form (the order choices that best fit data with the fewest coefficients) is $n = m$. Start with a simple empirical model and progressively increase the order until the fit to the transient data is satisfactory.

15.5 A Caution About Scale-Up

Model coefficients may come from lab- or pilot-scale data. The full-scale process may have different mechanisms than the unit that generated data to determine model coefficient values. For instance, a lab-scale reactor may be effectively ideally mixed. The contents are uniform. However, a large-scale reactor might not be instantly well mixed, and the contents touching the heat transfer surfaces might have a different temperature from the inside contents.

Also, for instance, a fluidized bed in a pilot-scale reactor might be well mixed, but in a large-area reactor, there may be individual cells of rotating fluidized particles.

Do not accept the lab- or pilot-scale models as true to the full-scale process. The models may need to be revised.

On the other hand, you will never obtain a perfect model, and a model with a grade of C may be fully adequate for control because of the frequency of control actions. So, seek perfection, but balance that with sufficiency.

15.6 Takeaway

- Document and explain how you derived, structured, and solved your models. It is important that the person who inherits your application can adapt the model to new situations.

Exercises

15.1 Defend using a third-order polynomial as an appropriate fit of $y = f(u)$ steady-state (SS) data, even if the relation is not a third-order polynomial.

15.2 Determine a and b values in Equation 15-9 from coefficients in the original ordinary differential Equation 15-6. If you did not have K and τ values, explain how to obtain them from empirical data that is noisy and subject to unmeasurable disturbances.

16 Equivalence to PID

16.1 Concept

When single-input, single-output (SISO), model-based control (MBC) uses simple linear models and is unconstrained, it reduces to conventional proportional-integral (PI) control. This is comforting considering the years of trust in PI.

Consider simple MBC with a first-order model and the desire to identify the manipulated variable (MV) value that begins moving the model to the biased set point.

The MBC diagram in Figure 16-1 can be rearranged. Consider that there is no override and that the Correct function is simply a pass-through of the process-model mismatch (pmm). Slide the process-to-model difference operation through the set point plus pmm summation. Figure 16-2 shows the controller acting on the actuating error within the dashed lines.

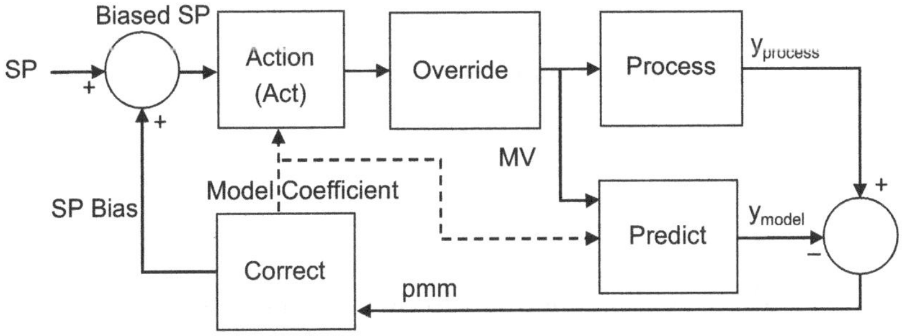

Figure 16-1. MBC representation.

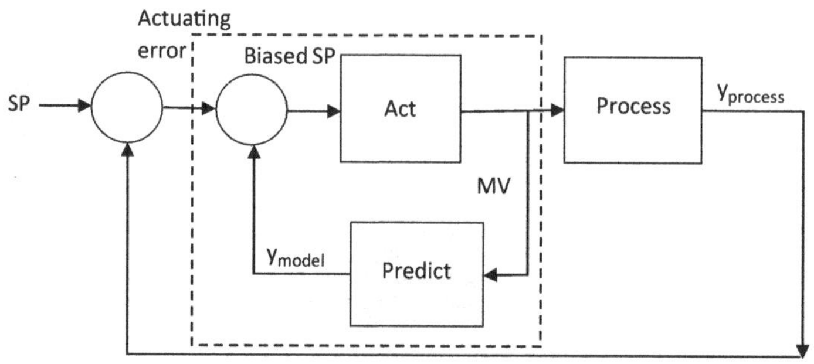

Figure 16-2. Figure 16-1 rearranged.

If the model is first-order

$$\tau_m \frac{dy_m}{dt} + y_m = K_m u \tag{16-1}$$

The desire is that the model changes at a rate proportional to the difference between the model set point and the model y value.

$$\left.\frac{dy_m}{dt}\right._{desired} = \frac{y_{mSP} - y_m}{\tau_{desired}} = \frac{y_{mSP} - y_m}{\tau_{desired}} = \frac{y_{SP} - y_p}{\tau_{desired}} = \frac{e}{\tau_{desired}} \tag{16-2}$$

Then, in the Act function

$$u = \left(\tau_m \frac{e}{\tau_{desired}} + y_m \right) / K_m \tag{16-3}$$

In Laplace notation, the Act function is

$$\hat{u} = \left(\tau_m \frac{\hat{e}}{\tau_{desired}} + \hat{y}_m \right) / K_m \tag{16-4}$$

The Laplace-transformed model function is

$$\hat{y}_m = \frac{K_m}{\tau_m s + 1} \hat{u} \tag{16-5}$$

Combining reveals the reset feedback version of the PI controller.

$$\hat{u} = \frac{1}{K_m} \frac{\tau_m}{\tau_{desired}} \hat{e} + \frac{1}{\tau_m s + 1} \hat{u} \tag{16-6}$$

By solving explicitly for $\hat{u}$, the control function becomes the conventional PI controller with a gain of $K_m = \frac{1}{K_m}\frac{\tau_m}{\tau_{desired}}$ and an integral time of $\tau_i = \tau_m$.

$$\hat{u} = \frac{1}{K_m}\frac{\tau_m}{\tau_{desired}}\left(1 + \frac{1}{\tau_m s}\right)\hat{e} \tag{16-7}$$

16.2 Takeaway

- Use conventional PI control if the process can be considered linear and stationary. Nonlinear controllers are needed only in processes with ill-behaved dynamics, constraints, and nonlinear behavior.

Exercises

16.1 Do the equivalence analysis in Section 16-1 using a second-order model. You will need to change the objective to follow a second-order reference trajectory. You cannot make the second-order model follow a first-order response. Choose the τ_{want} to match one of the model time constants. You should get that the controller is a proportional-integral (PI) controller acting on a filtered actuating error.

16.2 Do the equivalence analysis in Section 16-1 using a first-order plus dead time (FOPDT) model. You cannot invert the delay, so use an approximation for it. Commonly, $e^{-\theta s}$ is approximated with a first-order Pade approximation, $\frac{1-\theta s/2}{1+\theta s/2}$. Using some algebra, you should get that the controller is proportional-integral-derivative (PID) on a filtered actuating error.

Section 3

Supporting Techniques for Multivariable Constrained Control

17 Data and Action Filtering

17.1 Introduction

This chapter is extracted from the *InTech* article "Removing Unwanted Data Signal Components or Features" by Joseph S. Alford, Brian M. Hrankowsky, and R. Russell Rhinehart.[1]

In many situations, filtering is valuable to remove unwanted components or features from a data signal before the input is used in an automation and control application. The unwanted component can result from many situations, including random noise caused by mixing turbulence or mechanical vibrations, spurious outlier events such as missed data packets or isolated spikes, or periodic confusion, for instance, from power or rotational harmonics. Data filtering used properly can improve control, productivity, and decision-making.

Smoothing data and eliminating outliers have several advantages. For example, reducing noise in the derivative portion of a controller could lead to greater use of the derivative term in proportional-integral-derivative control (and, therefore, improved control). Data filtering also reduces distracting features in trend plots. In addition, less noise in a control loop can contribute to reduced variation in the controller output. However, data filtering can also have negative consequences, such as hiding or delaying the recognition of real problems occurring or developing in a process or its equipment.

---

1. Joseph S. Alford, Brian M. Hrankowsky, and R. Russell Rhinehart, "Removing Unwanted Data Signal Components or Features," *InTech* 65, no. 4 (July/August 2018): 14–19.

Filtering can also present a skewed (i.e., invalid) view of the magnitude and duration of real spikes occurring in the process. Data filtering generally causes a delay or a lag that can interfere with control. The engineer must understand that the real process value and the measured or displayed value are not the same.

In some applications, little or no data filtering is recommended. This is often the case with signals sent to alarm algorithms, sent to data collection systems, and presented on a human-machine interface for critical process parameters. In such applications, it is important to monitor and record true details of process excursions, not to reduce the perceived magnitude or extend the perceived duration of process spikes through filtering.

Several different data filtering technologies are utilized in industrial automation systems. This chapter discusses some of them, including comments on the pros and cons.

17.2 Noise Tempering Methods

In these noise tempering methods, it is assumed that the process holds a steady value in time and that random fluctuations corrupt the signal. The objective of the filter is to reveal the underlying value.

17.2.1 Moving Average Filter

A moving average (MA) filter reports the conventional average of data in a window.

$$\bar{X} = \frac{1}{N}\sum_{i=1}^{N} X_i \tag{17-1}$$

Here, $i = 1$ indicates the most recent data, and the average is over the past N data values. This is termed a *window* of the data. At each sampling, the window moves, the newest data enters, and the oldest leaves. The user must specify the window length as a time duration or as the number of data in the window. The more data, the lower the variability on the average. At a steady condition (the nominal value is not changing, but the measurements fluctuate randomly due to noise effects). Using σ the standard deviation to represent uncorrelated variability, the variability of the average is related to the variability of the individual data:

$$\sigma_{\bar{X}} = \sigma_{X_i}/\sqrt{N} \tag{17-2}$$

The $\sqrt{N}$ impact is true for any measure of variability, such as range.

Note: Variation cannot be eliminated by averaging, just attenuated. The user must choose the value of N. Larger N means less variation, but it also means that it will take longer for the average to move to the proximity of the new value when the data value changes.

Figure 17-1 reveals a process (data are the markers) that is initially at a noisy steady state at a nominal value of 2 and makes a step change to a new value of 5 at the 30th sampling. The data value is on the vertical axis, and the sample number is on the horizontal axis. A data spike, a spurious signal with a value of zero, occurs at sampling 120. The thin dashed line connects the data dots to help reveal the trend. Subsequent figures will show the results of diverse filtering methods on the same data.

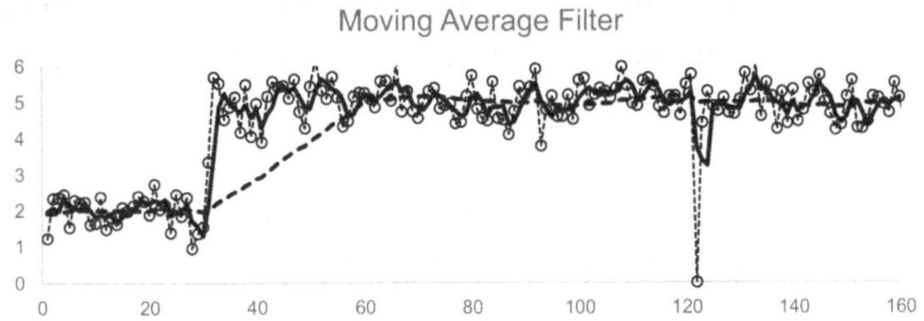

Figure 17-1. Characteristic performance of an MA filter to a process step change and spike.

In Figure 17-1, the solid line represents the MA filter with $N = 3$ and the dashed line with $N = 30$. Several features of the filter are revealed in the figure: (1) The MA with $N = 3$ rises more rapidly to the new data value, but its variation is larger than the MA with $N = 30$. (2) The character of an MA filter response to a step change in the signal is a linear ramp toward the new value, which lasts N samples. (3) The response to the one-time outlier is a pulse with a duration of N samples and a magnitude of $1/N$ of the outlier deviation.

An advantage of the MA filter is that the concept of averaging is commonly understood, and the tuning choice of window length N is easily understood. When periodic disturbances are present, a window length chosen to match the period will result in a signal that moves to the new average in one period. However, the algorithm response to a process change is delayed, which delays alarm and control application responses. The algorithm does not reject data outliers.

17.2.2 First-Order Filter

A first-order filter (FoF) has several alternate names, including an *exponentially weighted moving average* (EWMA) and an *autoregressive moving average* (ARMA). It represents the analog device methods for filtering an electronic resistor-capacitor or pneumatic restriction and bellows, but it can be performed digitally. The advantage is that it is computationally simpler than an MA filter, which must store and process all the data in the window. The FoF equation is

$$X_{f,i} = \lambda X_i + (1 - \lambda) X_{f,i-1} \tag{17-3}$$

Here, lambda, λ, is the filter factor (it is not related to lambda tuning). It ranges between 0 and 1, $0 < \lambda < 1$. X_i is the most recent data measurement, $X_{f,i-1}$ is the prior filtered value, and $X_{f,i}$ is the new filtered value. If one derives the equation as an approximation to the MA method, essentially $\lambda = \frac{1}{N}$. Some vendors use the symbol f for the filter factor, and some switch the λ and $(1 - \lambda)$ weighting. If one derives the FoF from the differential equation representing a first-order resistor-capacitor circuit, then $\lambda = 1 - e^{-\Delta t/\tau}$, where Δt is the sampling interval and τ is the time constant. With substantial filtering, the time constant is approximately related to the number of data, $\tau \cong \Delta t N$. The user must be aware of how the vendor presents the filter.

The FoF also tempers but does not remove noise. Here, if the process is steady with random fluctuations, the relation for variability attenuation is

$$\sigma_{X_f} = \sigma_{X_i} \sqrt{\frac{\lambda}{2-\lambda}} \tag{17-4}$$

The user chooses the value of λ (or alternately f or τ) to temper the noise. Although the benefit of an FoF over the MA filter is computational simplicity, the negative is the persisting influence of long-past data. In the MA, once a data value is out of the window, it no longer influences the average. However, in a FoF, the past values are exponentially weighted, fading in time but never totally leaving.

Figure 17-2 illustrates FoF performance. The filtered value makes a first-order response to a change in the process. The lambda values of 0.500 (solid line) and 0.0645 (dashed line) are chosen to represent the same noise reduction as obtained with the $N = 3$ and $N = 30$ values of the MA filter. Combining Equations 17-2 and 17-4 for equivalent variance reduction

$$\lambda = 2/(N+1) \tag{17-5}$$

Figure 17-2. Characteristic performance of an FoF to a process step change and spike.

Notes:

1. Both filtered values show an exponential rise to the new value, requiring three to four time constants, a period of $N \cong 3.5/\lambda$, to make about 97% of the change to reach the proximity of the new value.
2. After the spike at sample 120, there is a progressive relaxation back to the filtered value.

The key advantage of the FoF over the MA filter is that the algorithm is computationally simple. However, interpreting the filter factor or time constant is less intuitive than choosing N in a moving average. Also, the FoF does not reject outliers.

In the MA or FoF, the user must choose the filter coefficient value to best balance the lag or ramp period and desirably temper the variation. In either case, the user must realize that the lag can be detrimental to control if it is similar in magnitude to the process primary time constant or dead time.

17.2.3 Statistical Filter

One of several statistical approaches to filtering is based on a statistical process control desire to prevent tampering. I discussed the filters in articles in 1992[2] and published a follow-up discussion in 2002.[3]

Even at steady conditions, the filters discussed will continually report small deviations. If the filters are used in automatic control, the controller will seek to correct the deviations. The prevent-tampering concept is to hold a single filtered value until there is statistically confident evidence that the process value has changed and then change the filtered value. In one technique, the observed metric is the cumulative sum (CUSUM) of measurement deviations from the filtered value. If the CUSUM becomes statistically significant (perhaps at the 3-sigma level), there is adequate justification to change the filtered value.

$$IF\left(\frac{|CUSUM|}{\sigma\sqrt{N}} > 3\right) THEN\ (X_{SPC} = X_{SPC} + CUSUM/N) \tag{17-6}$$

Figure 17-3 illustrates the CUSUM filter on the same data. Note that the filtered signal does not change during the initial steady-state period and jumps quickly when there is a real change. It took about five samples to be statistically confident in the step change from 2 to 5. In this data set, the filter did not jump quite far enough but made

2. R. R. Rhinehart, "A CUSUM-Type On-Line Filter," *Process Control and Quality* 2, no. 2 (February 1992): 169–176.
3. R. R. Rhinehart, "A Statistically Based Filter," *ISA Transactions* 41, no. 2 (April 2002): 167–175.

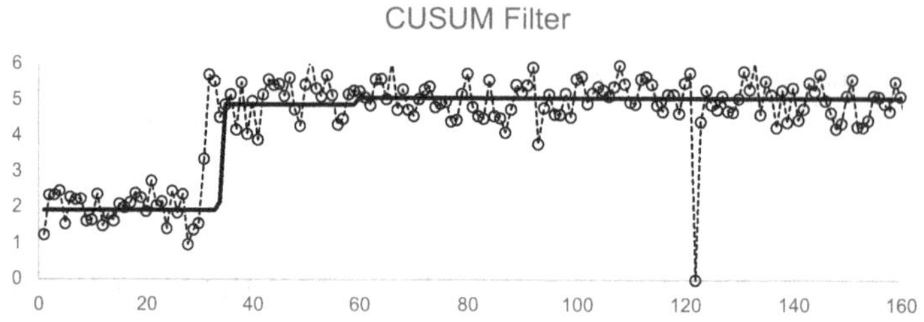

Figure 17-3. Characteristic performance of a statistical filter (3 sigma, 99% confidence) to a process step change and spike.

a correction after samples in the 40 to 60 sample period provided adequate confidence. Also, note that the outlier at sample 120 was rejected.

The statistical-based filters are scale-independent; they adapt to the noise amplitude.

A statistically based filter moves rapidly when a change is confidently detected and holds a constant value in between. When the process noise amplitude changes, the responsiveness automatically changes. However, the code for this CUSUM filter is about 10 lines and requires that the noise be relatively compliant with the basis of independent sample-to-sample fluctuations. The user interpretation of the step-and-hold behavior of the statistical filter will be unfamiliar to many.

17.3 Outlier Rejection Filters

Here, the concept is that the signal is steady, but an occasional event happens to provide a one-time (or brief) wholly uncharacteristic value. This is often called an *outlier, phantom,* or *spurious event*. The filter purpose is not to average but to ignore that outlier, which might be related to occasional dropped data or an electrically induced spike. Electrical sources include voltage or current surges from a nearby lightning strike, radio communications, nearby motors starting up, or electric floor scrubbers. Loose or corroding wiring connections, sensors that receive a mechanical shock, or dropped data packets in an overloaded communication system can also cause short-lived or one-sample outliers.

17.3.1 Median Filter

The median filter reports the middle of the most recent values—not the middle in chronological order but the middle in value. For instance, if the most recent values are 5, 6, and 3, the middle value 5 is reported. A median filter could be based on a larger number of sequential data, such as 5 or 7 values. If you suspect that two outliers could happen sequentially because of some common cause, then a median of five sequential values will reject them. The median filter does temper noise a bit, but the application intent should be to remove outliers.

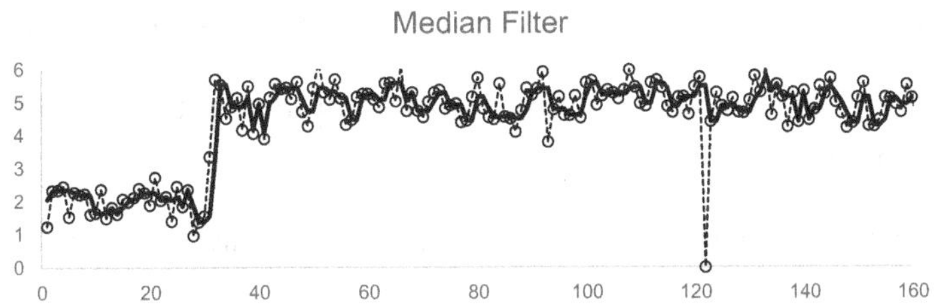

Figure 17-4. Characteristic performance of a median filter (middle of 3) to a process step change and spike.

Figure 17-4 illustrates the median filter (middle of 3) applied to the example data set in previous figures.

Notes:

1. When the signal makes a step change at the 30th sampling, the filter has a delay of about half the number of data.
2. The outlier at sample 120 is wholly ignored.
3. Throughout, the vagaries of the signal substantially mimic the measurement.
4. The median filter removes outliers and rapidly tracks real changes. However, the user must choose an N that is large enough to exclude persistent outliers. Masking outliers can misrepresent important features, and noise is not removed.

17.3.2 Heuristic Methods

These are user-defined, as appropriate, and could be based on any number of data consistency or validation checks that a human observer might use to judge veracity. For instance, if a sudden change in one measurement correlates to a simultaneous or prior change in another, the one-time effect may be interpreted as real, not an outlier. The logic is usually implemented as IF-THEN-ELSE rules that pass through valid data and may flag what appear to be outliers. Flagged data could then be "thrown out" or ignored by process control applications but may still be retained for historian recording purposes.

For instance, use a trip delay logic where a process variable (e.g., fired heater fuel gas pressure) must be above the interlock set point for more than 3 s (e.g.) for the interlock to activate. This is another method to prevent spuriously activating interlocks based on process noise.

17.4 Takeaways

- Filtering can be used to either temper noise or eliminate outliers. Use the right tool for disparate applications.
- If the signal is noiseless (and void of outliers), filtering is unnecessary.
- If the noise level changes, the user must adjust the filter factor to maintain the desired balance of noise attenuation to lag.

- Statistical filters automatically adapt to changes in noise amplitude.
- Filtering adds a lag or delay, which could impair control action and/or require alternate controller tuning.
- Diverse filtering methods can be used in combination, such as a median filter to reject outliers and an FoF to reduce noise.
- Filter effects and options are on nearly every device. Recognize where these might be.

Exercises

17.1 Propagate the variance on Equation 17-1 to obtain 17-2.

17.2 Propagate the variance on Equation 17-3 to obtain 17-4.

17.3 Explain the appropriate choices for the threshold in Equation 17-6. The value of 3 represents the 99.7% confidence. A value of 2 represents 95% confidence.

17.4 Quantify the computational burden improvement of the first-order filter (FoF) over the moving average filter.

18

Steady- and Transient-State Identification

18.1 Introduction

This chapter is extracted from Chapter 14 in *Applied Engineering Statistics*, Second Edition, by R. R. Rhinehart and R. M. Bethea.[1] Here, the methods are presented for the person seeking to apply steady-state (SS) or transient-state (TS) automated identification. For greater analysis and detail, see the original.

It is important to identify SS and TS in noisy signals. SS models are widely used in process control, online process analysis, and process optimization, and because manufacturing and chemical processes are inherently nonstationary, selected model coefficient values must be adjusted frequently to keep the models true to the process and functionally useful. Additionally, the detection of SS triggers collecting data for process fault detection, data reconciliation, neural network training, the end of an experimental trial (when you collect data and implement the next set of conditions), and other purposes. However, the use of SS models for process analysis or their data-based adjustment should only be triggered when the process is at SS.

---

1. R. R. Rhinehart and R. M. Bethea, *Applied Engineering Statistics*, 2nd ed. (Boca Raton, FL: Taylor & Francis CRC Press, 2022).

In contrast, transient, time-dependent, or dynamic models are also used in control, forecasting, and scheduling applications. Dynamic models have coefficients representing time constants and delays, which should only be adjusted to fit data from transient conditions. The detection of TS triggers collecting data for dynamic modeling. It also recognizes points of change, wake-up data recording, the beginning of a process response to an event, interruptions to the norm, and other events.

One characteristic of these online, real-time identification applications is that we only have past and current data. We do not have the future points. In many filtering applications, such as image and recording enhancement and detection, filtering is done offline after all the data has been collected. We have access to both the before and after data and can use kernel-type filtering approaches to use both sides to estimate the in-between value better. By contrast, we must decide now in real-time applications without seeing the future.

The methods presented here are for processes that asymptotically approach a steady state but where the actual signal is confounded by noise. These approaches are not applicable if rotating machinery creates a cyclic response as a background oscillation to the steady signal.

18.2 Approaches and Issues to SS and TS Identification

A conceptually simple approach to identify an SS would be to look at data values in a recent time window, and if the range between high and low is acceptably small, declare SS. This approach, however, requires the human to decide the acceptable range for each variable and the time window duration, and if the process noise level changes, the threshold should change. Although the approach is simple to understand, easy to implement, and often works acceptably, it is not universal.

Another straightforward implementation of a fully automated method for steady-state identification (SSID) would be a statistical test of the slope of a linear trend in the time series of a moving window of data. This technique is a natural outcome of traditional statistical regression training. Here, at each sampling, use linear regression to determine the best linear trend line for the past N data points. If the t-statistic for the slope exceeds the critical value, there is sufficient evidence to confidently reject the SS hypothesis and claim it is probably in a TS. A nice feature of this approach is that the determination is independent of the noise amplitude. However, at the crest or trough of an oscillation centered in the N data window, this slope would be nearly zero, and SS would be claimed during the TS. The approach is also somewhat of a computational burden.

Another straightforward approach is to evaluate the average value in successive data windows. Compute the average and standard deviation of the data in successive data sets of N samples, then compare the two averages with a t-test. If the process is at SS, ideally, the averages are equal, but noise will cause the sequential averages to fluctuate. If the fluctuation is excessive relative to the inherent data variability, the t-statistic (difference in average divided by the standard error of the average) will exceed the critical t-value, and the null hypothesis (process is at SS) can be confidently rejected to claim the process is probably in a TS. When windows are centered on either side of the crest or trough in an oscillation, the averages will be similar and SS will be falsely accepted. A solution could be to use three or four data windows, each with a unique N, to prevent possible matching to a periodic oscillation.

Many other approaches have been used, including statistical process control techniques, dual filters, runs tests, wavelets, polynomial interpolation, thresholds on variance or slope, closure of SS material and energy balances, and variance ratios in the data.

Such tests have a computational burden that does not make them practicable for applications that are online, in real time, and executed within most process control computers.

Notes:

1. A practicable method must be computationally simple, robust to the vagaries of process events, easily implemented, and easily interpreted.
2. Most methods reject the null hypothesis, which is a useful indicator that the process is confidently in a TS. However, not rejecting SS does not permit a confident statement that the process is at SS. A legal judgment of "not guilty" is not the same as a declaration of "innocent." "Not guilty" means there was insufficient evidence to confidently claim "guilty" without a doubt. Accordingly, there must be a dual approach that can confidently reject TS to claim probable SS and reject SS to claim probable TS. This chapter presents several dual approaches.

18.3 A Ratio of Variances Method

Von Neumann[2] and Crowe and colleagues[3] independently proposed an approach that calculates the variance on a data set by two approaches—the mean-square deviation from the average and the mean-square deviation between successive data. The ratio of variances is an F-like statistic, and assuming no autocorrelation in the data, it has an expected value of unity when the process is at SS.

---

2. J. von Neumann, "Distribution of the Ratio of the Mean Square Successive Difference to the Variance," *Annals of Mathematical Statistics* 12 (1941): 367–395.
3. E. L. Crowe, F. A. Davis, and M. W. Maxfield, *Statistics Manual* (New York: Dover Publications, 1955).

The filter approach to determine the value of the same ratio statistic published by Cao and Rhinehart[4] is similar but computationally simpler.

Begin with this conceptual model of the phenomena: The true process variable (PV) is at a constant value (at SS), and fluctuations in the measurement and signal transmission process create uncorrelated "noise," independently distributed fluctuations on the measurement.

If the noise distribution (mean and variance) were uniform in time, then statistics would call this time series stationary. However, for a process, the true value, nominal value, or average may be constant in time, but the noise distribution may change. So, SS does not necessarily mean stationary in the statistical sense of the term.

The first hypothesis of this analysis is the conventional null hypothesis that the process is at a steady state, H_0: SS. The statistic, a ratio of variances, will ideally have a value of unity, but due to the caprices of noise, it will have a distribution of values at SS. As long as the ratio statistic (R-statistic, or simply R) value is within the normal range of the distribution of SS values, the null hypothesis cannot be rejected. When the R-statistic has an extreme value (R-value), the null hypothesis can be rejected with a certain confidence level and probable TS claimed.

By contrast, there is no single conceptual model of a TS. A transient condition could be due to a ramp change in the true value, an oscillation, a first-order transient to a new value, a step change, or another event. Each is a unique type of transient. Further, each single transient event type has unique characteristics such as ramp rate, cycle amplitude and frequency, and time constant and magnitude of change. Further, a transient could be comprised of any combination or sequence of the not-at-SS events. Because there is no unique model for TS, there can be no null hypothesis or corresponding unique statistic that can be used to reject the TS hypothesis and claim probable SS. Accordingly, an alternate approach must be used to claim probable SS.

The alternate approach used here is to take a transient condition that is barely detectable or decidedly inconsequential (per human judgment) and set the probable SS threshold for the R-statistic as an improbable low value, but not so low as to be improbably encountered when the process is truly at SS.

So, there are two one-sided tests, one to reject SS and one to reject TS, and two critical values, as illustrated in Figure 18-1, where the vertical axis is the cumulative distribution

4. S. Cao and R. R. Rhinehart, "An Efficient Method for On-Line Identification of Steady-State," *Journal of Process Control* 5, no. 6 (1995): 363–374.

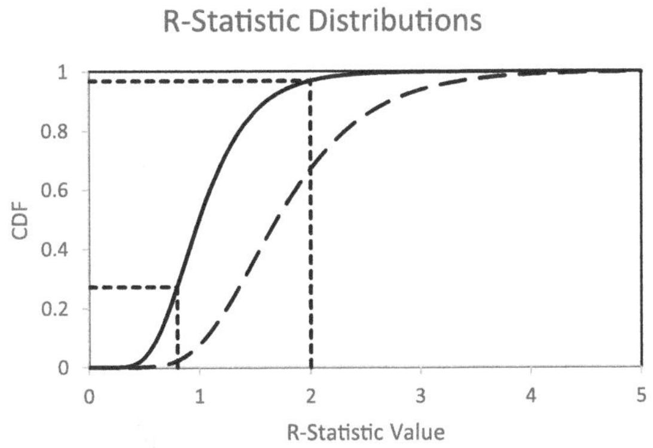

Figure 18-1. Illustration of dual rejection regions. The vertical axis is the CDF, and the horizontal axis is the SS ratio statistic. The solid curve is the F-like distribution of R-values when at SS. The dashed curve is the F-like distribution when not at SS.

Source: From *Applied Engineering Statistics* by R. R. Rhinehart and R. M. Bethea. Used with permission from Taylor & Francis Group LLC.

function (CDF) and the horizontal axis is the SS ratio statistic (R). The solid curve is the distribution of R-values at SS and is centered on a value of 1. The dashed curve is the distribution when the process is not at SS, but nearly so, and is centered on about 1.7.

The vertical dashed lines are the critical values. The line on the far right is the trigger to reject SS. It intersects the at-SS CDF at a value of about 0.95. There is a 5% chance that an at-SS process will generate an R-value greater than 2. However, there is about a 35% chance ($0.35 = 1 - 0.65$) that the nearly-but-not-at-SS process will generate an R-value greater than 2. If $R > 2$, bet on the TS condition.

The far-left vertical line is the trigger to reject TS. It intersects the at-SS CDF at a value of about 0.45. There is a 45% chance that an SS process will generate an R-value less than 0.9. However, there is only about a 5% chance that the nearly-but-not-at-SS process will generate an R-value less than 0.9. If $R < 0.9$, bet on the SS condition.

In this illustration, the odds of correctly rejecting SS are $0.35/0.05 = 7{:}1$. And the odds of correctly accepting SS are $0.45/0.05 = 9{:}1$. However, in most transient conditions, the dashed not-at-SS curve is further to the right, and both odds are better than illustrated.

18.3.1 Filter Method

Figure 18-2 illustrates the filter method concept to create an R-statistic. The markers represent process measurements over the 100 sequential samples indicated on the horizontal axis. The process value starts at about 10, ramps to about 15, and then holds steady. The true trend is unknowable; only the measurements can be known, and they are infected with noise-like fluctuations. The solid line is a first-order filtered value of

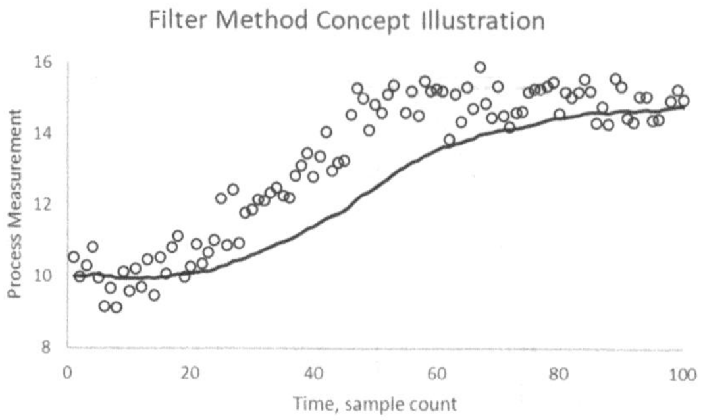

Figure 18-2. Filter method concepts from *Applied Engineering Statistics*, second edition, by R. R. Rhinehart and R. M. Bethea. Used with permission from Taylor & Francis Group LLC.

the data. It starts at about 10, then lags behind the process, and finally settles to a value representing the process level of 15 at the end of the chart. The filtered value is not smooth but reveals wiggles due to the high and low vagaries of the data.

The method first calculates a filtered value of the process measurements, then the variance in the data is measured by two similar methods. One is based on the difference between the measurement and the filtered trend. The other is based on deviations between sequential data measurements.

If the process is at SS, as illustrated in the 0 to 10 and 90 to 100 time periods, the filtered value, X_f, remains almost in the middle of the data. Then, a process variance, s_1^2, estimated by differences between the data and filtered value, will ideally be equal to the true value of σ^2. The variance can also be estimated by the data-to-data (not average-to-data) differences, s_2^2. Then the ratio of the variances, $r = \frac{s_1^2}{s_2^2}$, will be approximately equal to unity. Alternately, if the process is in a TS, such as in the 20 to 60 time period, then X_f is not in the middle of the data, the filtered value lags behind the process, the variance measured by the data-to-filter difference will be much larger than the variance estimated by sequential data differences, $s_1^2 \gg s_2^2$, and the ratio will be much greater than unity.

To minimize the computational burden, in this method, a filtered value (not an average) provides an estimate of the data mean:

$$X_{f,i} = \lambda_1 X_i + (1 - \lambda_1) X_{f,i-1} \tag{18-1}$$

where

X = the PV

X_f = filtered value of X

λ_1 = filter factor
i = time-sampling index

Note: Alternate terms for the filtered value are *first-order filter, first-order lag,* and *exponentially weighted moving average.*

The first method to obtain a measure of the variance uses an exponentially weighted moving "variance" (another first-order filter) based on the difference between the data and the filtered value, representing the average:

$$\upsilon^2{}_{f,i} = \lambda_2 (X_i - X_{f,i-1})^2 + (1 - \lambda_2)\upsilon^2{}_{f,i-1} \tag{18-2}$$

where

$\nu^2_{f,i}$ = filtered value of a measure of variance based on the differences between data and filtered values
$\nu^2_{f,i-1}$ = previous filtered value

In Equation 18-2, the symbol ν^2 is a measure of the variance to be used in the numerator of the ratio statistic. Because it is calculated from the filtered value, not the average, ν is slightly larger than σ. The previous value of the filtered measurement is used instead of the most recently updated value to prevent autocorrelation from biasing the variance estimate, $\nu^2_{f,i}$, keeping the equation for the ratio relatively simple. Equation 18-2 does not provide the true variance, even at SS, because using the filtered X_f, rather than the true average adds a small amount of variability to the difference $(X_i - X_{f,i-1})$.

The second method to obtain a measure of variance is an exponentially weighted moving "variance" (another filter) based on sequential data differences:

$$\delta^2_{f,i} = \lambda_3 (X_i - X_{i-1})^2 + (1 - \lambda_3)\delta^2{}_{f,i-1} \tag{18-3}$$

where

$\delta^2_{f,i}$ = filtered value of a measure of variance
$\delta^2_{f,i-1}$ = previous filtered value

This will be the denominator measure of the variance. Also, it is not the variance but is effectively twice the variance when the process is at SS. So, it uses the symbol δ instead of σ.

The ratio of variances, the R-statistic, may now be computed by the following simple equation:

$$R = \frac{(2-\lambda_1)v^2{}_{f,i}}{\delta^2{}_{f,i}} \tag{18-4}$$

Because Equations 18-2 and 18-3 compute a measure of the variance, not the true variance, the coefficient in Equation 18-4 $(2 - \lambda_1)$ is required to scale the ratio to represent the classic variance ratio. At a steady state, the R-value will fluctuate about a value of unity. However, during a TS, it will fluctuate at larger values. The calculated R-value must be compared to its two critical values to determine SS or TS. Complete executable code, including initializations, is presented in the software on the site *www.r3eda.com*. The essential assignment statements for Equations 18-1 to 18-4 are:

nu2f := l2 * (x - xf) ^ 2 + cl2 * nu2f	Equation 18-2
xf := l1 * x + cl1 * xf	Equation 18-1
delta2f := l3 * (x - x_old) ^ 2 + cl3 * delta2f	Equation 18-3
x_old := x	Update prior value
R_Filter := (2 - l1) * nu2f/delta2f	Equation 18-4

The coefficients l1, l2, and l3 represent the filter lambda values, and the coefficients cl1, cl2, and cl3 represent the complementary values. $cl1 = 1 - l1$. Equations 18-2 and 18-1 are calculated in reverse order so the prior xf value does not need to be stored.

This method's five computational lines of code require direct, no-logic, low storage, and low computational operation calculations. There are 4 variables and 7 coefficients to be stored, 10 multiplication or divisions, 5 additions, and 2 logical comparisons per observed variable.

Without prior knowledge of the value for σ, initialize filtered values with 0. This is more convenient than initializing them with a more representative value from recent past data. This leads to an initial not-at-SS value of the initial R-statistic, but after about 35 samples, the initial wrong values are incrementally updated with representative values.

Being a ratio of variances, the statistic is scaled by the inherent noise level in the data. It is also independent of the dimensional units chosen for the variable.

Critical values for the R-statistic, based on the process being at an SS with independent and identically distributed variation (white noise), were also developed by Cao and Rhinehart,[5] who suggest that filter values of $\lambda_1 = 0.2$ and $\lambda_2 = \lambda_3 = 0.1$ produce the best balance of Type I and Type II errors. The null hypothesis is that the process is at a steady state. If the computed R-statistic is greater than R-critical (a value of about 2.5), then we are confident that the process is not at an SS.

However, if the R-value is slightly less than the upper critical value, the process may be in a mild TS. Consequently, a value of R less than or equal to a lower R-critical value, $\sim$0.9, means the process may be at an SS.[6]

We often assign values of either "0" or "1" to a variable, SS, which represents the state of the process. If R-calculated > R-critical1 $\sim$ 2.5, "reject" SS and assign SS = 0. Alternately, if R-calculated < R-critical2 $\sim$ .9, "accept" that the process may be at an SS and assign SS = 1. If in-between values happen for R-calculated, hold the prior 0 or 1 (reject or accept) state because there is no confidence in changing the most recent declaration.

The method presumes no autocorrelation in the time series of the process measurement data at SS. This may require selecting a sampling time interval longer than the control interval. Running in real time, the identifier does not have to sample at the same rate as the controller.

18.3.2 Choice of Filter Factor Values

The filter factors in Equations 18-1, 18-2, and 18-3 can be related to the number of data (the length of the time window) effectively influencing the average or variance calculation. Simplistically, the effective number of data in the window is $N = 1/\lambda$. If $\lambda = 0.1$, then effectively, the method is observing $N = 10$ most recent data points. However, based on a first-order decay, long-past data retain some influence, and roughly, the number of data effectively influencing the window of observation is about $3.5/\lambda$. If $\lambda = 0.1$, then effectively, the method remembers the impact of $N \approx 35$ data points. However, not all data points have equal weighting. The filter is a first-order decay, an exponentially declining weighting N past data. The long-past data retain some influence, but the collective fractional influence is $1 - (1 - \lambda)^N$. With $\lambda = 0.1$ and $N = 35$, the old data only have a 0.025 fractional influence.

---

5. S. Cao and R. R. Rhinehart, "Critical Values for a Steady-State Identifier," *Journal of Process Control* 7, no. 2 (1997): 149–152.
6. N. Shrowti, K. Vilankar, and R. R. Rhinehart, "Type-II Critical Values for a Steady-State Identifier," *Journal of Process Control* 20, no. 7 (2010): 885–890.

Larger λ values mean that fewer data are involved in the analysis, which reduces the time for the identifier to catch up to a process change, reducing the average run length (ARL) to a decision. However, larger λ values have an undesired impact of increasing the variability on the statistic, confounding interpretation. The reverse is true: Lower λ values undesirably increase the ARL to detection but increase precision (minimizing statistical errors).

I recommend $\lambda_1 = \lambda_2 = \lambda_3 = 0.1$ for convenience and effectiveness. However, $\lambda_1 = \lambda_2 = \lambda_3 = 0.05$ will have less uncertainty but a bit longer ARL. This is not a critical choice.

18.3.3 Critical Values

If the supposition is that a process is at SS, then a claim of not-at-SS when the process is actually at SS is termed a *Type I (type one) error.* The concept is best understood by considering the distribution of the R-statistic of an SS process. The left curve in Figure 18-3 represents the statistical distribution of the R-statistic values at a steady state. The average value (the 50% value) is R = 1. Note that the distribution is not the symmetric, bell-shaped, normal distribution. It is an F-type distribution, skewed and with no values below zero. The curve on the right represents the R-distribution when a process is in the TS; here, the average ratio is 3. In either case, the R-statistic will have some variability because of the random fluctuations in the sequential measured data.

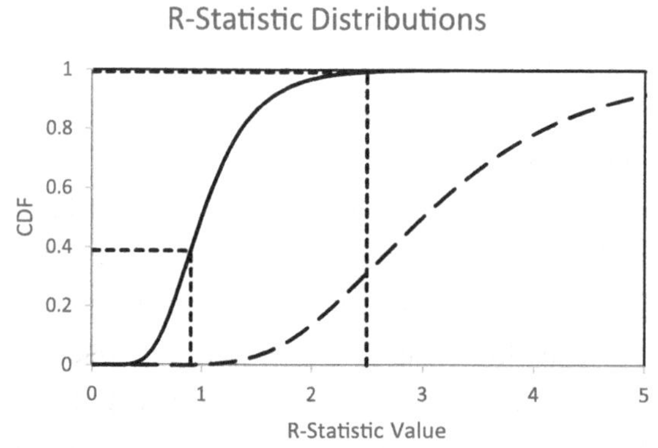

Figure 18-3. R-statistic distributions at an SS (left curve) and a TS (right curve).

Source: From *Applied Engineering Statistics*, second edition, by R. R. Rhinehart and R. M. Bethea. Used with permission from Taylor & Francis Group LLC.

If the value of R is larger than the upper 99% confidence value of about 2.5, illustrated by the vertical dashed line, there is about a 1% chance that the process could be at an SS but about a 70% chance that value would have come from a process in a TS. Bet on it being TS.

If the R-value is a bit less than 2.5, there is still a substantial probability (about 30%) that the process in the TS could have generated it. So, don't use a single critical value to reject and accept SS or TS.

There is also a lower critical value, about 0.9, indicated in the figure. If the process is in a TS, Figure 18-3 shows that the TS process has less than about a 0.01% chance that it will generate an R-value less than the lower critical value. Accepting SS when the process is in a TS could be classified as an alternate Type I (T-I) error. However, if the process is at an SS, the solid curve in Figure 18-3, there is about a 40% likelihood that it will generate such an R-value or lower. So, if $R < R_{\text{lower critical}}$, the odds are that the process is at an SS. Claim SS.

However, if the R-value is between the two critical values, the process could be in either an SS or a TS. There is no adequate justification to make either claim. So, retain the last highly probable claim.

Both T-I and alternate T-I errors are important. Critical values can be obtained from Cao and Rhinehart (1997) and from Shrowti and colleagues (2010). However, it is more convenient and less dependent on idealizations to visually select data from periods that represent a transient or steady period and to identify the R-critical values that make the algorithm agree with the user interpretation.

Experience recommends $R_{\text{upper}} \sim 3$ to 4 and $R_{\text{lower}} \sim 0.85$ to 1.0, chosen by visual inspection as definitive demarcations for a transient and steady process.

18.3.4 Steady-State and Transient-State Identification Illustration

Figure 18-4 illustrates this method. The PV is connected to the left-hand vertical axis (log10-scale) and is graphed with respect to the sample interval. Initially, it is at an SS with a value of about 5. At sample number 200, the PV begins a first-order rise to a value of about 36. At sample number 700, the PV makes a step rise to a value of about 40. The ratio statistic is attached to the same left-hand axis and shows an initial kick to a high value as variables are initialized, then relaxes to a value that wanders about the unity SS value. When the PV changes at sample 200, the R-statistic value jumps to values ranging between 4 and 11 and then relaxes back to unity as the trend hits a steady value at a time of 500. Then, when the small PV step occurs at sample 700, the R-value jumps to about 4 and then decays back to its nominal unity range. The SS value is connected to the right-hand vertical axis and has values of 0 to 1 that change when the R-value exceeds the two limits of $R_{\beta,TS}$ and $R_{1-\alpha,SS}$.

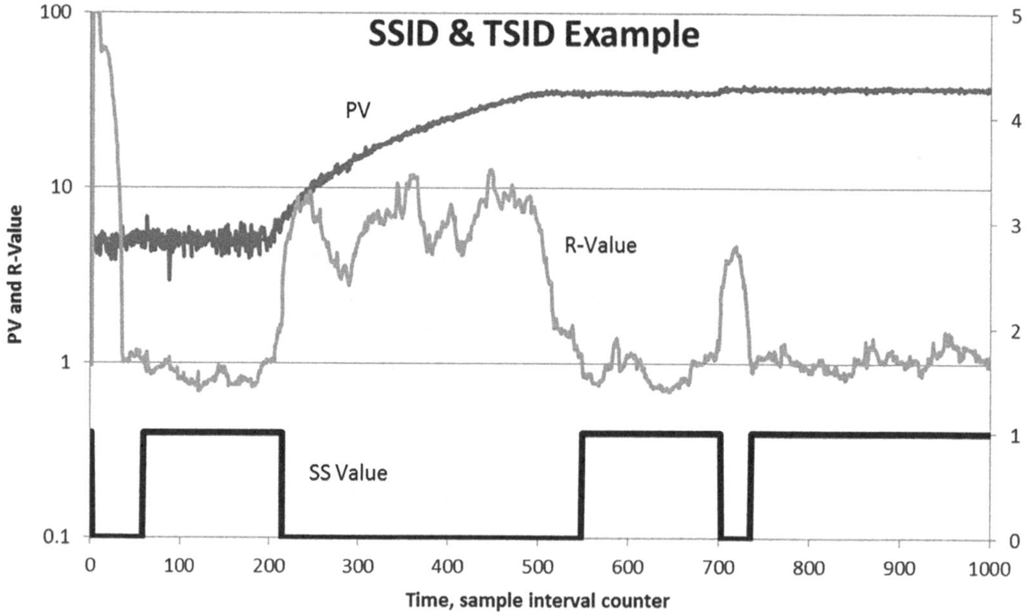

Figure 18-4. Illustration of the filter method to identify SS and TS.

Source: From *Applied Engineering Statistics*, second edition, by R. R. Rhinehart and R. M. Bethea. Used with permission from Taylor & Francis Group LLC.

18.3.5 Autocorrelation

The basis for the R-statistic method presumes that there is no autocorrelation in the time-series process data when at SS. Autocorrelation means that if a measurement is high (or low), the subsequent measurement will be related to it. For example, if a real process event causes a temperature measurement to be a bit high and the event is persistent, then the next measurement will also be influenced by the persisting event and will be a bit high. Autocorrelation could be related to factors such as control action, thermal inertia, data filters in sensors, or signal discretization. Autocorrelation would make the denominator estimate of the variance smaller, which would tend to make all R-statistic distributions shift to the right and require reinterpreting the critical values for each PV.

A convenient solution is to choose a sampling interval that eliminates autocorrelation. This can be done if the source of the autocorrelation is from random perturbations. Although they have some persistence, the influence fades, and new random perturbation values cause measurement fluctuation. This short-lived persistence from random perturbations differs from the regular cycling from a feedback mechanism or the flat-line constant value that arises when a signal has a coarse discretization.

A plot of the current process measurement versus the previous sampling of the process measurement over a sufficiently long period of time (equaling several time constants) at an SS is a simple visual method to establish the presence or absence of autocorrelation. To detect autocorrelation, visually choose a segment of data that is at an SS and plot the PV value with respect to (w.r.t.) its prior value.

Figure 18-5 plots data with a lag of one sample (a measurement w.r.t. the prior measurement) and shows autocorrelation. (In statistics jargon, lag means delay. In system dynamics and control jargon, lag means a time constant. Here, we are using the statistics jargon.) Figure 18-6 plots the same data with a lag of five samples and shows zero autocorrelation. This means that the SS or TS monitor should only be looking at every fifth data value from the process data stream to calculate δ_f^2.

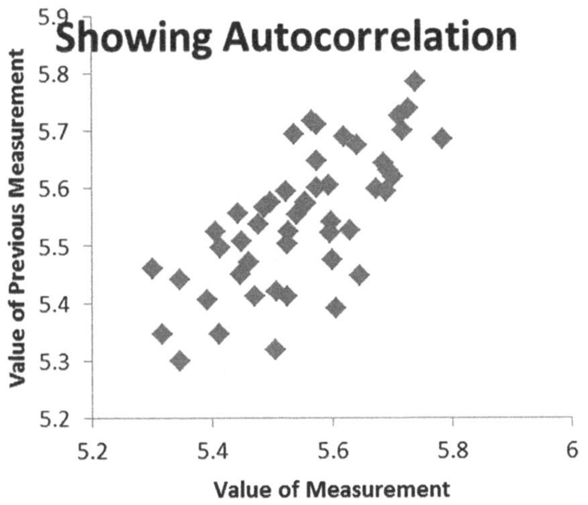

Figure 18-5. Data showing autocorrelation.

Source: From *Applied Engineering Statistics*, second edition, by R. R. Rhinehart and R. M. Bethea. Used with permission from Taylor & Francis Group LLC.

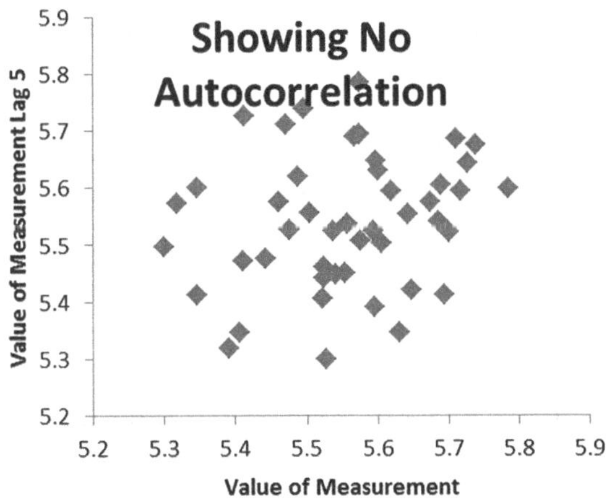

Figure 18-6. Data showing no autocorrelation when the interval is five samplings.

Source: From *Applied Engineering Statistics*, second edition, by R. R. Rhinehart and R. M. Bethea. Used with permission from Taylor & Francis Group LLC.

18.3.6 Signal Discretization

Every measuring device has a minimum increment that can be displayed or reported.

In a digital image, it is pixel size. On a digital clock, it is a 1 s interval; on a stopwatch, it is perhaps 1/100 of a second; on a digital speedometer display, it is 1 mph; and on a digital temperature display, it is 1°F or 0.5°C. As the time, speed, temperature, or other measurement progresses, the displayed value remains unchanged until the true value crosses the discretization threshold to the next interval. If a data transmission system is a 10-bit system, the binary number can have $2^{10} = 1024$ values. If the calibration uses the middle 80% of values (paralleling a standard 4–20 mA or 3–15 psig analog signal), the PV range is matched with about 820 binary counts. There are only 820 possible values, with a discretization interval of about 0.12% of full scale.

Discretization is visible as a signal that repeatedly jumps between only a few values.

Because many process measurements are nonlinear, such as orifice flow measurements based on orifice pressure drop, the discretization may not be visible in one data range but clearly distinguished in another.

Discretization confounds an estimate of data variance or noise amplitude. For instance, if a signal were at SS with a value of 53.5 and a noise sigma of 0.1, the noisy measurement would be within 53.5, ± 3 sigma, or about 53.2 to 53.8. However, if the digital discrimination interval was 1, all measurements would hold at the 53 value, and an estimate of the noise amplitude would, erroneously, be zero, not 0.1.

For most SSID or transient-state identification (TSID) methods to work properly, the signal discretization interval should be small relative to the signal noise variation.

18.3.7 Multivariable Extension

In a multivariable analysis, there is a greater chance that data vagaries will cause a false TS claim in any one monitored variable. Also, there could be an excessive wait for all PV monitors to return to SS.

One approach is to change the threshold to be more cautious about rejecting SS and less cautious about accepting SS. This analysis is predicated on the independence of all PVs.

In expanding the multivariable analysis technique, we choose to claim that a process is not at an SS if any PV is not at an SS and that the PV might be at an SS if all variables are. (However, because these are statistical tests, and "are" implies certainty, "might be" would be more precise.) This can be easily computed with a single statistic:

$$SS_{process} = \prod_{j=1}^{N} SS_j \tag{18-5}$$

where

N = total number of variables in the process

j = variable index

SS_j = 1 if the jth variable is at an SS, and $SS_j = 0$ if the individual PV is in a TS

$SS_{process}$ = 1 if all variables are at an SS, and $SS_{process} = 0$ if any one variable is in a TS

Assign a value of either 0 or 1 to a dichotomous variable, SSID(i), if the ith PV is at probable transient or probable steady conditions. At incidences of intermediate values of the R-statistic, the dichotomous SSID(i) variable retains its latest definitive value. Therefore, an $SS_{process}$ value of 1 means that the last definitive state of each variable was an SS; there has not been a definite transient since the last definite SS. Similarly, an $SS_{process}$ value of 0 means that the last definitive state of at least one variable was transient; there has not been a definite steady state since the last definite transient.

If the process is at an SS and variations and trends on each variable are independent of other variables, then

$$P(SS_{process} = 1) = \prod_{i=1}^{N} P(SS_i = 1) = \prod_{i=1}^{N} (1 - \alpha_i) \tag{18-6}$$

$$(1 - \alpha_{process}) = \prod_{i=1}^{N} (1 - \alpha_i) \tag{18-7}$$

Here, α represents the level of significance, the probability of making a Type I error. For example, if we want $\alpha_{process} = .05$, then to determine the required level of significance for the SSID test on each variable

$$\alpha_i = 1 - \sqrt[N]{(1 - \alpha_{process})} \tag{18-8}$$

where

N = number of variables observed in the analysis

$\alpha_{process}$ = desired overall level of significance for the process statement

α_i = required level of significance for each test

If, for example, there are $N = 10$ variables with independent behaviors and the desired level probability of a Type I error is 0.05 (95% confidence), then from Equation 18-8, the level of significance for each test would need to be about 0.005 (99.5% confidence).

In a multivariable analysis, there is a greater chance that data vagaries will cause a false TS claim, and there could be an excessive wait for all PV monitors to return to SS.

Further, the signals will probably not be independent. If an event makes one PV change or appear to change, it likely affects several PVs. So, as an alternate solution, the implementer might want to temper the monitors: Claim TS only if 10% or more of the PVs are at TS, and claim SS if 90% or more are at SS.

18.3.8 Cross-Correlation

In addition to the R-statistic requirement that there is no autocorrelation between data points during an SS, Equation 18-8 also requires that there is no cross-correlation between variables at an SS. When the process is at an SS, the noise on one variable is not correlated to the noise on another. Variables that are being observed should not have correlated noise at SS conditions. The investigator should identify the minimum set of independent PVs that would present a comprehensive view of the process state and have uncorrelated noise or responses.

For example, consider a tank with inlet flow, outlet flow, and level measurements. Only two of the three are necessary for observation. The third would be correlated to the other two.

As another example, consider that temperature is measured on Tray 7 and Tray 8 in a distillation column. If some event changes the composition on Tray 7, it will soon similarly affect Tray 8, so the temperatures will be cross-correlated. Only one of the two temperatures is needed to observe that section of the column.

18.4 Variable Selection

A process will likely have many variables that can be observed (e.g., temperatures on many trays in a distillation column), but if one is in a TS, whatever causes it will also be affecting other measurements, so not all variables must be observed to determine whether the process is at SS or in a TS. The user should consider which subset of all possible variables comprises the minimum number for detecting TS or SS and only set up the procedure on that minimal set. This minimizes computational burden. It also reduces the number of false alarms generated by normal vagaries in the data. To select the minimum number of variables, the user must understand the process to recognize redundancy in the variables. The user should also select variables that have the least autocorrelation. For instance, if there are two temperature measurements that could be used, one in a thermowell and the other not, the thermowell will create greater autocorrelation. Use the other. Also, select variables that show the largest change-to-noise ratio during a transient.

Some variables, such as flow rates, liquid inventory, or pressure, respond rapidly. Other variables respond much more slowly, such as composition or temperature in a distillation column. (However, temperature could respond rapidly in a heat exchanger, and the composition could change rapidly in a reactor.) Some variables may be grouped into a subcategory for rapidly responding phenomena such as "hydraulics." Other variables taking longer to approach SS might be grouped into subcategories such as "thermal" or "composition." This would reveal that one part of the process has settled, a comforting thing to know, but we must wait for the residual changes in temperature or compositions to come to SS.

18.5 Alternate R-Statistic Structure – Array

The rationale for the three first-order filters in Equations 18-1 through 18-3 is to reduce the computational burden. However, they have the disadvantage of permitting a past event to linger with an exponentially weighted diminishing factor. Also, the filter structure is not as comfortable for many who are familiar with conventional sums in calculating variances. Further, the window of fixed length N is easier to grasp than the exponential, weighted, infinite window or the interpretation of the three lambda values. Alternately, in using a window of N data, the event is totally out of the window and out of the analysis after N samples, which provides a clearer user understanding of data window length. Finally, using a single coefficient, N, is more convenient than specifying three separate λ values.

Returning to the base concept with the R-statistic as a ratio of variances, first, calculate the average from the N most recent observations.

$$\bar{X} = \frac{1}{N}\sum_{i=1}^{N} X_i \tag{18-9}$$

where i is the sample counter starting at the most recent sample and is counting back in time N data points in the window. Then, the conventional variance can be expanded:

$$s_1^2 = \frac{1}{N-1}\sum_{i=1}^{N}(X_i - \bar{X})^2 = \frac{1}{N-1}\left[\left(\sum_{i=1}^{N} X_i^2\right) - N\bar{X}^2\right] \tag{18-10}$$

and substituting for the average

$$s_1^2 = \frac{1}{N-1}\left[\left(\sum_{i=1}^{N} X_i^2\right) - \frac{1}{N}\left(\sum_{i=1}^{N} X_i\right)^2\right] \tag{18-11}$$

Assuming no autocorrelation (independent deviations) and at SS, the variance can also be estimated as the differences between successive data.

$$s^2 = \frac{1}{2(N-2)}\sum_{i=1}^{N-1}(X_i - X_{i-1})^2 \tag{18-12}$$

There are $N - 1$ terms in a set of N data because there are $N - 1$ differences in the data window and, therefore, $N - 2$ in the divisor. Then, the ratio of variances is

$$R = \frac{s_1^2}{s_2^2} = 2\frac{(N-2)\left[\sum_{i=1}^{N} X_i^2 - \left(\sum_{i=1}^{N} X_i\right)^2/N\right]}{(N-1)\sum_{i=1}^{N-1}(X_i - X_{i-1})^2} = 2\frac{(N-2)}{(N-1)}\frac{[Sum1 - Sum2^2/N]}{Sum3} \tag{18-13}$$

This is essentially the von Neumann (1941) approach, but what appears to be a computationally expensive online calculation can be substantially simplified by using an array structure to the window of data and a pointer that indicates the array element to be replaced with the most recent measurement. With this concept, the sums are incrementally updated at each sampling.

The method is as follows: First, increment the counter to the next data storage location. Then, decrement the sums with the data point to be removed. Next, replace the data array element with the new value. Increment the sums and recalculate the R-statistic. Compared to the filter method, this requires storing the N data, the N squared data differences, and the three sums. The pointer updating adds a bit of extra computation. Our investigations indicate that the array approach takes about twice the execution time as the filter approach. Also, because it stores $2N$ data values, it requires that much more computer memory. The increased computer burden over the filter method may not be excessive in today's online computing devices.

This approach calculates a true data variance from the average in the N samples data window. It has the additional benefit that the sensitivity of the R-statistic to changes is greater than with the filter method (about twice as sensitive from experience). The array approach can respond to step changes in the PV of about 1.5σ. Accordingly, this defined window array approach is somewhat better for detecting small deviations from SS.

Further, the value of window length, N, explicitly defines the number of data to fully pass through the data window after an event clears and no longer has an influence on the R-value. By contrast, the filter method exponentially weights the influence of past data.

In addition, the array method seems less conflicted by autocorrelation and discretization of flat spots in the time series of data. It also has the benefit of having only one adjustable variable, the data window length N, which is an advantage in comprehension.

However, the array method is computationally more burdensome than the filter method. The filter method must store and manipulate seven values at each sampling. With $\lambda = 0.1$, this characterizes a window of about 45 samples. For the same characterization, the array method must store and manipulate over 180 variables at each sampling.

My preference is to use the filter method for its computational simplicity. However, if data vagaries (changes in autocorrelation, flat spots, coarse discretization, or small shifts relative to noise) make it dysfunctional, then the array method is a bit better.

18.6 4-Points Method

This relatively simple approach is rooted in a statistical process control approach. Here, four points in the moving window are considered, and a filtered value is calculated for each. The first filtered value represents the average of the most recent data value (the first in the window), the second is that of a data point that is $(1 - \gamma)\, N$ samples old, the third is γN samples old, and the fourth is that of data that is N samples old (the last in the moving window). Gamma is the golden ratio, $\gamma = \frac{\sqrt{5}-1}{2} \cong 0.618$. The variance is calculated using the sequential data differences from the leading data. The t-statistic comprises the maximum difference between filtered values scaled by the filtered sigma.

Tests indicate that this is simple and effective.

A classic approach is to segregate the data into the older and recent halves of the window and then do a t-test on the differences in the averages in the window. However, in an oscillation, when the split is at the peak or valley, the averages will be the same and SS will be erroneously declared.

A solution could be to divide the entire data window into three subwindows and conduct the t-test on the two windows that have the largest maximum average difference. One option would be to vary the sizes of the subwindows. For example, place the most recent data in a narrow subwindow (to make recognition faster), the next most recent data in a wider subwindow, and the oldest data in the widest subwindow (to best establish a historical baseline). It might be best to use unequal spacing to prevent symmetry in a data pattern from confounding the test. Another option would be for the middle subwindow to be the widest, making the difference between the first and third subwindows the largest in a ramp or first-order change.

The number of points, 4, and the golden ratio spacing were chosen to eliminate the potential for missing a periodic signal.

Based on experience with the other techniques, we chose 0.1 for each data filter lambda and 0.05 for the variance filter. Larger filter values create variation that can mask smallish changes, and smaller filter values can lead to a delay in detecting a TS or a return to an SS.

In a t-test, the denominator is a pooled sigma from variance calculated by the number of items in the two averages in the numerator. Here, with filter factors of 0.1 for the data averages, the effective N for averaging is $10 = 1/0.1$. However, the sigma is calculated with a filter value of 0.05, making N effectively 20. It is simpler not to normalize the t-statistic by the degree of freedom.

The statistic is

$$T_i = \frac{x_{f_{largest},\,i} - x_{f_{smallest},\,i}}{\sqrt{\delta_{f,i}^2}} \tag{18-14}$$

where i is the sample index, x_f would be calculated by Equation 18-1, and δ_f^2 would be calculated from Equation 18-3. From experience $\lambda_1 = 0.1$ and $\lambda_2 = 0.05$, and the two critical values for T—2 to reject SS and 1 to reject TS—give good results. If autocorrelation in the data is significant, extend the sampling interval for δ_f^2.

In my investigation of diverse confounding effects, the 4-Points approach has been the best overall. However, when the noise is uncorrelated, the filter method is effective and remains the simplest.

- Overall, for Type I and Type II errors and performance speed, the 4-Points method is the best, and the array method is the second best.
- The filter method is the simplest and fully adequate to recognize SS when noise is uncorrelated.

18.7 Takeaways

- The filter technique is the simplest to implement, but autocorrelation and signal discretization can confound it in process monitoring.
- In a multivariable process, the user must select the variables to be monitored. The selection criteria should be that the variables that are observed should be the minimum number needed to monitor all aspects of the process. In a multivariable analysis, there is a greater chance that data vagaries will cause a false TS claim and an excessive wait for all PV monitors to return to SS. So, the implementer might want to temper the monitors: Claim TS only if 10% of the PVs are at TS, and claim SS if 90% or more are at SS.
- Unlike typical hypothesis-rejecting procedures with only one critical value, these tests have two for the single statistic. One critical value is used to reject the SS hypothesis and the other to reject the TS hypothesis.

Exercises

18.1 Explore the filter and 4-Points techniques on your data.

18.2 Access the steady-state identification simulator (SSID) at *www.r3eda.com* and explore the techniques on data generated with diverse attributes.

19 Data Validity

19.1 Introduction

Controllers make decisions on process data. If the data is in error, the control action will reflect that error. Most normal measurement errors, such as those resulting from noise or imperfect calibration, are inconsequential to control. However, there can be substantial errors when sensors are failing or when they are impacted by process conditions (orifice erosion, flashing, fouling, etc.). The control action may be degraded when there is a substantial input error.

This problem of erroneous data is no different for model-based control than for classic control. Single-input, single-output, proportional-integral-derivative control would have one controlled variable. Simple ratio, cascade, and feedforward strategies will have two measurements. Multivariable control with measured disturbance variables could have many more. When using inferential measurements of soft sensors, several primitive measurements are used to calculate a more complex variable that might require laboratory analysis, such as polymer molecular weight, product composition, or emission concentration. The greater the number of measurements involved in the control strategy, the greater the probability of a data fault.

Filtering can temper random noise but not correct a failing or biased sensor value. Further, measurements don't occur just once; process sensors are sampled once each minute, second, or whatever the sampling interval. When there is substantial measurement error, this produces a sequence of biased values.

However, in model-based control, the data must be consistent with how the model uses it. For example, if the inflow to a tank is 5 gpm and the outflow is also 5 gpm, the model will calculate that the level remains the same. However, if the inflow measurement has a calibration error, a systematic bias of 0.1 gpm, the model will be informed that the inflow rate is 5.1 gpm, and the modeled level will continually rise in time. Here, data reconciliation is needed to make the measurements consistent with the model.

Often, we use collections of online measurements and models as inferential sensors (soft sensors), impute data (fill in missing data with soft sensors), or override measurements when there is greater confidence in the model (Kalman filtering).

This chapter discusses several useful techniques for ensuring data validity.

19.2 Voting

The *voting* technique uses three redundant sensors and accepts the middle of the three values. If one sensor is failing, one of the remaining sensors will be in the middle of the three and provide a good measurement. A large range between values indicates maintenance is needed, but tripling sensors is a cost.

19.3 Kalman Filtering

This is a statistical technique. First, create an inferential measurement—a model that uses related measurements to forecast the measurement of interest that may become corrupted. Then, report the value (either from the model or the direct measurement) with greater statistical confidence. Although this technique is common in the electronics and aerospace industries where linear models are appropriate and powerful computers are justified, it is not common in the chemical process industry.

19.4 Heuristic Overrides

Human logic (intuition or sanity checks) can also be effective in data validation. Use any sanity checks on either the data or the output from a soft sensor. A check may be, "Is it within a normal range?" or "Does the material balance around a section have an acceptable error?" That logic can be programmed in the computer; it does not have to be performed by a human supervisor. If an error is sensed, you could take (or automate) any of many appropriate actions: Turn off the associated controller feature. Freeze the last seemingly acceptable value. Use a nominal value. Switch from a primary sensor to an inferential modeled value. Call instrument maintenance.

19.5 Data Reconciliation

This section is extracted from Chapter 14 in *Nonlinear Regression Modeling for Engineering Applications: Modeling, Model Validation, and Enabling Design of Experiments*[1] and Case Study 10 in *Engineering Optimization: Applications, Methods, and Analysis*,[2] both by R. R. Rhinehart.

Data reconciliation is a procedure to use online data and models of relations within the data to estimate the systematic errors of the same online data. In data reconciliation, the correction is based on redundant measurements (but not at the same location) and a model.

For example, if liquid flows into a tank and flows out, then the inflow minus the outflow must equal the accumulation in the tank. However, if the flowmeters are of the orifice type, then a 5% error is not unexpected. If the inflow meter reads 103 gpm, it could actually be 97 gpm. If the outflow meter reports 91 gpm, it could actually be 97 gpm. Even though the flow rates are identical and there is no accumulation, the readings would indicate an accumulation of 103 – 91 = 12 gpm. However, the level indicator might reflect the actual flow rates and report no change in level. If the level sensor is not calibrated exactly or the tank area does not account for ribbing, dents, or volume-immersed devices, then the reported inventory accumulation is also wrong.

One solution is to adjust each process measurement by a bias. The corrected value, y', is the measurement, y, corrected with the bias, b.

$$y' = y + b \tag{19-1}$$

There are many measurements, and each measurement is reported at each sampling. Below, the subscript i indicates the *ith* variable measurement, and t represents the sample time. With the condition that b_i does not change in time or changes very slowly over a time window

$$y'_{i,t} = y_{i,t} + b_i \tag{19-2}$$

---

1. R. R. Rhinehart, *Nonlinear Regression Modeling for Engineering Applications: Modeling, Model Validation, and Enabling Design of Experiments* (Hoboken, NJ: John Wiley & Sons, 2016).
2. R. R. Rhinehart, *Engineering Optimization: Applications, Methods, and Analysis* (Hoboken, NJ: John Wiley & Sons, 2018).

There are uncertainties in each measurement, the magnitude of which can be estimated as the standard deviation of the noisy value during a reasonably steady-state period, σ_i.

Desirably, the bias error of the sensor would be determined by a calibration. However, removing sensors from a process for offline laboratory calibration is often impractical, and later, they can become uncalibrated.

In data reconciliation, the b_i values are chosen to best reconcile all data and seek internal consistency in the data with respect to expectations or models. The models are typically material and energy balances. For example, assuming constant density, the material balance model about a vertical cylindrical tank is

$$A\frac{d\tilde{h}}{dt} = \tilde{F}_{in} - \tilde{F}_{out} \tag{19-3}$$

Where A is the tank cross-sectional area, $\tilde{h}$ is the modeled tank level, and $\tilde{F}$ is the measured flow rate.

Using the corrected measurements, the material balance model should have a zero error, e.

$$\begin{aligned} e = 0 &= (F_{in} + b_{in}) - (F_{out} + b_{out}) - A\frac{d(h + b_h)}{dt} = \\ &= (F_{in} + b_{in}) - (F_{out} + b_{out}) - A\frac{dh}{dt} \end{aligned} \tag{19-4}$$

The objective is to determine the bias values to minimize J, the square of the deviation of the model over the time duration of a data window.

$$\underset{\{b_{in}, b_{out}\}}{min} J = \Sigma_t [e_t]^2 \tag{19-5}$$

In this case, many combinations of the corrections, b_{in} *and* b_{out}, will make the error zero. So, we will add a penalty for large corrections relative to the normal standard deviation on the measurement.

$$\underset{\{b_{in}, b_{out}\}}{min} J = \Sigma_t [e_t]^2 + \left(\frac{b_{in}}{\sigma_{y-in}}\right)^2 + \left(\frac{b_{out}}{\sigma_{y-out}}\right)^2 \tag{19-6}$$

Recognizing that requiring $\sum_t [e_t]^2 = 0$ means that b_{out} can be directly calculated from b_{in}. This solves two problems. It eliminates the dimensional $\sum_t [e_t]^2$ term from the objective function, leaving only dimensionless terms, and it reduces the number of decision variables for the optimizer.

$$\min_{\{b_{in}\}} J = \left(\frac{b_{in}}{\sigma_{y-in}}\right)^2 + \left(\frac{b_{out}}{\sigma_{y-out}}\right)^2 \qquad (19\text{-}7)$$
$$S.T.: b_{out} = f(b_{in})$$

In general, multiple models are used to adjust each variable, and the data reconciliation method is a bit more complicated.

The data reconciliation method will seek independent b-values to minimize all b-value deviations tempered by the expected standard deviation (an equal concern factor).

$$\min_{\{\underline{b_x}\}} J = \frac{1}{N_x}\sum_i \left(\frac{b_{xi}}{\sigma_{xi}}\right)^2 + \frac{1}{N_y}\sum_j \left(\frac{b_{yj}}{\sigma_{yj}}\right)^2 \qquad (19\text{-}8)$$
$$S.T.: \underline{b_y} = f_j(\underline{b_x})$$

The S.T.: constraints represent the "closure" on the N_j number of models. The subscript x represents i number of biases chosen by the optimizer, and the subscript y represents j number of biases inferred by the constraints. The biases are summed over all the N_i decision variables and all the N_j models. The reciprocal N scaling normalizes for the number of variables in each sum. The terms in the optimization statement represent the value of the correction, scaled by the expected random uncertainty of the measurement.

The procedure to determine the b-values could be a classic optimization as indicated in Equation 19-8; however, I have seen it done manually and helped program the human heuristics to automate the procedure using fuzzy logic.

The data reconciliation approach was first published by R. S. Mah, G. M. Stanley, and D. M. Downing.[3]

---

3. R. S. Mah, G. M. Stanley, and D. M. Downing, "Reconciliation and Rectification of Process Flow and Inventory Data," *Industrial & Engineering Chemistry Process Design and Development* 15, no. 1 (1976): 175–183.

19.6 Steady-State (SS) Relations

The data reconciliation procedure is somewhat simpler if SS models are used, but only SS process data should be selected.

19.7 Takeaway

- The age-old computer mantra is garbage in, garbage out. For critical control applications, include procedures to ensure that input data is valid.

Exercises

19.1 Use the simple steady-state material balance of Equation 19-3 with two flow rate measurement biases. Fill in Equation 19-8 with that example.

19.2 Use your control simulator and show the impact of a bias on the controlled variable (CV) measurement and a disturbance or ratio measurement. You should determine that feedback control compensates for all errors except those on the CV maintaining control as long as constraints are not hit. However, measurement errors will affect the validity of model adjustment and the use of model coefficient values in process monitoring.

20 Optimization

20.1 Introduction

Optimization is a trial-and-error procedure that seeks the best possible solution (PS)[1]. It gets the results of that PS, termed the *objective function value*. With information about whether the objective function (OF) value is better or worse than that from a former PS, the procedure proposes a new PS. When the procedure appears to be close enough to the optimum PS, the procedure claims convergence, stops the search, and reports the result.

There are many ways to automate the sequential PS guesses. Gradient-based approaches (Newton's, Levenberg-Marquardt, and many others) use the sensitivity of the OF to the PS to define the best direction to move. Direct search approaches (Hook-Jeeves, Nelder-Meade, Leapfrogging, Particle Swarm, and many others) only use the OF value. Although gradient-based searches are faster on well-behaved functions, in my experience, direct search approaches are best for constrained nonlinear searches because they can cope with discontinuities, flat spots, constraints, and many other OF-to-PS responses that misdirect gradient-based searches.

If you have convenient access to an optimization procedure that is good and functional for your application, use it. This chapter discusses two direct search optimization techniques chosen for their balance of simplicity of understanding and implementation, robustness, and efficiency in identifying the optimum. One, Cyclic Heuristic

1. TS (trial solution) is the standard nomenclature for a possible solution, and TP (trial point) is sometimes used; but in this book, PS is chosen so that the single acronym TS (transient state) does not have dual meanings.

Direct (CHD), uses a single possible solution and incrementally moves it toward the optimum. The other, Leapfrogging (LF), uses many possible solutions (players) scattered within the feasible range, and the player with the worst OF value leaps over the player with the best OF value. I have found that CHD and LF work well in pilot-scale, real, control implementations.

To reference and compare many search types, I suggest *Engineering Optimization: Applications, Methods, and Analysis*[2] by R. R. Rhinehart. The book also has more details on CHD and LF.

20.2 Optimization Terminology

Consider that you wish to determine the value of a variable, x, that minimizes a function, $f(x)$. In predictive functional control (PFC), the function would be the deviation between the biased set point and the now-to-future (yN2F) model forecast value at the coincidence time. In horizon predictive control (HPC), the function would be the sum of squared differences over the horizon between the forecast reference trajectory and the yN2F model projections. The function being minimized is the OF, and its value is the OF value. The variable being changed to minimize the function value is the decision variable (DV). The DV and the manipulated variable (MV) are the same for control, but optimization and control give them different names.

In PFC, the possible solution (PS) is the possible MV value sent to the model. The single MV is termed the DV in an optimization context. In HPC, the PS is a sequence of MV values. If there is only one variable in the PS, the search for the best is termed a *univariate search*. If the PS is a choice of three values in the MV sequence, then the PS has three DV values. One set of three DVs is still termed the PS. A multivariate search has more than one DV.

For example, in HPC with three future u-values being decided, one PS might be $\underline{u} = \{46\%, 32\%, 37\%\}$. This means first, set the output to 46%; one-quarter of the way through the horizon, set it to 32%; halfway through the horizon, set it to 37%. Using this sequence to move the yN2F model for temperature control, the OF value might be 17 (°F)^2. The former PS might have been $\underline{u} = \{52\%, 50\%, 35\%\}$, generating an OF value of 26 (°F)^2. The new PS is better.

The optimum PS is identified as PS*, and the corresponding OF value is OF*. The optimum is not the true perfect optimum but is close enough to be a fully adequate

2. R. R. Rhinehart, *Engineering Optimization: Applications, Methods, and Analysis* (Hoboken, NJ: John Wiley & Sons, 2018).

representation. The optimizer could continue the search and get closer to perfection; however, this requires more computational time, and perfection to the 100th decimal would not be possible to implement. Accordingly, the convergence criterion is set to provide a balance of utility in the answer.

Each successive "guess" of the PS by the optimizer is an *iteration*. After each iteration, there is a test for convergence to stop continued work if the PS is close enough to PS*. Because PS* is unknown, the convergence criterion might be the difference between successive DV values. When the successive difference is very small relative to any practical impact on control, claim convergence. The value used for very small would be the convergence criterion threshold; for example, stop iterations and state that the current PS is close enough to PS* when the sequential change in the MV is less than 0.1%.

Constraints are also a part of optimization. Sometimes the constraints are on the DVs, such as rational limits of 0% to 100%, or a rate of change limit between sequential MV values. Sometimes, they are on the controlled variable or auxiliary variables that the process model predicts. Sometimes, they are on relations between model input or output variables. Sometimes, they are on internal variables to prevent execution errors in the model (divide by zero, square root of a negative number, etc.).

Some constraints are "hard." They cannot be violated. For instance, the DV must be between 0% and 100%. Do not exceed a 5%/s rate of change in the MV. Do not create an infeasible operation. Here, after making a PS guess, the optimizer is told, "No, you cannot do that. There is no OF value. Guess again."

Some constraints are "soft," meaning some violation would be permitted. Here, the penalty for the constraint violation is added to the OF value. The optimizer does not know why an OF value gets better or worse, but it uses that information to make a better PS choice.

The optimization statement has a common formulation.

$$\begin{aligned} &\underset{\{\underline{u}\}}{min}\; J = f(\underline{u}, \underline{y}, y_0, \underline{d}, \underline{p}) \\ &\text{S.T.}: g(\underline{u}, \underline{y}, y_0, \underline{d}, \underline{p}) = 0 \end{aligned} \tag{20-1}$$

In the top line, the word *min* or *max* indicates whether the optimization is to minimize or maximize the function. $f(\underline{u}, \underline{y}, \underline{d}, \underline{p})$ is the OF (including any constraint violation penalties), and J is its value. The DVs are elements of the vector $\underline{u}$, within braces under the operation type. The function being optimized has input variables $\underline{u}$, the DVs, and $\underline{d}$, disturbances (external inputs). In control, the model predicts state variables, $\underline{y}$, and a dynamic model must be initialized with initial state variable values from the past-to-now (yP2N) model, $\underline{y}_0 = \underline{y\text{P2N}}$. Finally, $\underline{p}$ represents parameter values in the model.

In the second line, "S.T.:" means "subject to:" and what follows is a listing of all the constraints.

In this generic presentation, neither $f(\underline{u}, \underline{y}, \underline{y}_0, \underline{d}, \underline{p})$ nor $g(\underline{u}, \underline{y}, \underline{d}, \underline{p})$ provide any substantive information. In a real application, neither would likely be one line equations. They would probably be somewhat complicated functions or subroutines. You could use human language descriptions instead of a generic function. For instance, $f(\underline{u}, \underline{y}, \underline{y}_0, \underline{d}, \underline{p})$ could be replaced with the phrase "squared deviation between the biased set point and the yN2F modeled value at the coincidence time."

Equation 20-1 cannot be solved directly, like a normal algebraic equation. Instead, it summarizes all the essential elements of the optimization.

20.3 Cyclic Heuristic Direct (CHD)

The term *direct search* means the procedure only uses information from function evaluations and constraint violations. There is no derivative or second derivative (no gradient, no Hessian) in the algorithm. No mathematical surrogate model of the surface is formulated. The term *"cyclic"* means that the procedure cycles through the list of DVs, making an incremental change in each, one at a time. Then, it repeats a cycle. Each complete cycle of exploring a change in each DV ends a search iteration. The term *heuristic* means that human logical rules guide the search increment expansion or its reversal and contraction. Cyclic Heuristic Direct is abbreviated CHD.

This is my second favorite algorithm for moving DV values toward the minimum (or maximum). It is high on my list of favorites because of its simplicity, robustness, broad applicability, and effectiveness. It can be implemented in a human, mentally directed, trial-and-error search and does not need a computer. It is my second favorite algorithm because it is a single-trial solution approach that can get stuck in a local optimum if the surface has multiple minima and the CHD is not initialized near the global optimum. However, in control applications, initializing the PS at the prior MV values seems to avoid any such local traps.

The procedure pseudo-code is:

1. Initialize.

   i. DV base (the initial possible solution in a feasible region; a best initialization may be the former optimum PS) and evaluate the OF

ii. ΔDV_i to be used for exploratory changes

iii. Convergence method and threshold

2. Start the cycle for one iteration (alternately, this may be called an *epoch* or a *stage*).

3. Taking each DV, one at a time, sequentially, set the new possible DV value.

$$DV_{i,possible} := DV_{i,base} + \Delta DV_i$$

Note $DV_{possible}$ and DV_{base} are vectors. Only add ΔDV_i to the *i*th element. All other $DV_{possible}$ elements remain at the base value.

i. Evaluate the function at the possible solution value.

1. If worse or "FAIL," keep *DV base* and set ΔDV_i to a smaller change in the opposite direction: $\Delta DV_i := -contract * \Delta DV_i$. This means that each dimension has its own ΔDV_i value.
2. Otherwise, keep the better solution (make it the base point) and accelerate future moves in what seems to be the right search direction.

$$OF_{base} := OF_{possible}$$
$$DV_{i,base} := DV_{i,possible}$$
$$\Delta DV_i := expand * \Delta DV_i$$

4. At the end of the cycle, check for stopping criteria (excessive iterations or another indication of futility).

i. If stopping, exit with a "Did Not Converge" message.

ii. Otherwise, go to Step 5.

5. Check for convergence criteria.

i. If converged, exit with an appropriate message.

ii. Otherwise, return to Step 3.

This algorithm is simple to understand and implement. Although other search algorithms may be more efficient (use fewer iterations or fewer function evaluations) for compliant functions, to my taste, the simplicity, robustness, general applicability, and constraint accommodation of CHD are substantial advantages.

I believe that the *expand* coefficient in the range of 1.1 to 1.2 is good. Larger numbers can make the jump to new DV values excessively large after a few moves in the right direction. I like the *contract* coefficient to be in the 0.5 to 0.8 range. Smaller values quickly lead to tiny steps, slowing the search. If the prior step had an expansion factor of 1.2 but led to a worse place than the prior OF, then a contraction of $\frac{1}{1.2} = 0.8333$

would return the next DV value to the prior one. So, I keep the *contract* value less than the reciprocal of the *expand* value. My default values are *expand* = 1.1 and *contract* = 0.5.

In the initialization step, the best initialization for the DV values would likely be the prior optimum values. These would not be a good initial guess only if a significant disturbance or set point happened within the last control interval.

20.4 Leapfrogging (LF)

LF is an optimization algorithm in the contemporary multiplayer genre, and it now has many supporting publications. The key reference is "Leapfrogging and Synoptic Leapfrogging: A New Optimization Approach" by R. R. Rhinehart, M. Su, and U. Manimegalai-Sridhar."[3]

LF is a direct search algorithm that only uses function values, not derivative (gradient or Hessian) information. This makes it robust to many surface features that confound or misdirect derivative-based searches (saddle, inflection, ridges, discontinuities, numerical striations, flat spots, integers, classifications, stochastic, etc.).

Start with a randomly placed team of players (alternately termed *swarm of particles, individuals,* or *trial solutions*) in feasible spots in the range of DV space thought to be of interest. For control applications, initialize one player with the former optimum DV values.

Evaluate the OF value for each player and determine which has the best and worst OF values.

Move the worst player to the other side of the best player in the DV space. The worst player jumps over the best player and lands in a random spot in the projected DV "area" on the other side, which explains the name *Leapfrogging.*

The leap-over may place the player at an OF value that is still the worst or into an infeasible region (where it cannot generate an OF value because it violates constraints, leads to an execution error in the OF calculation, etc.). If so, it remains the worst player and is selected for the next leap-over.

3. R. R. Rhinehart, M. Su, and U. Manimegalai-Sridhar, "Leapfrogging and Synoptic Leapfrogging: A New Optimization Approach," *Computers & Chemical Engineering* 40 (May 11, 2012): 67–81.

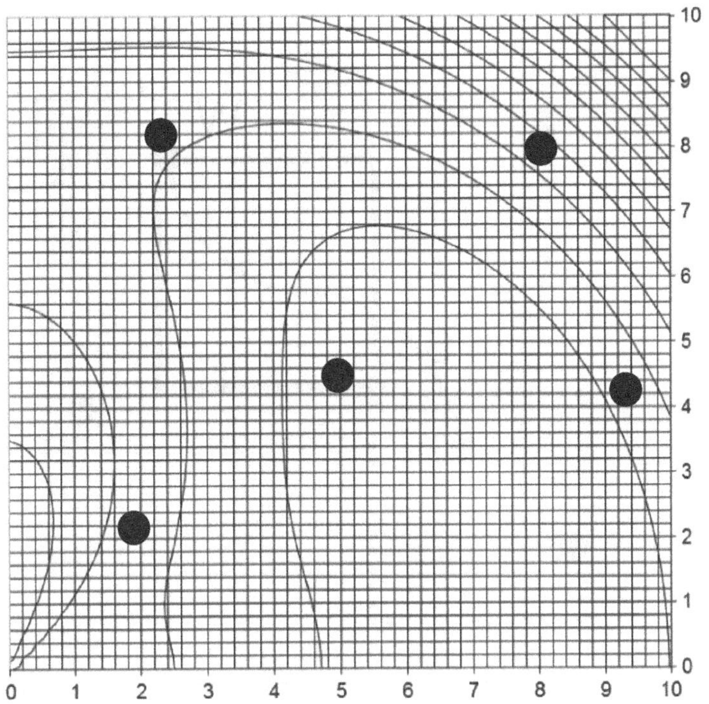

Figure 20-1. LF explanation step 1, initialization.

Source: From *Engineering Optimization: Applications, Methods, and Analysis* by R. R. Rhinehart.[4] Used with permission from John Wiley & Sons.

Figure 20-1 illustrates a 2-D contour, with a minimum at about $x = 7.5$, $y = 2$. Five trial solutions, five players, are illustrated with markers and are randomly placed over feasible DV space.

Note: Five players per dimension with a minimum of 20 players is recommended. Here, there should be 20 players, but only five are shown to simplify the presentation.

Calculate the OF for each player and identify the worst and best. These are labeled W and B in Figure 20-2, which also shows the leap-from DV area (the window with the solid border) and the leap-into area (dashed border). The leap-to area is illustrated with the same size in DV space but is reflected on the other side of B from W. Both window borders are aligned with the DV axes. As the arrow indicates, W leaps to a random spot inside the reflected DV space window.

Calculating the new DV position is very simple. If there are two DVs, x and y, the leap-to location is calculated as

4. R. R. Rhinehart, *Engineering Optimization: Applications, Methods, and Analysis* (Hoboken, NJ: John Wiley & Sons, 2018).

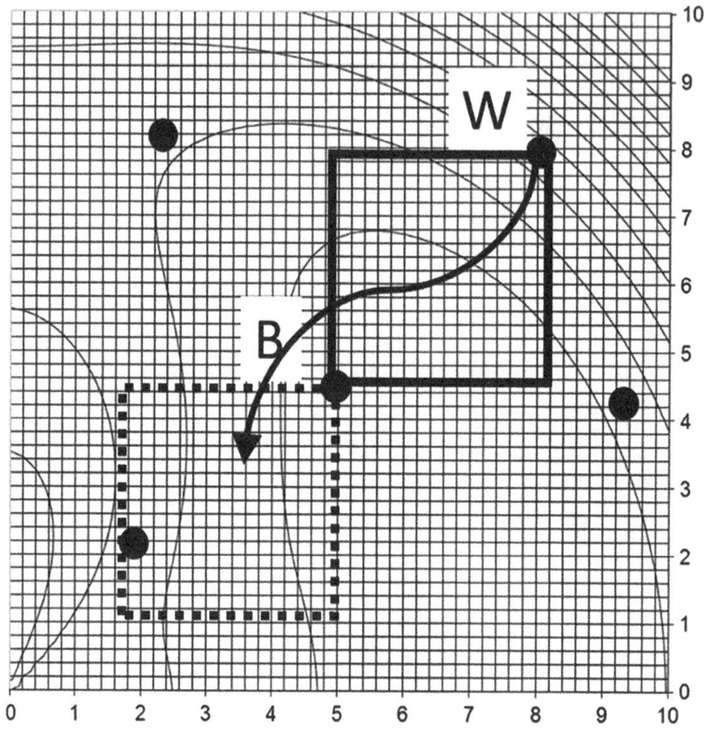

Figure 20-2. LF explanation step 2, choosing the leap-to area.

Source: From *Engineering Optimization: Applications, Methods, and Analysis* by R. R. Rhinehart. Used with permission from John Wiley & Sons.

$$\left.\begin{aligned} x_{w,\,new} &= x_b - r_x(x_w - x_b) \\ y_{w,\,new} &= y_b - r_y(y_w - y_b) \end{aligned}\right\} \tag{20-2}$$

Here, r_i are independent random numbers with a uniform distribution and range of $0 < r \leq 1$. These random perturbations are independent for x and y.

Note: The procedure scales to higher-order or to a univariate search.

As Figure 20-3 reveals, W moved to a new position, and its prior location was vacated. The algorithm now must determine the OF value at the new player position.

Note several features: The new position is not in the center of the leap-into window; it is placed between the center and the player labeled B in this leap-over. There are still five players. There is one function evaluation per leap-over.

Test for convergence. As Figure 20-4 shows, the players are still far apart in DV space, as indicated by the size of the ellipse that contains them. They have not converged yet. Many metrics can be used to determine when the players are all close to the optimum. I often use the average or root-mean-square deviation of all players from the

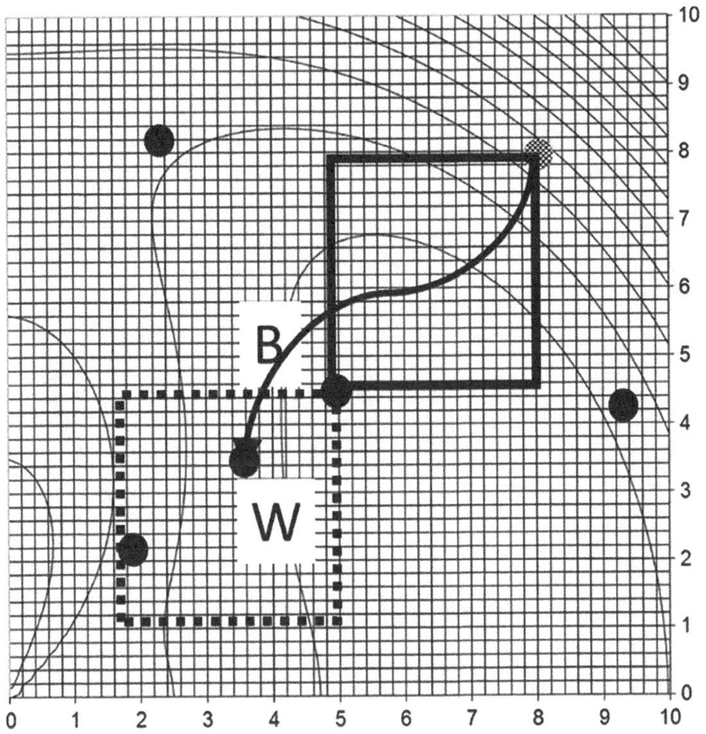

Figure 20-3. LF explanation step 3, the leap-over.

Source: From *Engineering Optimization: Applications, Methods, and Analysis* by R. R. Rhinehart. Used with permission from John Wiley & Sons.

best represented in DV space. This distance represents the MV ranges, which I think is easy for a user to specify. For control, if the range is less than 1%, there is little practical justification to seek something more perfect.

Because the team has not converged, repeat the procedure. Identify B and W as shown in Figure 20-5. The best player now is still the former best. The former worst now has a mid-contour value, so a new worst is identified.

Notes:

1. There is no need to search through all player OF values to identify the best. If the new leap-to-player OF value is better than the former best, it is the new best; otherwise, the old best remains the lead.
2. If the leap-to spot was either infeasible (for any reason, an explicit DV constraint, an execution error in the OF evaluation, etc.) or had a worse OF value than the leap-from point, that player remains the worst. There is no need to search for the best, and the algorithm must only search for the worst if the OF value of the leaping player improves. This aspect reduces the computational burden.

In Figure 20-5, the new W leaps over B. One player is moved. Now, the OF of the new location must be determined. In this illustration, the randomized leap-to position is

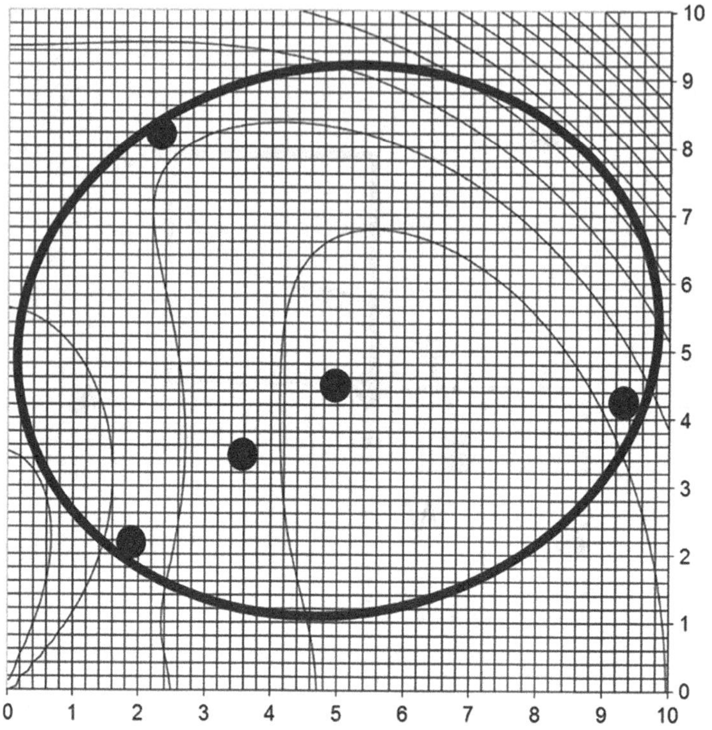

Figure 20-4. LF explanation step 4, test for convergence.

Source: From *Engineering Optimization: Applications, Methods, and Analysis* by R. R. Rhinehart. Used with permission from John Wiley & Sons.

on the far side of the leap-to window. Because the player is closer to the optimum, the former W will become the new B.

> Note: The team of players is both contracting and moving toward the optimum in Figure 20-6. The team is converging.

After repeating the steps a few times, the players converge on the optimum, as Figure 20-7 reveals, and in a few more leap-overs, convergence will be declared. When converged, report the DV and OF of the best player in the cluster as DV* and OF*.

If there are M players (trial solutions), there are M initial OF evaluations. Then, each leap-over only costs one additional OF evaluation. I recommend that the number of players is five times the number of DVs, $M = 5N$, but with a minimum of 20.

$$M = \max\{20, 5N\} \tag{20-3}$$

However, this number is from experience gleaned when seeking the number of players to maximize the probability of identifying the global optimum and minimizing

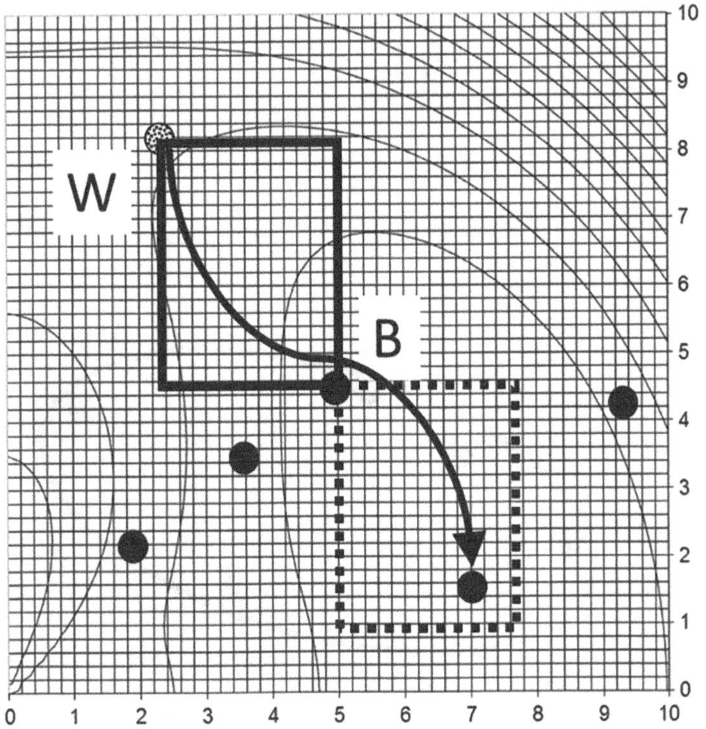

Figure 20-5. LF explanation step 5, next leap-over.

Source: From *Engineering Optimization: Applications, Methods, and Analysis* by R. R. Rhinehart. Used with permission from John Wiley & Sons.

the number of function evaluations to converge. It seems to partly depend on the leap-over amplification factor, which I recommend as unity for simplicity and convenience.

Note: If there are N dimensions, each leap-over changes all N DV values for the jumping player.

I term an iteration in LF as N leap-overs, one leap-over per dimension, representing N OF evaluations. This is roughly equivalent to the work per iteration of other optimizers. Not all players will move in an iteration. There are two reasons: (1) $M > N$, and (2) the worst could jump to a new spot, remain the worst, and jump again.

As a direct search, LF does not require surfaces to be continuum-valued. It can handle cliffs and discontinuities, flat spots, and ridges that would result from the discretization of the DV, time or space discretization in numerical methods to determine the OF value, conditionals that switch models and, therefore, create discontinuities, or sampling discretization in time series models. Also, LF can handle constraints of any sort (explicit or implicit DV equations or as surprise, nonexecutable encounters on variables within the OF).

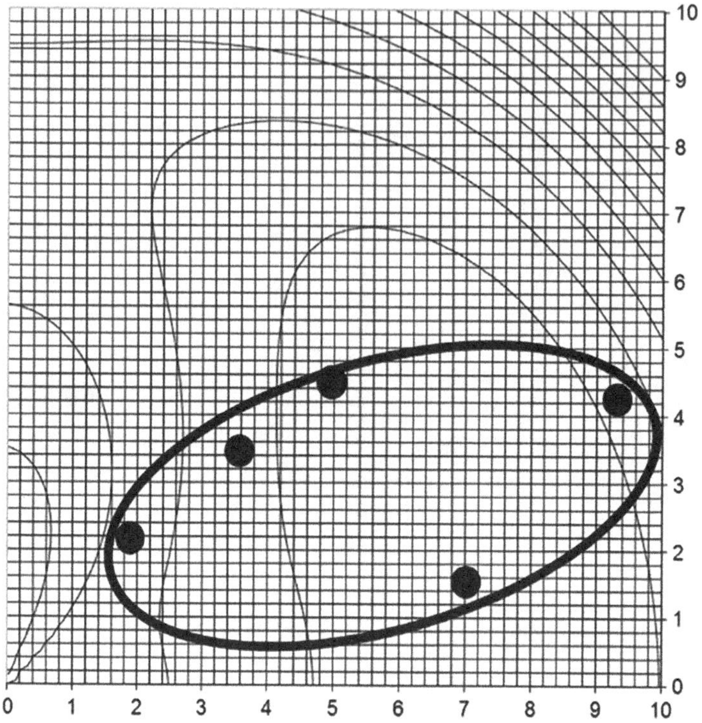

Figure 20-6. LF explanation step 6, test for convergence.

Source: From *Engineering Optimization: Applications, Methods, and Analysis* by R. R. Rhinehart. Used with permission from John Wiley & Sons.

On average, each leap-over moves the player with the worst OF half of the distance closer to the player with the best OF. If the best player remains the best, all players converge to it. However, if the leaping player becomes the best, the solution moves toward the optimum. On the side of a hill, the worst player, from the back of the team, leaps a long distance over the lead player, and this expands the team, accelerating motion and widening exploration as it leapfrogs downhill.

20.5 Takeaways

- Some optimizers are faster than others on ideal surfaces. However, speed does not seem to be the primary issue with today's computers. Use any convenient optimizer that can cope with infeasible spots, numerical striation, discontinuities, nonlinearities, and multiple optima.
- Do not make convergence too tight. 'Do not seek perfect solutions. The model is imperfect, and a perfect solution is not perfect for the process. Make convergence tight enough so the error is negligible relative to noise and device resolution.

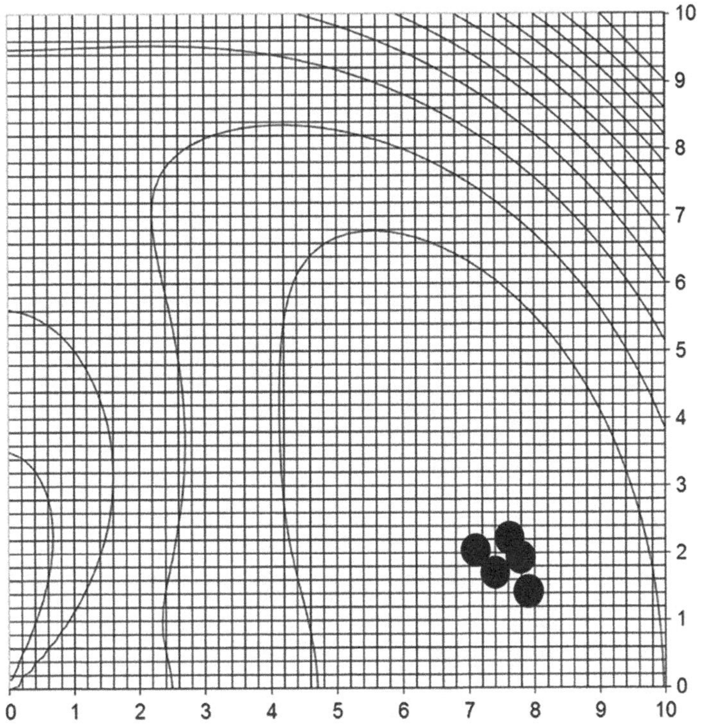

Figure 20-7. LF explanation step 7, players are converging.

Source: From *Engineering Optimization: Applications, Methods, and Analysis* by R. R. Rhinehart. Used with permission from John Wiley & Sons.

Exercise

20.1 Software to illustrate and compare Cyclic Heuristic Direct and Leapfrogging to many other optimization algorithms is available on my website, *www.r3eda.com*. Access it and explore various techniques on diverse applications.

Section 4

Constrained MIMO Control

21 Constraints and Balancing Desirables

21.1 Introduction

This chapter introduces the diverse categories of constraints and how they can be handled as "hard" or "soft." It also discusses methods to accommodate cases with extra degrees of freedom.

Constraints mean that you cannot do what you want. For instance, the objective might be to increase the throughput to a set-point value, but doing so might lead to cavitation, an undesirable consequence. Lowering throughput could avoid cavitation, but it has the undesirable aspect of missing a production target. Should you choose the manipulated variable (MV) to meet the production target or to get as close as possible without creating cavitation? Should you choose a controlled variable (CV) value that gets closer to the production target and causes only mild cavitation? This chapter offers a method for the controller to choose an MV that balances the penalties for missing the set point and accommodating constraints.

21.2 Constraint Categories

Constraints may be on any variable: the MV, the CV, or an auxiliary variable (AuxV). There may also be several types of constraints on each. If the process is multiple-input, multiple-output (MIMO), there may be several MVs and CVs, each with constraints.

In addition, there can be constrained combinations of variables. For instance, if the vapor boil-up and reflux rate are considered the MVs for a distillation column, then, simplistically, the reflux rate is limited by the vapor boil-up rate because you cannot put more back into the column than what comes out. Finally, the constraints may be "hard" and cannot be violated, or they may be "soft," permitting some level of violation.

21.2.1 Value Limits

The MV cannot exceed its range limits. Usually, this range is 0% to 100%, but some control devices have a range of –6% to 106%, accounting for calibration errors in the transmission devices that may not let a valve stroke fully close or fully open at the 0% to 100% limits. For some other reason, perhaps cavitation, the user might limit the output to a 70% maximum, or, to ensure there is flow to prevent gelling, limit it to a minimum of 10%.

To correct an upset, the optimizer in a controller might want to investigate a value that exceeds such limits. If the optimizer is told, "No, you cannot choose a value that exceeds the limits. Try again," this would be a hard constraint.

There could also be limits on the future value of the modeled CV or an AuxV (another state in the model): Do not let the temperature exceed a value that leads to charring, do not let the level get too low, and do not let the pressure exceed the vacuum or the high limit. Conceptually, these cases should use a hard limit, but it is possible that near a limit, noise or a disturbance causes the CV or AuxV to exceed the limit. Consequently, nothing the controller can do in the future can alleviate the constraint violation now. So, for CV and AuxV limits, it is best to use either a soft constraint or a barrier function as penalty additions to the objective function.

21.2.2 Rate-of-Change Limits

Aggressive movement of the MV, such as large immediate changes, may stress equipment (thermal stresses, water hammer, etc.) and cause utility upsets to other processes on the same utility lines. So, a user might want to impose rate-of-change (RoC) constraints on the MV. The rule might be, "Do not make sequential MV changes that are 5% from the prior value." Or, the limit might be stated as a rate, for example, "Do not exceed an MV rate of change of $\pm 1\%/s$."

Because these are MV constraints, they can be treated as hard. If the optimizer asks to explore an MV value that exceeds an RoC limit, it is told, "No, you cannot choose a value that exceeds the limits. Try again."

There may also be RoC constraints on the CV or AuxV. But again, because a disturbance may be causing a RoC violation, there is nothing the controller can do to fix the current state. So, use a soft constraint penalty approach on CV and AuxV to temper any violation.

21.2.3 Soft Penalty

A soft penalty approach permits some violation, even if undesired. Although we want our children to behave politely, we often accept a bit of violation, but not excessive violation. The same logic applies to posted speed limits on the road. We want the process to follow a first-order reference trajectory, but knowing this might not be possible, we accept some deviation.

In a soft penalty approach, there is a desired limit for some modeled variable values. If the target is "Keep this value," such as a set point or a resting value, then any deviation is undesirable. In the soft penalty approach, the measure of undesirability is the square of the deviation. For small deviations, the squared value is very small, but for large deviations, the squared value amplifies the impact. This deviation-squared evaluation matches the least-squares regression approach to determine model coefficient values that minimize the sum of squared deviations between the model and process. It also seems to match the human acceptance of small offenses but not large ones.

$$penalty = (x - x_{target})^2 \tag{21-1}$$

The soft penalty would be added to the objective function (OF) the controller seeks to minimize. Scaling is discussed in Section 21.4.

For many variables, there is a one-sided upper or lower limit like these examples: Keep the temperature below some value. Keep the impurity below some value. Keep the pressure above some value. In this case, there is no penalty as long as the limit is not exceeded, but there is a quadratic penalty if it is exceeded.

$$penalty = \begin{Bmatrix} 0, & if\ x \le x_{limit} \\ (x - x_{limit})^2, & if\ x > x_{limit} \end{Bmatrix} \tag{21-2}$$

The soft penalty approach can also be applied to MV values and RoC limits. Some optimizers can cope with hard constraints; if you use one that cannot, convert the MV limits to soft penalties.

21.2.4 Shifted One-Sided Soft Penalty

An undesirable aspect of the soft penalty approach is that it permits some violation. One solution is to shift the target so the penalty starts prior to the violation.

$$penalty = \begin{cases} 0, & \text{if } x \le x_{limit} - \epsilon \\ (x - (x_{limit} - \epsilon))^2, & \text{if } x > x_{limit} - \epsilon \end{cases} \tag{21-3}$$

This approach requires a choice of the ϵ value that shifts the limit, which adds a burden to the designer; being a soft penalty, it still will permit a bit of deviation.

21.2.5 Barrier Function for One-Sided Limits

The barrier function approach adds a penalty for proximity to a limit and amplifies the penalty toward infinity when on the constraint. An example is a reciprocal model for the penalty.

$$penalty = \begin{cases} \dfrac{1}{(x - x_{limit})}, & \text{if } x < x_{limit} \\ large\ number, & \text{if } x \ge x_{limit} \end{cases} \tag{21-4}$$

The advantage of this approach is that the variable x will not be permitted to exceed the limit. One disadvantage is that the value of the *large number* penalty must be defined.

This may seem like a good approach if the limit is related to safety. For instance, keep the concentration below the lower explosive limit. However, one would not want to use the actual limit because of measurement error, nonideal mixing in a process, uncertainty on the true value, other possible materials in the gas phases, and other possible issues. Therefore, you might also want to shift the limit.

$$penalty = \begin{cases} \dfrac{1}{(x - (x_{limit} - \epsilon))}, & \text{if } x < x_{limit} - \epsilon \\ large\ number, & \text{if } x \ge x_{limit} - \epsilon \end{cases} \tag{21-5}$$

Some people use a log-barrier function, the logarithm of the variable deviation from a limit, rather than the reciprocal.

If you want to shift the limit, the one-sided soft penalty method of Section 21.2.4 may be more convenient.

21.3 Extra Degrees of Freedom

There are extra degrees of freedom if there are more MVs than needed. Here are a few options for treating extra MVs:

- Have the optimizer decide on an energy rate, not a particular MV. Then, use the least expensive MV to meet that desired energy rate; when the MV meets a limit, add the next most expensive MV.
- Include a penalty in the OF for using the most expensive MV.
- Include a penalty in the OF for an MV deviation from a desired resting value.
- Include the cost of all MVs in the OF.

21.4 Combining Undesirables

The OF for control might be to minimize the deviation from a target (a CV RoC or a CV coincidence point), or it might be to minimize the sum of squared deviations between the model and reference trajectory over a future horizon.

The classic approach to including undesirable actions or consequences in determining control action is to add a penalty to the OF.

$$\begin{aligned} &\min_{\{\underline{u}\}} J = \sum_{k=1}^{m}\sum_{i=1}^{H_k}(\tilde{y}_{i,k} - \tilde{r}_{i,k})^2 + \sum_{j=1}^{n}\sum_{i=1}^{H_j} penalty_{i,j} \\ &\text{S.T.: } 0 \le \underline{u} \le 100 \\ &\qquad \Delta\underline{u} \le RoC \end{aligned} \tag{21-6}$$

Here, penalties are added to the nominal OF, and hard constraints are listed in the S.T.: section. $\underline{u}$ represents the future possible MV values, H is the future time horizon, i is the time interval counter between now and the horizon, m is the number of CVs included, and k is the CV counter. This would be for a multivariable controller. Also, n is the number of penalty types included, and j is the penalty counter.

One issue is that the OF may not be dimensionally consistent. If one CV is flow rate and another is composition, the units on the deviation squared might be $(L/m)^2$ and $(kg/L)^2$ which should not be added in the first OF term. Further, if a penalty is on a temperature violation, the units on a penalty term might be $(^0C)^2$ or $(^0C)^{-1}$ depending on the form of the penalty being applied. The disparate units on the OF cannot be added.

In addition, one CV might be much more important than another, or a penalty may be of greater or lesser impact. There is no relative weighting in Equation 21-6. A solution for relative weighting and dimensional consistency is using equal concern (EC) weighting factors for each term.

Finally, the summations may not have the same number of terms over the horizons, as indicated by the subscripts on the H_k and H_j counters. In this case, one deviation may be five times more important to the optimizer just because the number of terms in the sum is five times more than that of another.

A solution is to normalize each term by the number of items in the sum.

The scaled and normalized optimization statement is

$$\begin{aligned} &\min_{\{\underline{u}\}} J = \sum_{k=1}^{m} \frac{1}{H_k} \sum_{i=1}^{H_k} \left(\frac{\tilde{y}_{i,k} - \tilde{r}_{i,k}}{EC_k} \right)^2 + \sum_{j=1}^{n} \frac{1}{H_j} \sum_{i=1}^{H_j} penalty'_{i,j} \\ &\text{S.T.: } 0 \le \underline{u} \le 100 \\ &\qquad \underline{\Delta u} \le RoC \end{aligned} \tag{21-7}$$

where $penalty'$ represents the EC-scaled penalty.

For example, Equations 21-2 and 21-5 become

$$penalty = \begin{Bmatrix} 0, & if\ x \le x_{limit} \\ \left(\dfrac{x - x_{limit}}{EC} \right)^2, & if\ x > x_{limit} \end{Bmatrix} \tag{21-8}$$

$$penalty = \begin{Bmatrix} \dfrac{EC}{(x - (x_{limit} - \epsilon))}, & if\ x < x_{limit} - \epsilon \\ large\ number, & if\ x \ge x_{limit} - \epsilon \end{Bmatrix} \tag{21-9}$$

21.5 Setting EC Factor Values

Here is a procedure for setting the EC factor values.

1. Choose the most critical term to minimize. One option is the term $\sum_{i=1}^{H}(r_{1_i} - \tilde{y}_{1_i})^2$ if it is of greatest concern.
2. Choose a moderate $\tilde{y}_1$ deviation from perfection. It might be the deviation from the reference trajectory along the future time or the deviation from the biased

set point at a steady state. This value is an EC factor; EC_1 is equal to the $\tilde{y}_1$ deviation considered to be moderate.

3. Feel the concern this deviation creates. Use both qualitative and quantitative aspects. It may be helpful to rate the concern level on a 1 to 10 scale.

4. Assign EC values to each penalty term so that each deviation from perfection has the same concern level. As an example:

   - Perhaps $EC_1 = 0.5wt\%\ product\ deviation\ from\ SP$ has a 1 to 10 concern rating of 3.

   - To have the same concern rating of 3, perhaps the deviation from a target flow rate is 15 lb/min. Then, $EC_2 = 15\ lb/min$.

   The EC values are situation-dependent, meaning they change with the situation.

Notes:

1. Traditional optimization uses Lagrange-type multipliers for the terms in a sum. Here, those values are the reciprocal of the denominator values. I think the weighting method using EC factors is intuitive and makes it easier to assign weighting values that provide the relative importance.

2. Except for the barrier function, larger EC values mean it takes a larger deviation from the desired value to have an equivalent concern. By contrast, larger values in the barrier function mean the variable must be farther from the limit to have an equivalent concern.

21.6 The Impact of Penalties

The optimization statement for horizon predictive control (HPC) is

$$\min_{\{U\}} J = \frac{\sum_{i=1}^{H} (r_i - \tilde{Y}_i)^2}{EC_Y^2} + \sum_{k-1}^{M} \frac{\sum_{i=1}^{H} Penalty_{k,i}}{EC_k^2} + \frac{\Delta t \sum_{l=1}^{L} \sum_{t=now}^{t=H} c_l \cdot MV_{l,t}}{EC_{cost}} \tag{21-10}$$

Subject to $0 \le U_i \le 100\%,\ |\Delta U_{i,j}| \le threshold_i,\ \ldots$

If a penalty is "active," which means it is limiting MV values, and the system is "square" (has the same number of CVs as MVs), you will not be able to keep all CVs at their set points. If the number of CVs and active constraints is greater than the number of MVs, the degree of freedom is less than zero. In this case, optimization will balance the deviation from the desired value, but it cannot make a process with inadequate degrees of freedom match all performance desirables. In such a case, the EC factors would indicate which CVs get more attention.

For simplicity in analyzing this situation, consider $r_i = Y'_{SP}$ with the process at a steady state and a single $\underline{U}$ element.

Figure 21-1 illustrates the objective for the CV. At $\tilde{Y} = Y'_{SP}$, the penalty is zero, and that U value is $U_{desired}$. The three curves represent three EC values. The tighter, steeper upper curve has the lower EC value.

Consider that Z is the constrained AuxV. When $U \geq U_P$, Z does not violate Z_c. When $U < U_P$, there is a violation. This is illustrated in Figure 21-2. Again, the three curves represent three EC values. The tighter, steeper upper curve has the lower EC value.

Figure 21-3 illustrates the sum of the two penalties, showing U^* at the OF optimum. Notice that the penalty has shifted U^* to the right of U_d.

To determine the steady-state deviation from the set point, recognize that, at the minimum, the slope of the OF with respect to U is zero. At U^*

$$\frac{dJ}{dU} = 0 = \frac{d\,Penalty_Y}{dU} + \frac{d\,Penalty_Z}{dU} = 0 \tag{21-11}$$

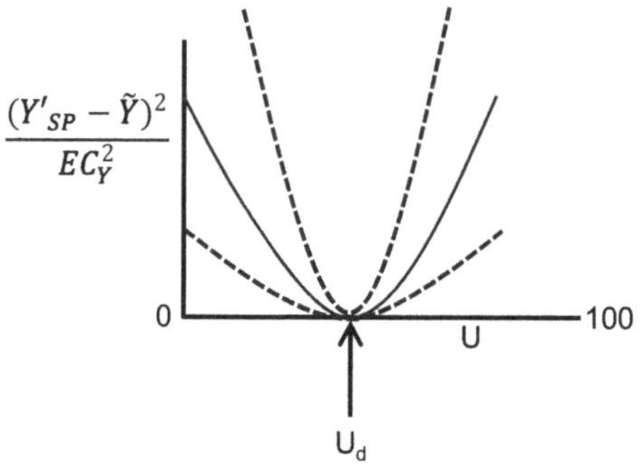

Figure 21-1. Illustrating the OF value for a CV.

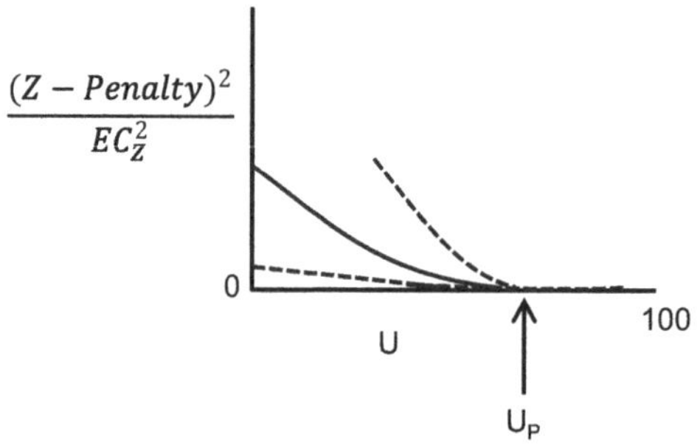

Figure 21-2. Illustrating a penalty for a deviation.

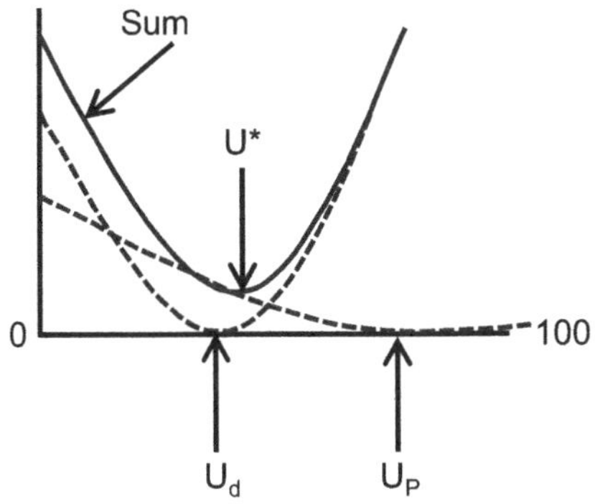

Figure 21-3. Illustrating the combined OF.

Substituting formulas for the penalties

$$0 = \frac{1}{EC_Y^2}\frac{d(Y'_{SP} - \tilde{Y})^2}{dU} + \frac{1}{EC_Z^2}\frac{d(Z_C - \tilde{Z})^2}{dU} \tag{21-12}$$

and taking the derivatives of the binomials

$$0 = -\frac{2}{EC_Y^2}(Y'_{SP} - \tilde{Y})\frac{d\tilde{Y}}{dU} - \frac{2}{EC_Z^2}(Z_C - \tilde{Z})\frac{d\tilde{Z}}{dU} \tag{21-13}$$

Recognizing that at a steady state, with locally linear models

$$\frac{d\tilde{Y}}{dU} = K_Y; \qquad \frac{d\tilde{Z}}{dU} = K_Z \tag{21-14}$$

and using $(Y'_{SP} - \tilde{Y}) = \Delta Y$ as the steady-state offset from the set point and $(Z'_C - \widetilde{Z}) = \Delta Z$ as the steady-state offset from the constraint

$$0 = \frac{\Delta Y K_Y}{EC_Y^2} + \frac{\Delta Z K_Z}{EC_Z^2} \tag{21-15}$$

Under the conditions of steady state, one-step-and-hold U, and the penalty on deviation from Y'_{SP} not r, gives

$$\Delta Y = -\Delta Z\left(\frac{EC_Y}{EC_Z}\right)^2 \frac{K_Z}{K_Y} \tag{21-16}$$

The relationship between the Y and Z offsets depends on the EC factors (user choice) and the process sensitivity to U (the process gain).

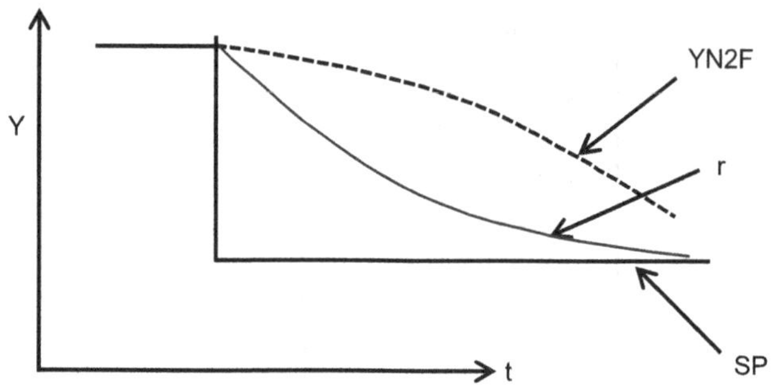

Figure 21-4. Illustrating deviation during a transient.

Further, because the controller OF is not based on $(Y'_{SP} - \tilde{Y})^2$ but on $(r - \tilde{Y})^2$, as Figure 21-4 shows, deviations from r are less than those from Y'_{SP}. So, during a transient, Y_{offset} will be larger than predicted above.

Choosing the economic EC factor may be difficult because it must balance the consequential steady-state offset. To help see the way to include this aspect, consider a single CV with one MV and the OF value $J = \left(\frac{y'_{SP} - \tilde{y}}{EC_y}\right)^2 + \frac{c_l \cdot \Delta u}{EC_{cost}}$. Ideally, the deviation from the set point is zero with some particular MV value. If the optimizer chooses a less expensive MV value, perturbed by Δu, there will be a corresponding deviation from the set point. If the CV is locally linear to the MV, then $\Delta \tilde{y} = K_m \Delta u$, and the steady-state CV deviation factor will be $\left(\frac{\Delta \tilde{y}}{EC_y}\right)^2 = \left(\frac{K_m \Delta u}{EC_y}\right)^2$. The cost term improvement will be $-\frac{c_l \cdot \Delta u}{EC_{cost}}$. At an optimum, the derivative $\frac{dJ}{du} = 0$. This means $2\frac{\Delta \tilde{y}}{EC_y}\frac{K_m}{EC_y} - \frac{c_l}{EC_{cost}} = 0$. We would like the CV deviation to be imperceptible, perhaps small relative to the CV noise or the width of the image on the display. We desire $\Delta \tilde{y} \le \varepsilon$. Solving for EC_{cost}, $EC_{cost} > \frac{c_l \cdot EC_y^{\,2}}{2\varepsilon \cdot K_m}$.

21.7 Takeaway

- Choosing approaches to manage constraints, extra degrees of freedom, and the relative weighting used in the optimization can substantially impact the control.

Exercises

22.1 Confirm the analysis for the EC_{cost} factor value at the end of Section 21.6.

22.2 Extend the analysis to a system with two manipulated variables (MVs) and one controlled variable (CV). Presume the CV can be locally linearized.

22 Multivariable Processes

22.1 Introduction

Multivariable processes have two or more controlled variables (CVs) and two or more manipulated variables (MVs). They may also have auxiliary variables (AuxVs) that must be kept within limits. Interaction, nonlinearity, and constraints combine to create control difficulty.

As an example, consider the simple binary distillation column illustrated in Figure 22-1. Feed to the column (F) has several components with compositions (z) that are separated by boiling. *Binary* means that there are two product streams or, more simply, two key feed components. Exiting at the top is the distillate (D) with key component composition y, and exiting at the bottom is the bottom product (B) with key product composition x. Steam enters a heat exchanger (the reboiler) at the bottom, which boils the liquid. The steam does not mix with the boiling fluid, and condensate from the steam recycles to the boiler to make more steam.

Boiling the liquid creates vapor that rises through the column and exits at the top. The vapor is mostly composed of the more volatile components in the feed, but it does contain some of the less volatile components as an impurity. Exiting vapor is then condensed, perhaps by plant process water from an evaporative cooler, which is recycled to the cooler. The vapor condensate is collected in the condensate receiver. Some of the condensate is returned to the column as reflux (R), washing impurity back down the column and putting the top product back into the column. In the column, the rising

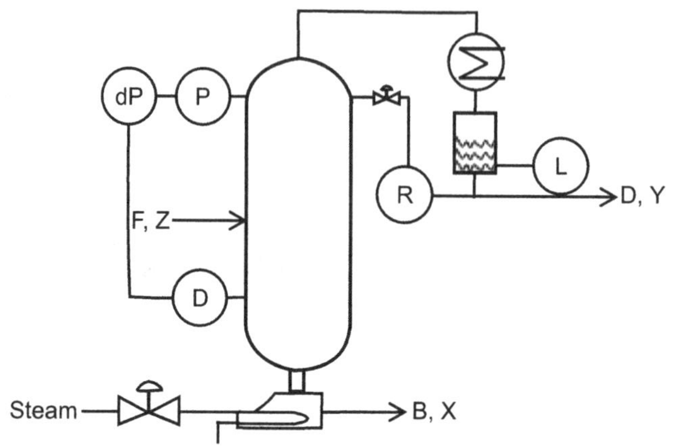

Figure 22-1. A simple illustration of distillation as a multiple-input, multiple-output (MIMO) process example.

vapor exchanges material with the falling liquid, and in the many stages within the column, the less volatile and more volatile products are purified.

22.2 Interaction

The role of a distillation column is to separate the feed products, meeting purity specifications on the top and bottom product compositions. Basically, and for this discussion, the two CVs are the top and bottom product compositions, y and x; however, pressure, differential pressure, and levels could also be CVs. The column must respond to the feed rate, F, and feed composition, z, it is given, as well as to feed temperature and other ambient disturbances such as environmental heat losses. Also, basically, from a process view, the two MVs are the steam flow rate, S, and the reflux rate, R; however, from a process control view, the MVs may be the controller signals to the steam and reflux valves. Nominally, this is a two CV, two MV system. It is termed *"square"* because the number of CVs is the same as the number of MVs.

As a complication, each MV affects both CVs. If the steam is increased to boil up more impurity, it also boils up more bottom product, affecting top product purity, which requires the reflux to change to counter it. The term is *interacting*, meaning that each MV will affect many CVs. A portion of the interaction is illustrated in Figure 22-2.

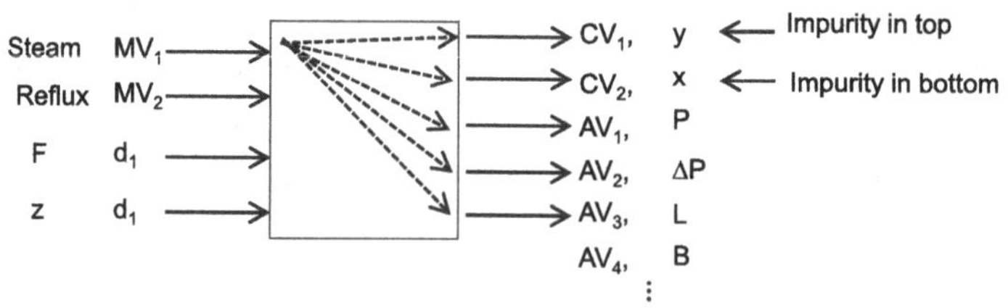

Figure 22-2. A subsample of relations indicating interaction.

The dashed lines within the process box indicate that MV_1 affects each response variable. The dashed lines from each other input to each response variable are not shown.

In Figure 22-2, the auxiliary variables (AVs) are column pressure, P; column differential pressure, ΔP; and reflux and bottoms product flow rates, L and B. The input disturbances, d, are input flow rate, F, and inflow composition, z.

Also, as a complication, compositions are difficult to measure and are probably implied by tray temperatures near the top and bottom of the column. These are termed *inferential measurements*.

The composition controller will likely send flow rate set points to lower-level controllers that send signals to the valves that adjust the steam and reflux flow rates. The output of the composition controller would probably not be a direct signal to the valve. Accordingly, the output of the primary controller would be the process engineer's viewpoint of the MVs, *S* and *R*.

22.3 Secondary MVs and CVs

However, there is more to consider. The reboiler and the condensate receiver need level control to prevent them from going empty or overflowing. The valves on the distillate and bottom product lines adjust the flow rates to keep the levels within specification. The level controllers will be continually confounded by the composition controller action. If there is an impurity upset in the top product, the reflux flow rate will respond to fix it, but this upsets the level in the condensate receiver. Eventually, the change in liquid falling through the column upsets the reboiler level.

Secondary CVs are not the primary reason to operate the process, but they must be kept under control to make the process operate. For example, you don't buy a car to keep the engine coolant at 180°F or the oil pressure at its set point, but temperature and pressure control are required secondary objectives in operating an engine.

Further, the column pressure must be controlled so that overpressure does not cause a safety vent to release vapor or so that vacuum does not cause the vessel to collapse, suck in air through seals, or create low suction pressure on the fluid pumps. Column pressure could be controlled by adjusting the cooling water rate to the condenser. Slower cooling water flow would condense less vapor, keeping the pressure up. However, changes in the steam or reflux rates will affect column pressure, requiring the pressure controller to continually change the condenser water flow rate.

Therefore, there could be at least five CVs and five MVs, not just the nominal two for the binary column.

22.4 Constraints

Some CVs have constraint limits. Column pressure, for instance, must be maintained between high and low values. The compositions will have specification limits.

The MVs have constraints. Controller outputs are limited to 0% to 100%, and if the valves are fully open, the MVs, *S* and *R* will be limited by the process's maximum capability. Further, the reflux cannot exceed the rate of supply of condensate from the vapor leaving the top of the column.

There is substantial interaction between the MVs.

There are also AuxVs that must be kept within limits. Differential pressure is an AuxV with constraints. If it is too low, the trays will weep (lose liquid through the holes designed to let the vapor pass through) and run dry, eliminating a separation stage. On the other hand, if the differential pressure is too high, the trays will flood (retain liquid and fill up). Here, column ΔP is an AuxV with greater than and less than constraints.

22.5 Degrees of Freedom

There are also extra MVs in this example. As long as the column *P* is between the high and low limits, the column can operate. But is there a best *P* value? Pressure affects the separation, and changing the *P* to maximize separation will reduce the steam required to do the boiling. It would be nice to use this additional degree of freedom (DOF) to minimize the energy required to perform the separation.

In some applications, the extra MVs provide additional DOF that can be used to optimize the operation; for others, the extra MVs have resting values, nominal values that keep them ready for immediate action when their help is needed.

On the other hand, when an MV, CV, or AuxV hits a constraint, you lose a DOF. The constraint limits the MVs from keeping the CVs under control. In this case, you could give up control of one CV and use the reduced set of MVs to control the others, or you could share the burden among all CVs, keeping them near but not at the set point.

Or, if there are extra MVs, the extra DOF could move them off their resting values to help the constrained primary MVs keep all CVs at their set points.

All in all, MIMO processes have multiple CVs and multiple MVs. Often there are extra MVs that provide a DOF to optimize operation. Constraints remove MVs, and when there is an inadequate number for control, one must balance the deviation of each CV from its set point.

It is possible to have a switching logic that assigns extra DOF variables to a limit when they are not needed for control and then permits those MVs to change when a primary MV hits a constraint. However, when a second or third variable hits a constraint, choosing which remaining MVs to engage quickly becomes a complicated decision tree diagram.

22.6 Ill-Behaved Dynamics and Nonlinearity

Ill-behaved dynamics are typical of MIMO processes. This is illustrated in Figure 22-3. The top two sketches indicate steps in the two primary MVs (steam and reflux), and the lower four sketches illustrate responses for the two primary CVs (top and bottom composition). Notable are delays of varying degrees and a diverse order of responses. The significant interaction is also notable. If one MV changes to make one CV move to a new value, it upsets the other CV, requiring the other MV to change. However, that second change to fix the second CV upsets the first CV.

Not indicated in Figure 22-3, the gains and dynamics also change with operating conditions. This includes the MVs as well as the influences (F and z), AuxVs (ΔP), and set points.

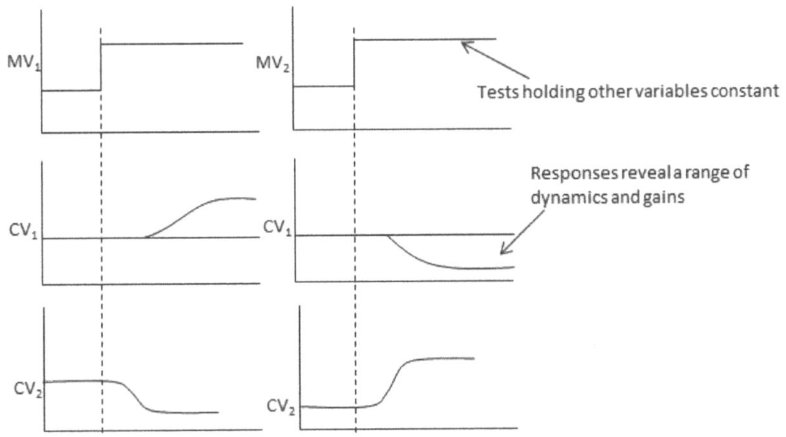

Figure 22-3. A subsample of relations indicating dynamic responses.

22.7 What to Do?

The issues are multiple CVs, extra MVs, changes in DOF, constraints, nonlinearity, interaction, and ill-behaved dynamics. See Chapter 23 for simple model-based control techniques and Chapter 24 for the horizon predictive control approach.

22.8 Takeaways

- Much of model-based control takes the process engineer's view of process influences and determines the set points for lower-level, proportional-integral-derivative control.
- Modeling is important. There will be inferential variables as well as process modeling.
- Data values must be consistent with their use in the models. The models, such as a material balance, may be needed to reconcile data inconsistencies, such as off-calibration flow rate measurements.
- The MIMO process is likely more complicated than it appears at first. For control, be sure to understand the diverse primary, secondary, and auxiliary variables and their diverse constraints and flexibilities.

Exercise

22.1 Consider a process that is familiar to you. List the primary controlled variables (CVs) and manipulated variables (MVs). State the secondary CVs and MVs. State the auxiliary variables that are observed. State the constraints and limits for all variables. If one MV is constrained, what else could be used as a compensating MV?

23

MIMO Process-Model-Based Control (PMBC)

23.1 Introduction

In addition to nonlinearity, multiple-input, multiple-output (MIMO) systems express interaction, constraint situations of degrees of freedom (DOF < 0), and extra manipulated variable (MV) situations of DOF > 0.

The first aspects, nonlinearity and interaction, are each revealed in the example of mixing hot and cold water in a continuous flow stirred tank.

$$F = F_h + F_c \tag{23-1}$$

$$\frac{V}{F_h + F_c}\frac{dT}{dt} + T = \frac{T_h F_h + T_c F_c}{F_h + F_c} \tag{23-2}$$

Here, F is flow rate, V represents volume, and T is temperature. Subscripts h and c refer to hot and cold.

Nonlinearity means that a process output is not a linear response to a process input. For example, in Equation 23-2, T is a nonlinear response to both F_h and F_c (or to the signals to the respective flow control valves, u_1 and u_2). *Interaction* means that process inputs affect more than one process output. If either u_1 or u_2 change, both F and T respond.

A process-model-based controller would specify desired individual rates of change for the modeled T and F; with the modeled values of T and F updated, this would leave Equations 23-1 and 23-2 as two equations and two unknowns. For control, solve the equations for the two unknowns, both F_h and F_c (or generically u_1 and u_2). In this specific case, it is possible to solve for u_1 and u_2 explicitly. However, in certain cases, it may not be possible to solve for the inverse explicitly, in which case a root-finding approach could be used. However, instead of root-finding, I recommend optimization because it can handle hard constraints on excessive MV values, the cases of DOF > 0 and DOF < 0, and a penalty when auxiliary variables exceed limits.

For MIMO systems with simple model-based control (MBC) (one-step calculations not intended to manage ill-behaved dynamics), the basic optimization statement is as follows:

$$\begin{aligned} &\underset{\{u_k\}}{\text{Min}}\, J = +\sum_i (h_i/EC_i)^2 + \sum_j B_j/EC_j + \sum_k c_k u_k/EC_k \\ &S.T.: 0 \le u_k \le 100, \text{and other hard constraints such as rate of change in the MV.} \end{aligned} \tag{23-3}$$

In this statement, h is the contribution to the objective function, J that represents the controlled process performance desire, such as the deviation between the desired and model-predicted outcomes due to u choices. The sum in Equation 23-3 includes all controlled variable (CV) objectives.

The symbol B represents a penalty for "badness," an undesirable excess or constraint violation. Constraints could be on the CV, the MV, or any of many auxiliary variables. B would typically represent the square of the deviation from a constraint when there is an undesired deviation and zero otherwise. This is a soft constraint, or penalty method. The sum is for all variables that have constraints, including MVs. For example, if there are extra MVs, optimization may want to set one to zero to minimize cost. But Operations may want to keep it at a low value to ensure that that part of the process is working and ready to move when needed.

The coefficient c represents the cost associated with each MV. If there are extra MVs and multiple ways that the flexibility could get all h values to zero with no constraint violations, optimization will identify the set of MV values that minimizes operating cost.

The equal concern (EC) factors are user-defined factors that balance the importance of each item. Some people prefer to show the normalizing coefficients as multipliers (often called *Lagrange coefficients* or *weighting factors*). I prefer using EC factors because I think the concept is easier to use to define weighting factors that appropriately balance all issues within the operational context. To set the EC factors, choose one of the most important of the h or B terms and choose a moderate deviation from the ideal. This deviation will be the *EC* value for that variable. Consider all the related issues caused by that deviation and rank the level of concern that it raises on a 0 to 10 scale. The issues might be related to economics, safety, your boss's reaction, or other quantitative and qualitative issues. Then, choose a deviation from the ideal for each variable that generates an equivalent concern. The *EC* factors are the reciprocal of the λ weighting factors you might otherwise use.

Note: Although some people think the sum of the λ weighting factors should be unity, $\sum_k \lambda_k = 1$, there is no need. You can multiply the objective function by any positive number and the optimum is unchanged. There is no rule for the sum of the EC factors. The only requirement is that EC factors represent equivalent concerns for a deviation from ideal.

The S.T.: part means "Subject To" and lists the hard constraints. Hard constraints cannot be violated. Apply these to the MVs. These could be rate-of-change limits or nonzero minimum values.

In contrast, soft constraints permit a bit of violation. One might prefer that there is no violation of a safety-related variable (e.g., pressure, temperature, or explosive mixture) and desire a hard constraint. However, if a disturbance or some past conditions have evolved responses causing a constraint violation now, there might not be any feasible MV action now that can immediately remove the constraint condition. It may take several future control actions to have the MV impact ease the constraint condition. So, on soft constraints, do not make the true disastrous limit the constraint. Include a safety factor and make the constraint value comfortably before the disastrous limit, as in Equation 21-3.

23.2 Multivariable Model Adjustment

The process parameter adjustment approach can also be extended to multivariable processes. If there are the same number of measured process outputs and adjustable parameters, Equation 12-9 becomes a vector of operations. However, if there are fewer adjustable parameters than process measurements, Equation 12-9 can be reformulated as a least-squares minimization. Also, there are many options for model adaptation and filtering that the user might prefer.

23.3 Illustration

This example involves in-line mixing hot and cold water. Process details and equations are presented in Appendix B. Hot and cold water flow rates are controlled by signals to valves. The flows are mixed at a piping confluence point, and the mixed warm water flows downstream to flow rate and temperature transmitters. The valves are air-actuated and have a response time constant. Opening one valve increases the total flow rate, which increases back pressure on the other influent, causing it to slow, representing an extra level of interaction. Flow rate measurements of the incompressible fluids are essentially instantaneous, but they are noisy and will be filtered. There is a transport delay on the mixed stream temperature and then a thermowell sensor lag, but the measured temperature is relatively noiseless.

Figure 23-1 indicates the temperature and flow rate responses to the controllers in manual (MAN) mode and without environmental effects (drifts and noise). As Figure 23-1a

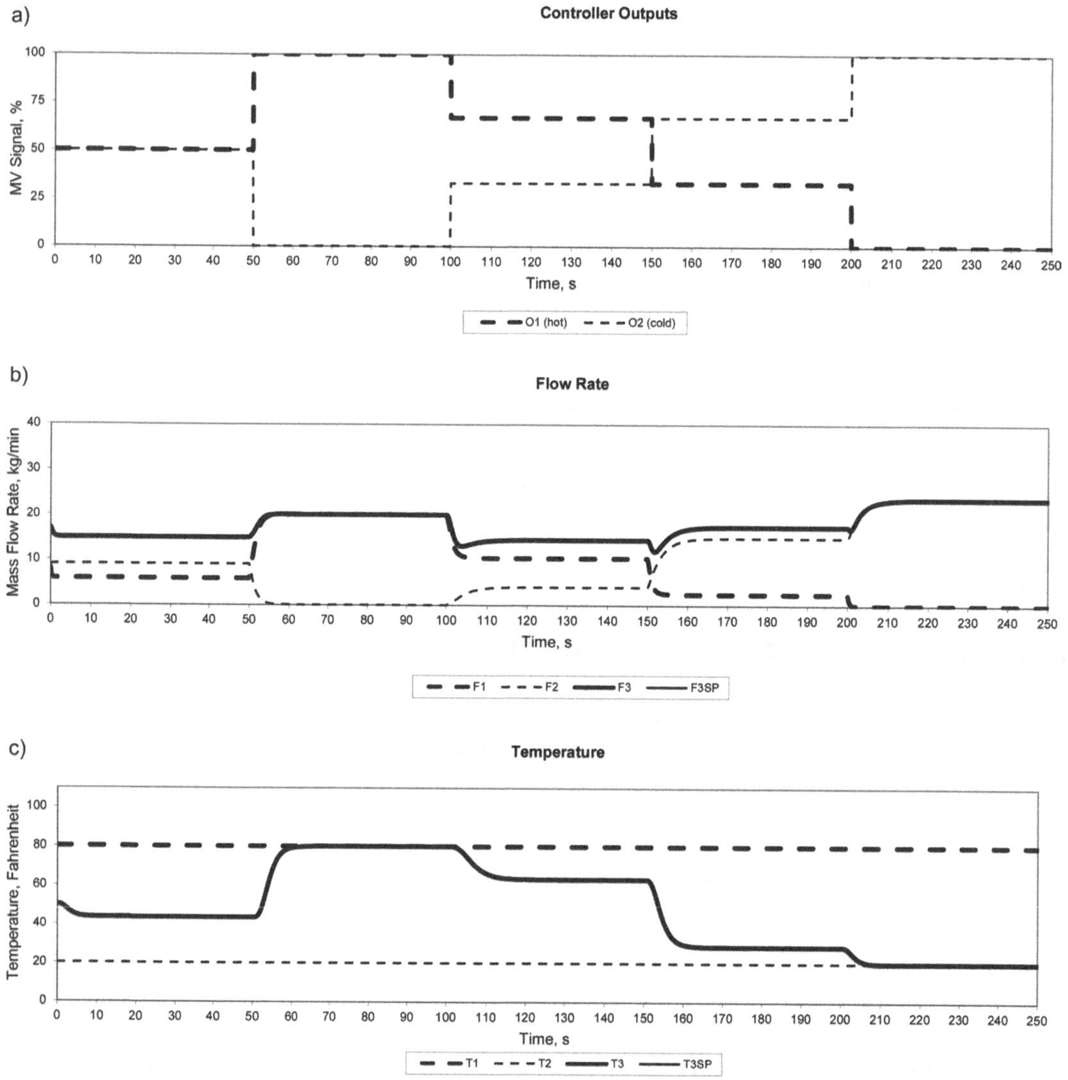

Figure 23-1. Controller output signals (a), flow rate responses (b), and temperature responses (c).

indicates, the controller outputs are stepped between 5%, 50%, and 95%. O1 represents the signal to the hot water valve, and O2 is the signal to the cold water valve. Figure 23-1b indicates the flow rates (hot, cold, and mixed). At times of 100, 150, and 200 s, the control signals made compensating 45% steps. One indication of nonlinearity is the difference in the change of the flow rates. Note also the inverse response of the total flow rate due to unequal valve dynamics. Figure 23-1c shows the mixed temperature response. Note the approximate 5 s transport delay, which also depends on the total flow rate. Observe the difference in temperature changes despite equal controller output changes, indicating nonlinearity.

The process is interactive, nonlinear, and has disparate dynamics. The dead time depends on the operating conditions.

Figure 23-2 illustrates simple MBC with optimization solving for the pair of MVs and balancing constraints. Because this is in automatic (AUTO) mode with both temperature and flow rate set points, the sequence of graphs has been shuffled. Graph (a) illustrates the temperature and graph (c) reveals the controller outputs. In this simulation, environmental effects include drifts in the inlet temperatures, drifts in the pressure driving the fluids, and noise on the flow rate measurement. The measurements are filtered values. Initially, the controller is in MAN mode, and at 30 s, it is switched to AUTO.

In the temperature graph, Figure 23-2a, the set point is the thin line, visible after the MAN-AUTO transition and making step changes. Notice that it tracks the mixed temperature while in MAN, then makes steps in the set point in the AUTO mode. The thicker line that follows the set point is the mixed T, and the upper and lower dashed lines are the inlet temperatures for the hot and cold water, which independently drift during the run.

The set point is the thin solid line in the flow rate graph, Figure 23-2b. Notice that it tracks the mixed flow rate while in MAN and then makes steps in AUTO. The thicker line that follows the set point is the mixed total flow rate, and the lower and middle dashed lines are the flow rates for the hot and cold water, which are influenced by supply pressure drifts during the run and the controller outputs.

Figure 23-2c illustrates the controller outputs. Note that at each temperature set-point change in the 70 to 160 s period (with a constant flow rate set point), the two controller outputs simultaneously coordinate their changes to maintain the flow rate while effecting a temperature change.

At a time of 190 s, the flow rate set point makes an extreme rise, and the process cannot simultaneously reach the two set points. The signal to the hot water valve goes to

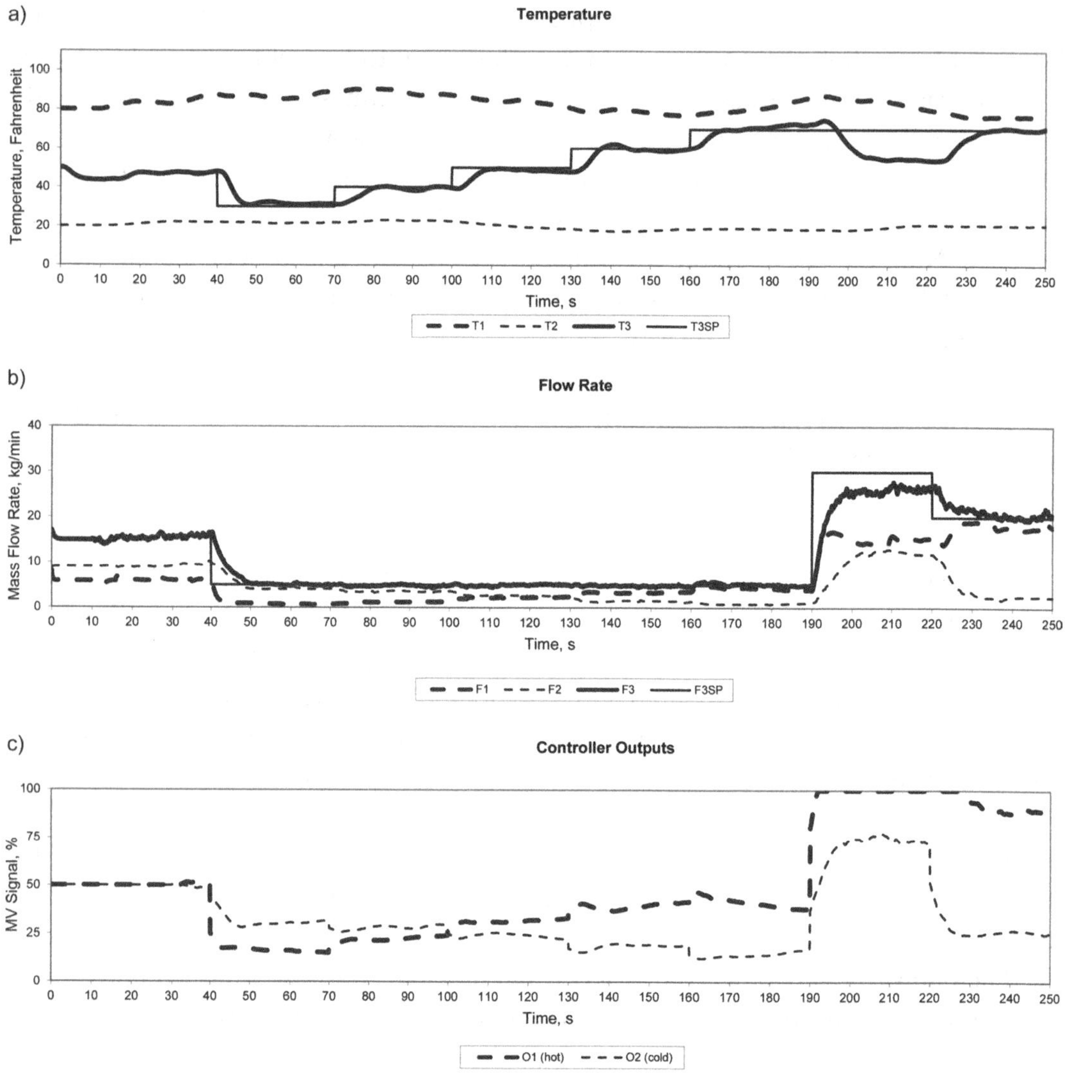

Figure 23-2. AUTO mode temperatures (a), flow rates (b), and controller outputs (c).

the 100% constraint, leaving the cold water signal to balance the mixed temperature and total flow rate. Note that the temperature and flow rate deviate from the set point in Figures 23.2 a and b. The deviation magnitudes are balanced by the equal concern choices, here chosen to cause approximately visually equal deviations on the graphs. Note that when the flow rate set point is reduced, removing the constrained condition, two control signals coordinate their action to return both CVs to their set points immediately (no windup).

23.4 Takeaway

- This is the same as the Chapter 12 takeaway, except there are more variables. The control benefit of incremental model adjustment is inconsequential. The important benefit of model adjustment is real-time, online monitoring of the process characteristics for forecasting maintenance needs, unit switching, and constraint operation. Optimization balances constrained conditions. The model handles nonlinearity and interaction. If the model dynamics are relatively well-behaved (low-order, little delay, no inverse action), multivariable PMBC is a good choice.

Exercises

Use the hot and cold mixing simulator to work each exercise.

23.1 Choose Trials 1 and 2 (steps in one manipulated variable [MV]) and describe nonlinear and interaction.

23.2 Choose Trials 3 and 4 (steps in both MVs) and describe inverse action and nonlinearity.

23.3 Choose Trial 5 (steps in controlled variables [CVs]) with simple model-based control and describe bumpless transfer, the ability to control both CVs uniformly for set-point changes, the ability to coordinate control for interaction, the balance of objectives for constraint, and no windup when the constraint is removed.

23.4 Choose Trial 6 (the same as Trial 5 but with environmental effects included).

23.5 Choose Trial 5 or 6 with proportional-integral-derivative control to see that tuning for one region does not make it good in another, one set-point change upsets the other, and the inability to balance offsets during the constraint.

24 MIMO Horizon Predictive Control

24.1 Multiple-Input, Multiple-Output Horizon Predictive Control

How can you solve the challenges of multivariable coupled processes with constraints on future variable values and ill-behaved dynamics?

First, structure the controller.

Create a model that represents the dynamic response of controlled variables (CVs) and auxiliary variables (AuxVs) to each manipulated variable (MV). Options include the following:

- Weiner
- Hammerstein
- First principles (my preference, but not the only effective way)
- Neural networks
- Linear ordinary differential equations
- Finite impulse response

Define a method to adjust for process-to-model mismatch, pmm. Again, there are many ways to accomplish this, including the following:

- Bias the set point by the associated pmm, $Y'_{SP_i} = Y_{SP_i} - pmm_i$. (This is my preference, but not the only right way.) Also, bias the limit (or constraint) for each predicted state or AuxV by its associated pmm.
- Bias the set point by the associated $\int e_i$.
- Bias the model by the associated pmm (residual).
- Adjust the model coefficients by pmm.

Define the control horizon (how far into the future to run the now-to-future model) and the number and location of forecast MV moves. I prefer three MV moves per MV and forecast over a horizon that is about 85% of the settling time and at 0 to 1/4, 1/4 to 1/2, and 1/2 to horizon intervals. Again, this is not the only way.

Define the procedures the ACT function is to use in the automatic (AUTO) mode:

1. The optimizer "guesses" all future MV moves (six are illustrated in Figure 24-1 in this case, three for each MV). Then, the N2F model forecasts the result over the control horizon on all CVs (two in this case) and all AuxVs (three in this case) (labeled as AVs).
2. The Objective function (OF) value is calculated. The calculation includes deviations from the set point or reference trajectory; penalties for high or low values or excessive rates of change on the MV, CV, and AuxV; a FAIL status for hard constraints; and economic penalties for expensive MVs when the degree of freedom (DOF) > 0.
3. Next, the optimizer re-guesses at MV sequences until the best OF value is achieved.

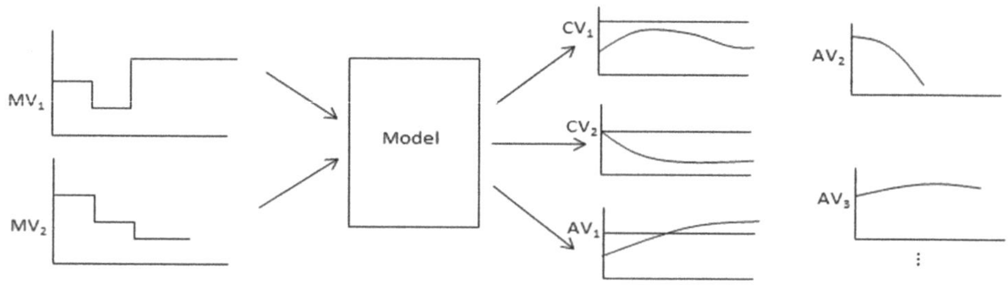

Figure 24-1. A multivariable illustration.

4. The controller then implements the first control action in the sequence for each MV. Further calculations are not performed until the control interval ends..

The optimization statement could be represented as

$$\min_{\{\underline{U}_1,\underline{U}_2\}} J = \frac{\sum_{i=1}^{H}(r_{1_i} - \tilde{Y}_{1_i})^2}{EC_1^2} + \frac{\sum_{i=1}^{H}(r_{2_i} - \tilde{Y}_{2_i})^2}{EC_2^2} + \cdots + \frac{\sum_{i=1}^{H} Penalty_{1,i}}{EC_3^2} + \frac{\sum_{i=1}^{H} Penalty_{2,i}}{EC_4^2} + \cdots \tag{24-1}$$

where J is the Objective function, H is the number of intervals in the control horizon, i is the interval counter, U is the MV, $\tilde{Y}$ is the modeled CV, $\frac{\sum_{i=1}^{H}(r_{1_i} - \tilde{Y}_{1_i})^2}{EC_1^2}$ represents CV_1 performance, $\frac{\sum_{i=1}^{H}(r_{2_i} - \tilde{Y}_{2_i})^2}{EC_2^2}$ represents CV_2 performance, and $\frac{\sum_{i=1}^{H} Penalty_{1,i}}{EC_3^2} + \frac{\sum_{i=1}^{H} Penalty_{2,i}}{EC_4^2}$ represent the penalties (a squared deviation from a limit, a soft penalty) for constraint violations. (There may be more than the two illustrated.)

The equal concern, *EC* values must be chosen to weight each factor by its concern for the process owner.

If the reference trajectory, *r*, is first-order (which is my preference for simplicity and effectiveness), then

$$\tau \frac{dr}{dt} + r = SP', \qquad for\ now \le t \le H, \qquad r_{t=now} = \tilde{Y}_{P2N,\,now} \tag{24-2}$$

Tau, τ, the time constant for the path, is the tuning coefficient for the controller aggressiveness. There will be independent τ values and reference trajectories, *r*, for each CV. SP' is the biased set point, and t is the time from now to the control horizon, H.

Numerically, Euler's explicit method can be used to calculate the reference trajectory:

$$r_{t+\Delta t} = \frac{\Delta t}{\tau} SP' + \left(1 - \frac{\Delta t}{\tau}\right) r_t \tag{24-3}$$

Note that Δt is the controller time interval. This method is predicated on $\frac{\Delta t}{\tau} \ll 1$. $\Delta t \le 0.1\tau$ would provide good values. In the extreme, if $\Delta t > \tau$, the reference trajectory can oscillate. The developer must add a code feature to inform users if the choices for Δt and τ violate the condition or to remove the $\frac{\Delta t}{\tau} \ll 1$ constraint, use $1-e^{-\frac{\Delta t}{\tau}}$ instead of $\frac{\Delta t}{\tau}$ in Equation 23-3. e is the base of the natural logarithm

The soft penalty is based on a deviation of a limit, L_1.

$$Penalty_1 = \begin{Bmatrix} 0, & if\ AuxV_1\ is\ within\ the\ limit \\ (AV_1 - L_1)^2, & if\ AuxV_1\ exceeds\ the\ limit \end{Bmatrix} \tag{24-4}$$

If there are extra MVs, one should choose the combination that minimizes utility or other operational costs, c. A simple cost model for one MV is that the cost rate is $\dot{C} = c \bullet MV$. Then, the cost over a future horizon where the plan is to change the MV value is $\sum_{t=now}^{t=H} \Delta t \bullet \dot{C}_t = \Delta t \sum_{t=now}^{t=H} c \bullet MV_t$. For several MVs, the projected cost over the horizon is

$$\Delta t \sum_{l=1}^{L} \sum_{t=now}^{t=H} c_l \bullet MV_{l,t} \tag{24-5}$$

Generically, the OF becomes

$$\min_{\{U_{l,m}\}} J = \sum_{j=1}^{N} \frac{\sum_{i=1}^{H} (r_{ji} - \tilde{Y}_{ji})^2}{EC_j^2} + \sum_{k=1}^{M} \frac{\sum_{i=1}^{H} Penalty_{k,i}}{EC_k^2} + \frac{\Delta t \sum_{l=1}^{L} \sum_{t=now}^{t=H} c_l \bullet MV_{l,t}}{EC_{cost}} \tag{24-6}$$

$$S.T.: list\ hard\ constraints$$

In Equation 24-6, i and t are time counters from the current time into the future to the control horizon H. N, M, and L, are the number of variables in the category, and j, k, and l, are their indices.

Notes:

1. With ill-behaved dynamics, near-term $\tilde{Y}$ deviations are expected, so the CV performance terms might be weighted to make long-term deviations of greater importance.
2. The EC_{cost} value must be large enough so that including the optimization of operating cost does not preclude keeping the CV close enough to the set point.

Alternately, the desire could be to keep the CV at the set point and sum the squared deviations as the factor to minimize. Then, realizing that this leads to extreme control action, seek to minimize a penalty for large control changes.

$$J = \sum (Y_{SP_{i,j}} - \tilde{Y}_{i,j})^2 + \sum_j^M \lambda \sum_i^H (\Delta U_i)_j^2 \tag{24-7}$$

Note: When the controller is in manual mode, the model should be running, and the set points should follow the CVs to prevent bumpless transfer when switched to AUTO.

24.2 Takeaway

- This is the same as the Chapter 14 takeaway but with more variables. As a strategy that combines solutions to all control-related issues, normalized horizon predictive control (nHPC) could be considered the best strategy. However, it is also the most difficult to understand and complex to program. Some people recommend using the simplest strategy that addresses the needs of the application. Others recommend using horizon predictive control (HPC) for all applications, unifying all control implementations to one best overall strategy (except for keeping it simple and safe).

Exercise

24.1 Implement multiple-input, multiple-output (MIMO) normalized horizon predictive control (nHPC), preferably with more manipulated variables (MVs) than controlled variables (CVs). Test for individual CV tuning, coordinated action, and MV constraints and recovery. This includes the 0%, 100%, and rate-of-change limits. Test for the situation of an operator placing one MV in manual (MAN) mode and leaving the rest in automatic (AUTO) mode. Investigate the role of the equal concern factor values. Test for the constraint impact of the auxiliary variable. Include measurement noise, drifting disturbances, and drifting process characteristics.

Section 5

Ending

25 Insight

25.1 Introduction

Model-based controllers have been used successfully in commercial-scale, pilot-scale, and lab-scale processes. According to my research investigations, single-input, single-output (SISO) applications (not simply simulations) include fluid flow rate control, heat exchanger temperature, distillation bottoms composition, plasma reactor pressure, and pH. Multiple-input, multiple-output (MIMO) applications include distillation dual-end composition control and fluidized bed gasification. Cascade applications provide differential pressure control for packed tower absorbers. The processes are nonlinear, control is effective, and model parameter adjustment provides legitimate and reproducible insight into the process behavior.

Many have explored and demonstrated using nonlinear model-based controllers in commercial-scale processes and equipment.

To my knowledge, applications have been limited to processes with just a few controlled variables (CVs) and auxiliary variables (AuxVs).

There have been many simulations to test and explore model-based controllers, much more extensively than experimental time and safety would permit on real processes.

I believe there is credible evidence that nonlinear model-based control is practicable (including theoretical simulations, credible simulations, and actual full-scale applications) and is now ready for others to implement.

25.2 Sample Simulation Demonstrations

This book does not report on published results from real-scale implementations. Instead, it provides links to simulated processes that can be used for desktop investigations. Anyone wishing to investigate nonlinear model-based control should use simulations to convince themselves of its applicability. I chose four processes that are fairly simple to understand and model yet also express behaviors that are difficult for controllers—nonlinearity, nonstationarity, continual disturbances, interaction, and ill-behaved dynamics.

Researchers commonly use simulators to guide developing new automation and data processing techniques and then to demonstrate their techniques and benchmark them to others. Instructors use simulators to aid learning. The simulators are often conveniently simple, which is important for ease of understanding, ease of coding, and immunity from execution errors. However, most simulations are too idealized to establish the credibility of the technique for practitioner acceptance. When phenomenological models are replaced with transfer functions, undesirably, mechanistic understanding is masked, not enhanced. Real processes are nonlinear and nonstationary, have measurement noise, and are subject to continually varying inputs and disturbances. The following simulators balance the diverse issues related to simplicity and credibility.

25.2.1 Car Speed

The SISO simulation in Appendix A is perhaps the simplest to understand. It is nonlinear with several disturbance sources and is used to demonstrate nonlinear control, online model adjustment, and constraint avoidance.

25.2.2 Hot and Cold In-Line Mixing

The hot and cold water mixing process in Appendix B is a MIMO simulation. Here is a summary of its features: First, the flow rates in the simulator are not linear functions of the controller outputs; the valves have a parabolic inherent characteristic (modified equal percentage) installed in a process line. Further, because the system pressure drop, ΔP, experienced by either fluid depends on the flow rate of the other fluid through the common discharge pipe, the ΔP for either fluid depends on the other fluid's flow rate. This also makes the total flow rate a nonlinear and interactive response. In addition, as the flow rate changes, the residence time in the discharge pipe from the mix point to the temperature sensor changes, causing the transport delay (dead time) to change. Further, the dynamics of the heat transfer from the flowing fluid to the thermowell depend on the heat transfer coefficient, which is modeled as a 0.8^{th} power law function of the fluid flow rate. The process is fifth-order plus delay, with dynamics associated with valve response, fluid acceleration, transport delay, and sensor response. Finally, the process includes environmental disturbances in several forms. There are drifts in the input values for the two fluid temperatures and the two inlet pressures. Also, temperature and flow rate measurement devices include calibration drifts and measurement noise.

By contrast, the simple first-principles (elementary) model used for control in the simulation is the first-order idealization—linear and not interactive in the flow rate response to controller outputs and simplistically nonlinear in the temperature response to controller outputs.

Although the controller model can be explicitly rearranged to solve for u_1 and u_2, given the desired rates of change and current modeled values for T and F, the simulator uses optimization. This provides the exact values as an explicit solution when the solution is feasible but can use the unconstrained manipulated variable (MV) to balance deviations from the two CVs when a constraint is encountered.

25.2.3 pH

Because pH expresses "extreme nonlinearity," it is commonly chosen as a test case for controllers. Based on a full-scale, nonideal, continuous stirred tank reactor (CSTR), the simulator in Appendix C is nonlinear with third-order lags and subject to unmeasurable and measurable disturbances. Although rich in expressing process features, the simulator is relatively simple to code.

Many commercially important chemical reactions depend on acidity, as do biological and food processes. Because water is about neutral, we do not want to discharge acids or bases and ruin the environment for the natural life forms in rivers, lakes, or underground geology. Therefore, pH control is important and ubiquitous. Because pH control is also difficult, it has become a classic test case for control algorithms.

Many researchers use clever approximations to simplify the pH calculation and often use very simple pH process representations, which are too trivialized to represent a legitimate and credible demonstration basis. Accordingly, the approach in Appendix C is recommended as a benchmark challenge application for those who wish to explore pH control and related data analysis.

25.2.4 Heat Exchanger

The SISO simulation model of a process heat exchanger in Appendix D is based on a full-scale exchanger. The simulator is nonlinear with variable transport delay and third-order lags and is subject to unmeasurable and measurable disturbances. Although rich in expressing process features, the simulator is relatively simple to code and can legitimately guide technique development and establish credibility.

25.3 Insight

Here are guides to help those implementing nonlinear model-based control.

- In tuning, τ_w (tau-want, the desired time constant for the return path to the set point) should be a bit less (80% perhaps) than the dominant process-model time constant. This makes the controller push the process slightly faster than the process would normally respond.
- When adjusting model parameter values, τ_{pmm}, the time constant for the model parameter adjustment should be about five times the dominant model time constant. This tempers noise and reduces any interaction that might arise between the controller and the model adaptation. However, it makes model adjustment slightly slower than the speed at which the process might change.
- Model parameter adjustment can provide insight into the process behavior for maintenance, health, condition, future constraints, and other factors. This secondary benefit from a control viewpoint might be the primary justification for implementing first-principles, model-based control from a maintenance, operations, or supervisory optimization viewpoint.
- Model parameter adjustment is not necessary for control. The feedback mechanism for control is to bias the set point for the model with the process-model mismatch (pmm).
- Supervisory optimizers use steady-state process models to determine the best economic operating regions. The optimizer must understand constraint possibilities and the temporal state of the process. If the controller models are adjusted in real time to reflect the temporal process conditions, integrating the controller models (and model parameter values) within the plant-wide optimizer will provide better set-point values and constraint recognition.
- Using process models in the controllers preserves and propagates process knowledge among the operational staff. This enhances process understanding, rational troubleshooting, and process improvement, and provides another benefit to managing the process.
- The case-by-case model development and controller design, unique for each implementation, is a disadvantage. Each case introduces unique challenges in

structuring and implementing the equations to present execution errors and guarantee a solution. Often, this is easy, but it requires engineering attention.

- Although, in my experience, local or temporal linearizing of these controllers permits mathematical proof that they are stable (with appropriate tuning), nonlinearity means an inability to globally guarantee characteristics such as convergence and stability. Some will object to the absence of a universal mathematical proof of stability. However, in my experience, the nonlinear controllers have been tunable and stable over the entire operating range. "Proof" might not be mathematical. Rather, enough successful experimental evidence, with tests on a wide variety of processes that create challenges, could provide sufficient confidence. For the practitioner, this type of testing may even be stronger than any mathematical "proof" that is always based on many assumptions.
- The inverse relation in the Act function might not provide an explicit solution. Use optimization (as opposed to root-finding) when the controller has multiple MVs or the solution for the MV value cannot be from an explicit rearrangement. Optimization is the best way to handle the degree of freedom (DOF) > 0 and DOF < 0 situations.
- You can use model parameter tracking independent of process control for process monitoring and constraint prediction even when the controller is in manual (MAN) mode.
- Nonlinear control ability relieves a constraint on process design, permitting design to focus on more economically competitive outcomes.
- In MAN mode, the controller must be running the model, updating pmm to bias set points and limits and tracking the CVs as set points to have bumpless transfer to automatic (AUTO) mode.
- Implement the controller in stages. Be sure that each stage works before letting it reveal its control decisions. Start with exploring the model offline to test it for compliance with the process. Then, simulate the controller offline to ensure all execution errors are compensated in MAN-to-AUTO switches, start-up, and other conditions. Then, run the model in MAN mode to see if it adequately tracks the process. If the model needs fixing, fix it. Fix the model before it is used to suggest control action. Do not let a bad model lead to control action that creates a bad reputation for the project. After there is confidence in the model, operate the controller in AUTO mode and suggest a control action. You must use the implemented MV values, not the controller-suggested output, as controller model inputs. After there is confidence in the controller, place it online.

- The model for the controller must be developed and validated. This takes more time than retuning a controller. However, it may eliminate many retuning instances and much cumulative retuning time in the long run.
- After the first-principles model is developed and the functional relations are defined, calibrating the model (matching it to the process) requires much less process testing and, with only a few adjustable coefficients, much less time and effort.
- If the process changes significantly (reconfiguration, product type), the model may have to be changed. Characteristic of linear model predictive control (MPC) and advanced process control (APC) applications, process changes over a year or so require model recalibration. However, because the first-principles models are nonlinear and phenomenologically based, the nonlinear model-based controllers will have an extended operating range where they remain valid.
- Implementing and diagnosing such controllers requires a process engineer with several years of control experience.
- The controller model does not have to get a grade of A when rating its fidelity to the process. A model with just a grade of C will be fully adequate for control.
- If the control problem is related to process faults (dropped signals, stiction, analyzer delay, undersized pumps, ill-sized orifices, etc.), nonlinear control is not the solution.
- Use the actual MV (the overridden MV value in safety or select control, not the prior calculated MV) to calculate the modeled value in the Predict function.
- Operations will remove MVs for maintenance. The optimizer needs a switch to not use that MV. Similarly, if Maintenance overrides an MV (to repair a valve) or drops a CV (to repair an orifice), the optimizer needs a switch to remove those variables.
- Regardless of controller strategy (advanced regulatory control [ARC], APC, fuzzy, nonlinear model predictive control [NMPC], etc.), if the controller has equivalent feedforward, decoupling, nonlinear, and other features, it has equivalent performance in reducing variance and permitting higher throughputs, lower costs, reduced waste, higher quality, and other benefits. I suspect that NMPC will have the same process economic benefits as APC, be easier to adjust to the process (calibrate), and last longer.
- In some situations, data inconsistencies could make a model state value diverge. This would include a model of liquid level in a tank if the inflow and outflow measurements have errors that do not make the material balance true. In some situations, data reconciliation (adjusting measurements so that they are mechanistically compatible with the model) will be necessary.

25.4 Candidate Processes for Nonlinear Control

Candidate processes could have any of these features:

- **Nonlinear or nonstationary** – If tuning the controller is a continual problem because the process gain and/or time constant and/or interactions change with the operating region and/or they change in time due to process reconfigurations, product type, flow rate, or input material properties, then this process may be a candidate for nonlinear control.
- **Condition monitoring** – If this is important, use the simple models with model parameter adjustment. This provides information related to process conditions and constraints.
- **First-order-like dynamics** – If the process response is first-order-ish, the one-step-ahead control action methods—process-model-based control (PMBC), generic model control (GMC), and internal model control (IMC)—are appropriate.
- **Ill-behaved dynamics** – For processes with inverse action or significant dead time, processes in which an action now could lead to a future constraint violation, or MIMO interactive processes, use predictive functional control (PFC) or horizon predictive control (HPC).
- **MIMO** – PMBC, PFC, and HPC can be applied within multivariable situations. GMC and IMC would get complicated.
- **Supervisory benefits** – If the modeling effort has additional perceived benefits (process knowledge retention, process analysis/monitoring, constraint forecasting, maintenance planning, training, etc.), these might provide complementary incentives to implement nonlinear model-based control (NMBC).
- **Extra degrees of freedom** – If the process has extra MVs, economic optimization can determine the resting values for the more expensive MVs and make them active only when needed, when the other MVs cannot handle the load or constraints.
- **Quality giveaway** – Experience has shown that each advance in control algorithms has led to about a halving of variability. Each advance permits operating closer to constraints or specification limits, permitting increased throughput or reduced utility and waste expenses. If the process is operating near constraints/limits, consider an advanced control strategy.

By contrast, traditional proportional-integral-derivative (PID) and advanced regulatory control (ratio, cascade, override, feedforward, and decouplers) are proven and readily accessible solutions for processes that are reasonably linear and have dynamics that are compatible with first-order plus deadtime (FOPDT) models.

25.5 Cautions

When you are considering nonlinear control solutions, take the following actions:

- Include MAN and AUTO modes with bumpless transfer features—set point following and model updating in MAN.
- Initialize all coefficients and variables in the first call of the subroutine.
- Identify and isolate execution errors. Look for errors such as divide by zero, overflow (real or integer), square root or log of a negative number, type mismatch "O" entered for "0," and missing input. Rearrange the equations to prevent the errors or bypass the calculation if such an event happens. The logical option should be clear. Be sure that variables are appropriately initialized in a power-up. There are many forms for nonlinear equations, which create many, and not obvious, possibilities for execution errors.
- Check for rational values. For instance, the value should not be negative when a model parameter is adjusted (for efficiency, reactivity, fouling, resistance, inlet composition, reactivity, or yield).
- If there is an external override, use the overridden MV value within the controller model (kin to external reset feedback).
- Prior to implementation, keep it simple and safe (KISS) so that you do not chastise yourself with the alternate admonition after implementation. Use a more complicated model only after determining that the simpler one will not work. For instance, explore pseudo-steady-state conditions for models with states that change relatively fast. The objective is sufficiency, not perfection.
- Ensure that data is valid. The greater the number of process measurements used within the model, the greater the likelihood of a sensor or data transmission fault. Some data faults can cause execution errors. Minimize the reliance on possibly faulty data. Ensure data integrity. Include fault detection or error isolation in the code.
- To permit Euler's forward finite difference approximation to be valid, set the control interval (and simulation interval) to less than one-tenth of the smallest time constant in the process or controller.
- Test the controller. First, use model shadowing in the MAN mode. Make sure the model is working as desired. Second, use the AUTO mode as a suggestion. Do not connect to the process. Be sure that the controller-suggested MV

action is reasonable and the execution time is much less than the scan time. Only use AUTO mode when you are fully confident that the model, as implemented, is acceptable. Murphy's Law applies to control. If it does not work at 3:00 a.m., you will get a call and may be unable to implement another model-based controller.

- Preferentially, solve for the inverse by rearranging the model equations to provide explicit values for the MV.
- Ensure the model inverse and the model have identical steady-state gains; otherwise, control will result in a steady-state offset. It is possible to obtain separate empirical SS relations for the model and the inverse with one data set. However, the regression slope of y versus x is not usually the exact reciprocal of the regression slope of x versus y.
- If an optimizer solves the inverse, it must handle a nonlinear constrained case with the possibility of local optima traps. Use a direct search approach to avoid misdirection from gradient-based methods and to accommodate hard constraints. Thoroughly explore the optimization surface and, where possible, structure the equations so there is only one minimum. If local minima are possible, use a multiplayer (multiparticle, multi-individual) optimizer to increase the probability of identifying the global minimum. Apply convergence criteria to the decision variable and set it to about one-tenth of the smallest detectable MV change.
- Develop and validate the controller model. This takes more time than retuning a controller. However, it may eliminate many retuning instances and much cumulative retuning time in the long run.
- You may have to change the model if there are significant changes in the process (reconfiguration, product type).
- Implementing and diagnosing model-based controllers requires a process engineer with several years of control experience.
- Recognize that multivariable applications may be a big step in complexity over SISO applications if they use either (1) optimization to handle constrained (DOF < 0) or excess MV (DOF > 0) situations or (2) data validation/correction techniques to handle data faults. Be prepared to invest the effort.
- Do not use nonlinear control if the control problem is related to process faults (dropped signals, stiction, analyzer delay, undersized pumps, ill-sized orifices, etc.).

25.6 Q&A

In presenting these techniques over the years, I have received recurring questions from audiences. These questions and answers distilled from my former presentations may be useful.

Q: How does the PMBC technique differ from other model-based control techniques that readers may already be familiar with (e.g., model reference adaptive control, Smith predictors, etc.)?

A: In the 1970s and 1980s, many people realized the potential for computers to do more than proportional-integral-derivative (PID) control and independently conceived of "hundreds" of model-based control approaches. Basic concept elements were independently conceived but with many nuances for each stage in the control algorithm function. In my experience, with the right limiting assumptions, one controller becomes another.

- GMC with a steady-state nonlinear model is the same as output characterization from a PI controller. With a linear model, the output is simply PI with a process gain for scaling.
- IMC with a second-order model and a desire to track a first-order response to a set-point change leads to a PID controller.
- If the process were linear, first-order, and with a fixed delay, then PMBC would reduce to a Smith predictor, as does IMC.
- The functions within APC—or the academic community term for it, model predictive control (MPC)—can be expressed as shown in Figure 1-1.
- In a limiting case of a multiple-input, single-output (MISO) linear model and a single step-ahead ACT function, the MBC algorithm reduces to IMC.
- Classically, the Smith predictor uses a first-order plus dead time (FOPDT) process model to account for the dead time and the pmm to bias the set point for a proportional-integral (PI) controller. PMBC, IMC, and other model-based controllers would have a similar structure, but the Act function in Figure 1-1 would be the model inverse, not PI.
- I have not done much with model reference adaptive control (MRAC). I have been more interested in using nonlinear models in control, and all my touches on MRAC showed people using linear models. However, I suspect one can be converted to another with the right choices.

Q: The examples focus on nonlinear processes, where the nonlinearity is in the process itself. Would PMBC also be used when the process is linear and only the control valve is nonlinear? Or are simpler methods OK for this situation?

A: I promote using the best tool for the situation. *Best* could mean simplest, most convenient, standard, conventional, least expensive, and so on. PMBC preserves process knowledge and has only one tuning coefficient. Yes, PMBC could handle a linear process with a nonlinear valve. But, if PMBC is low on the learning curve and gain scheduling or output characterization is expedient or understood, then those classic advanced control approaches would be the best choice. A nonlinear valve could be handled by either gain scheduling or output characterization. So could many nonlinear processes.

Q: Suppose dead time is the problem. Will nonlinear control work?

A: Yes, if dead time is substantial, then either PFC or HPC will work. However, if the process is linear, either IMC or a Smith predictor may also work and might have the familiarity and comfort that would make them preferred. As with those classic methods, if the process transport delays vary and the controller dead time is fixed, control can go "bad." With a process-based model, the PMBC approach would base the delay on the process conditions. If a process has dead time, it is not feasible to expect to be able to move it instantaneously. Following the IMC strategy, the appropriate controller desire would be to start the process moving at the desired rate after the delay.

Q: Can PMBC be used in a cascade, ratio, or override situation?

A: Yes. The PMBC approach discussed in this book is meant to replace SISO PID controllers that must be continually retuned because of process nonlinearity. PMBC could be used as primary or secondary controllers, either as both or mixed with a PID controller. I recommend keeping a stage-by-stage design of simple controllers, each with the same Figure 1-1 form, rather than trying to generate a single code that does it all in one. The signal to the model in PMBC must be the actual, selected, or overridden value, kin to external reset feedback used to prevent windup when a PID controller is in an override or select situation.

Q: Is it really necessary to use the set point biasing concept? Pushing the model toward a biased set point at some rate is the same as pushing the CV toward its set point at the same rate.

A: Yes, the two rates are the same thing. One could calculate the desired rate using the true set point and the measured CV. There is no difference in the numerical value.

Either way, if the measurement is noisy, the noise confounds the desired rate. If using the measurement and true set point, the measurement is noisy. If using the biased set point and the model, the biased set point is as noisy as the measurement because the pmm reflects the measurement.

Here is why I chose to use the biased set point: (1) Because the model is used to calculate the control action and it does not exactly match the process, you must use the biased set point in the inverse of the model to calculate the MV value. Because the model must be driven to the biased set point, I like calculating the rate on the model and biased set point values for consistency. (2) Where should one filter the noise? If the noisy CV is filtered, that creates a dynamic lag, which increases model mismatch. However, if one filters the pmm value, it might slow the correction, but it does not create an additional mismatch.

Q: I am confused about the benefit of horizon predictive control. Why do we predict the MV out into the future if we are only going to use the first step of it? Is there another value to predicting the response? It seems like a lot of computation that is not adding value to the result.

A: Good question. If the simple model-based control that started the book (identify the MV that makes the model move toward the biased set point at a desired rate) works, use it.

The horizon optimization that we are doing with HPC is much more complicated. However, it has many advantages that make it desirable, and this approach has led to many commercial software packages. Considering several MV moves into the future provides the benefit of (1) constraint avoidance (constraints on the MV, CV, or auxiliary variable, whether value or rate of change) and (2) shaping overall action to best match a desired reference trajectory. Further, (3) it solves the problem with ill-behaved dynamics where calculating a now-MV value may have no impact (if the model has a delay) or inverse impact (if the model is inverse-acting). Additionally, it is (4) a robust universal solution, which works whether the inverse has an explicit solution or not. Also, (5) the past-to-now (P2N) and now-to-future (N2F) models have exactly the same equations (N2F is not the inverse or a truncated variation of P2N). Finally, (6) this approach leads us to a solution to a nonsquare multivariable approach with extra degrees of freedom (extra MVs).

Q: Can function blocks be used in a distributed control system (DCS) or programmable logic controller (PLC)?

A: For generic model control with a steady-state model (GMC-SS), which is effectively output characterization, yes, a function block that has the model inverse could be used. However, desirably, the model inverse should be a simple direct calculation from an explicit equation. If it requires an iterative solution procedure, the function block may not have the necessary tools.

All my implementations have been in executable code. Some commercial control systems permit user-written subroutines or functions. For others, we have had to interface data exchange with a computer.

Q: Where do these applications fall within the typical automation hierarchy, the range of typical problem sizes (number of controlled and manipulated variables), and the technical reasons to consider their use over existing approaches?

A: These nonlinear control applications would replace the low-level PID controllers and combinations that are considered to be advanced regulatory control (ratio, feedforward, decouplers, etc.). They would be below the supervisory real-time optimization that determines process set points. Although direct control from these controllers to the final element is possible and constituted most of my investigations, commercially, one might prefer to have the model-based controllers determine flow rate and temperature set points to the standard PIDs of the DCS or PLC. Here, the structure would be like APC or MPC.

All my application investigations have been with just a few CVs and MVs. However, I do not see any reason why there might be a scale limit. It seems reasonable that as long as the models are valid and the computer can execute them fast enough, the methods can cope with processes with many inputs and outputs.

The technical reasons are to improve control and constraint handling. However, contrasting classical advanced regulatory control techniques, these techniques are simple to formulate, have few tuning coefficients, reinforce first-principles modeling in the staff, and provide process condition monitoring, multivariable decoupling, and nonlinear and nonstationary compensation.

Q: Can these techniques be integrated with regulatory control (PID, feedforward) strategies? Or with linear MPC, such as cascade?

A: Yes. We have demonstrated cascaded MBC in both the primary and secondary controller, but there is no reason that a lower- or upper-level controller could not be linear. Steady-state models could use classic lead-lag-delay dynamic compensations (e.g., feedforward) but use the nonlinear model (either first-principles or empirical) as the gain. Process Perfecter from Pavilion Technologies (now part of Rockwell Automation) used a neural network, empirically trained, to determine the nonlinear model gains.

Q: Could one model some relationships phenomenologically and, where other relationships are unknown, use an empirical model, that is, a hybrid model?

A: Yes. In my experience, any modeling combinations work as long as the model matches the process. One could start with a simple phenomenological model and then adjust its coefficient values to match process data. One could use a linear empirical model for some process attributes, such as dynamics. One could use a local linearized model (like we use in distillation) to model how pressure-adjusted tray temperature is used to infer composition or how a log transformation of composition is used as a linear approximation. One could use neural networks for gains and train the neural network (NN) from empirical data or a rigorous design software model. One could use fuzzy logic based on human judgment of gains or sensitivities. If the model is only locally valid, then when process conditions change (throughput, composition set points, process configuration, etc.), the model may not match the process well and may need adjusting.

Q: How long will a nonlinear model-based controller remain effective?

A: How long does any control scheme remain effective, or how often should it be recalibrated? In my experience with low numbers of CVs and MVs, classic ARC tools (ratio, feedforward, override, gain scheduling, etc.) can be as effective for control as a model-based controller when the ARC has all the relevant features. However, when the process makes substantial changes because of piping reconfiguration, a rebuild, or an alternate process unit, then the process gains, time constants, and delays may be significant enough to require recalibrating or restructuring the controller. Substantial changes in throughput or product specification could require the same. Substantial changes in the business objective (maximize throughput or minimize energy consumption) may also require restructuring the controller.

APC (alternately termed MBC, linear empirical model-based controllers) lasts about 6 months to 2 years. They are locally valid to the initial operation, and when the process changes, control degrades. I expect nonlinear model-based controllers to last longer because they are not just locally valid. However, substantial changes in the process configuration or objectives may require restructuring any model-based controller.

Q: Can phenomenological model-based control be used in batch processes?

A: Yes. In batch control, the process states change in time. These states could be degree of polymerization, reaction composition, fraction crystallization, etch depth, and so on. We often have phenomenological models relating the batch evolution of state values over time to influences. We use such models to define optimum recipes for the batch. However, each batch is unique, and its progression could be monitored in time and the model used to adjust the batch recipe in real time. Some have adjusted the recipe to return the batch states to the nominal trajectory. In several lab-scale experimental cases, I have used batch information to adapt the model, then used the revised model to adjust the recipe to optimize annual productivity.

25.7 When to Use Which

- Use PI(D) controllers for these reasons:
    - They are used 95% of the time.
    - They are well-understood.
    - They are inexpensive.
    - They are generally functional.
    - Techniques are available for gain scheduling, overrides, feedforward, ratio, decoupling, and more.
    - They are widely available in nearly all hardware or software products.
- If the process is SISO with nonlinear and well-behaved dynamics, the following apply:
    - GMC or PMBC are my recommended control algorithms.
    - PMBC would have the advantage of continuously adjusting the mechanistic model.
    - You could use a steady-state process model with a Weiner or Hammerstein (linear dynamics) structure.
    - The steady-state model could be empirical, such as a NN.

- If the process is SISO with linear responses and ill-behaved dynamics, the following apply:
  - You should use IMC.
  - This requires a model of the ill-behaved dynamics.
  - If IMC is based on FOPDT models, you could use the nonlinear model to generate gains, delay, and time constants for a linear surrogate model. This would be gain scheduling IMC.
- If the process is SISO and nonlinear, but not mathematically understood (if qualitatively understood), the following apply:
  - You should use fuzzy logic control or natural language processing.
  - Gain scheduling PI may be simpler than fuzzy logic control.
- If the process is MIMO and nonlinear with well-behaved dynamics, the following apply:
  - You should use PFC, simple MPC, or PMBC.
  - You must take care in accommodating constraints and faults.
  - There are no commercial products (to my knowledge).
  - You will have to structure it yourself.
- If the process is MIMO, linear, interactive, and subject to constraints and ill-behaved dynamics, the following apply:
  - You should use the horizon predictive structure (MPC, HPC, APC, dynamic matrix control [DMC], etc.).
  - Consider the time to create and validate models; modeling could be expensive.
  - It would be nice if a vendor provided a package that executes several functions and permits a user to insert function codes for the model.
- If the process is MIMO and nonlinear with interactions, constraints, and ill-behaved dynamics, the following apply:
  - You should use the nonlinear models in the horizon predictive structure or PFC.
  - You should use a grouped neural network MPC.

- This is the most expensive control approach.
- Again, it might be nice if vendors could provide a package that executes several functions and permits a user to insert function code for the model.

25.8 Takeaway

- Practice on SISO simulations and then SISO applications. With experience and confidence, move toward MISO and then MIMO applications. Keep it simple and safe.

Section 6

Appendixes

A

Car Speed Simulator

A.1 Model

This model starts with Newton's second law of motion, informally stated as $F = ma$, but with the g_c dimensional unifying factor required for dimensionless consistency.

$$F = ma/g_c \tag{A-1}$$

The forces on the car would be, from the engine, drag from the velocity through the air, and gravitational pull if on an incline. Primitively modeling the engine force as proportional to the pedal position, u, and the air drag force as proportional to velocity, v, squared, and using the velocity change rate as acceleration, the model becomes

$$Ku - dv^2 - \frac{mg}{g_c} \bullet sin(\theta) = \frac{m}{g_c}\frac{dv}{dt} \tag{A-2}$$

Here, θ is the angle from horizontal, K is the proportionality factor to convert pedal position, u, to engine accelerating force, and d is the factor to convert velocity squared to drag force.

In a conventional formulation for control, the model representation is

$$\frac{dv}{dt} = \frac{g_c}{m}\left(Ku - dv^2 - \frac{mg}{g_c} \bullet sin(\theta)\right) \tag{A-3}$$

This simplistic model might be the first-principles model useful for control.

However, reality is a bit more complicated. There is rolling friction (losses in the drive-train and from tire to road), which is ideally proportional to velocity. In addition, there is air drag because of the difference between the air and the car speed, and it may be an accelerating force if the air is pushing from behind. The ideal squared power may be more likely a 1.8 power.

$$\frac{dv}{dt} = \frac{g_c}{m}\left(Ku - d\,|(v - v_{air})|^{0.8}\,(v - v_{air}) - cv - \frac{mg}{g_c} \bullet \sin(\theta) \right) \tag{A-4}$$

Recognize that car mass, m, changes with the number of passengers, amount of luggage, and fuel tank contents. The air drag coefficient, d, changes with window positions and other factors that affect surface roughness, such as the collection of bugs on the car. The drag force depends on air attributes such as density (which is affected by humidity, temperature, pressure, and rain). The rolling friction coefficient, c, changes with road roughness, the angle of the wind pushing sideways on the car, lubrication, internal loads on the engine such as air conditioning, and steering through road curves. Finally, the road angle continually changes, so the coefficients normally considered to be fixed (m, d, v_{air}, c, θ) change in time in unexpected patterns.

However, there is more complexity. The force the engine puts on the car depends on the engine speed; it does not have a constant gain as the Ku model indicates. If the car's transmission is automatic, even if $u = 0\%$, when the engine is idling, it wants to move the car forward, contrasting the $F = Ku$ model that suggests there would be no force if $u = 0\%$. I do not know if this is a customary engine response model, but it makes sense to me.

$$F_{engine} = a + be^{-u/20} + ce^{-u/30} \tag{A-5}$$

where model coefficients $a = 25{,}000$, $b = 44{,}000$, and $c = 200 - a - b$.

So, the car simulator is

$$\frac{dv}{dt} = \frac{g_c}{m}\left(a + be^{-u/20} + ce^{-u/30} - d\,|(v - v_{air})|^{0.8}\,(v - v_{air}) - cv - \frac{mg}{g_c} \bullet \sin(\theta) \right) \tag{A-6}$$

There are also lags in the real process, for instance, the lag for the carburetor to increase the fuel/air to the engine. I have $u(t)$ in the simulator as a first-order response to the controller output.

Contrasting the simulator model of Equation A-6, the controller model is Equation A-3.

There are functionality mismatches and phenomenon mismatches. The coefficient values are not the same, and the simulator (m, d, v_{air}, c, θ) values change over time. The values are not knowable by the controller model. With this process, the $F = Ku$ relation for the model is not very good, but with $K = 900$, the velocity model is not a bad match for the process.

The controller model could be adjusted to better match the process in many ways. The road angle is not measured, so θ could be adjusted as a fictitious road angle needed to make the modeled speed match the process speed. Alternately, there are many uncertainties in the drag model, so the d coefficient could be adjusted. Or, one could attribute all forms of mismatch to a fictitious force as an equivalent to any of the mismatch sources. Then, the controller model would be

$$\frac{dv}{dt} = \frac{g_c}{m}\left(Ku - d_{adjust}v^2 - \frac{mg}{g_c} \bullet sin(\theta_{adjust}) - F_{adjust}\right) \tag{A-7}$$

Here, d_{adjust}, θ_{adjust}, and F_{adjust} represent the choices mentioned. Only one of the three would be used to adjust the model because there is only one process-model mismatch (*pmm*). You could choose several other model coefficients to adjust to remove pmm. It could be the K value that represents the quality of the fuel, the pedal-to-carburetor linkage calibration, piston ring wear, and so on. It could be the air drag coefficient that currently has the ideal value of 2. I chose to explore d_{adjust}, θ_{adjust}, and F_{adjust} in my simulator.

From the car model, the steady-state velocity, v_{ss}, can be explicitly solved by

$$v_{SS} = \sqrt{\frac{Ku - \frac{mg}{g_c} \bullet sin(\theta_{adjust}) - F_{adjust}}{d_{adjust}}} \tag{A-8}$$

Then, the coefficient adjustment technique from Section 12.2 can be used to adjust the coefficient at each sampling incrementally.

Exercises

Access the simulator from *www.r3eda.com* to perform these exercises.

A.1 Use the simulator Trial #1 (with Tau Adjust = 0, Enviro = N, and yp sigma = 0) to see the nonlinearity and process-model mismatch. Comment on the nonlinearity and mismatch.

A.2 Use the simulator Trial #2 (with Tau Adjust = 0, Enviro = N, and yp sigma = 0) to see the nonlinearity and process-model mismatch with large unmeasured disturbances. Comment on the magnitude of the impact.

A.3 Use the simulator Trial #3 (with Tau Adjust = 0, Enviro = N, and yp sigma = 0) to see simple model-based control. Vary the Tauw value between 1 and 20 to determine the impact on control. Comment on the time-trend of the process approach to the set point and zero steady-state offset, and the biased set point for the model and the model approach to the biased set point. Comment on the similarity of the approach to set point regardless of the nonlinearity. Comment on nISE, nTravel, and the aggressiveness of control. What tuning coefficient would you desire for your car? Comment on anticipatory action (the controller begins to change direction before the controlled variable overshoots the set point). Comment on the initial kick of the manipulated variable when there is a set-point change. In proportional-integral-derivative (PID) control, the P action provides the initial kick, the I action removes the steady-state offset, and the D action provides an anticipatory response. PID control requires three separate tuning coefficient values. Model-based control does it all with one aggressiveness factor.

A.4 Use the simulator Trial #4 (with Tau Adjust = 0, Enviro = N, and yp sigma = 0) to determine the constraint-handling ability of simple model-based control. Notice the immediate control action change when the constraint is removed. There is no windup. For aggressive control, there will be both upper and lower constraints.

A.5 Use the simulator Trial #7 (with Tau Adjust = 0, Enviro = N, and yp sigma = 0) to determine the regulatory control with large unmeasured disturbances. The first disturbance is a step-and-hold for a duration, then a pulse in the other direction that relaxes in time. Notice the magnitude of the upsets indicated by the required change in the manipulated variables. Comment on the change in the process-model mismatch (pmm) because of the unmeasured disturbances.

A.6 Repeat Exercises A.1 through A.5 but with the environmental effects set to "Y." This will impose autoregressive moving average (ARMA [1,1]) drifts on the road angle, wind velocity, and drag coefficient. Comment on the goodness of the control. In addition, Trial choices 5 and 6 provide extended regulatory periods at medium- and high-speed set points.

A.7 Repeat Exercises A.1 through A.5, but with Tau Adjust set to values between 10 and 100. In the Excel Cells(3,2), enter "a," "d," or "f" to choose which model variable is adjusted. Comment on the removal of pmm and the model adjustment speed. Also, note how the adjustable model coefficient value changes in time and how the model-predicted maximum speed tracks the actual process maximum speed. The model adjustment time constant must be larger than the controller time constant.

A.8 Repeat Exercises A.1 through A.5 but with y_p sigma (measurement noise) set to values between 0.2 and 1 km/h. Change the pmm filter time constant from 1 to 5 to determine the impact of noise filtering on the control action. Filtering the pmm also will lag the biased set point for the model, and you might explore the impact of excessive filtering on the quality of control metrics. The controller-desired time constant must be slower than the pmm filter time constant.

A.9 I concocted the $F_{engine} = a + be^{-u/20} + ce^{-u/30}$ relation for the simulator. If you can identify a more legitimate one, you might also want to change the controller $F = Ku$ model.

A.10 This car simulator does not include the automatic transmission changing gears. Explore including that feature.

A.11 When coasting down (perhaps a pedal change from 50% to 0%), the engine decelerating force on the car may be less than the accelerating force (if u changes from 0% to 50%). Explore the relationship between the pedal position and the engine's accelerating/decelerating force on a car.

A.12 Explore an alternate variable as the adjustable coefficient to make the model match the process.

B

In-Line Hot and Cold Water Mixing Simulator

B.1 Process

This appendix offers a simulation model of a process of hot and cold water mixing, which can be used to test and demonstrate automation algorithms, such as control techniques, controller tuning, data reconciliation, steady-state detection, model adaptation, noise filtering, constraint prediction, fault detection, and neural network and fuzzy modeling. Based on a full-scale process, the simulator is nonlinear, with variable transport delay and fourth-order lags, and subject to unmeasurable and measurable disturbances. Although the simulator is rich in process features, it is relatively simple to code. A VBA version and user guide are accessible to the public at *www.r3eda.com.*

Researchers commonly use simulators to guide developing novel automation techniques, demonstrate their techniques, and benchmark their techniques to other techniques. Instructors use simulators to aid learning related to process dynamics and standard control practices. Their simulator is often conveniently simple, which is important for ease of understanding, ease of coding, and immunity from execution errors. However, most simulations are too idealized to establish the technique's credibility for the practitioner. Real processes are nonlinear and nonstationary, have measurement noise, and are subject to continually varying inputs and disturbances. For instruction, simulators should reinforce process principles, not mask them with transfer function models. This appendix offers a simulator of in-line hot and cold water mixing, which is convenient for use (coding, understanding, etc.) yet also reasonably grounded in process reality to legitimately guide technique development and establish credibility.

The simulated process is hot and cold water mixing in a pipeline, illustrated in Figure B-1. The hot and cold water flow through separate valves to a common in-line mixing point, then through a common discharge line to the exit. The 2 × 2 process control challenge is to adjust the signals to the hot and cold water valves to achieve desired values for the measured total flow rate and mixed fluid temperature.

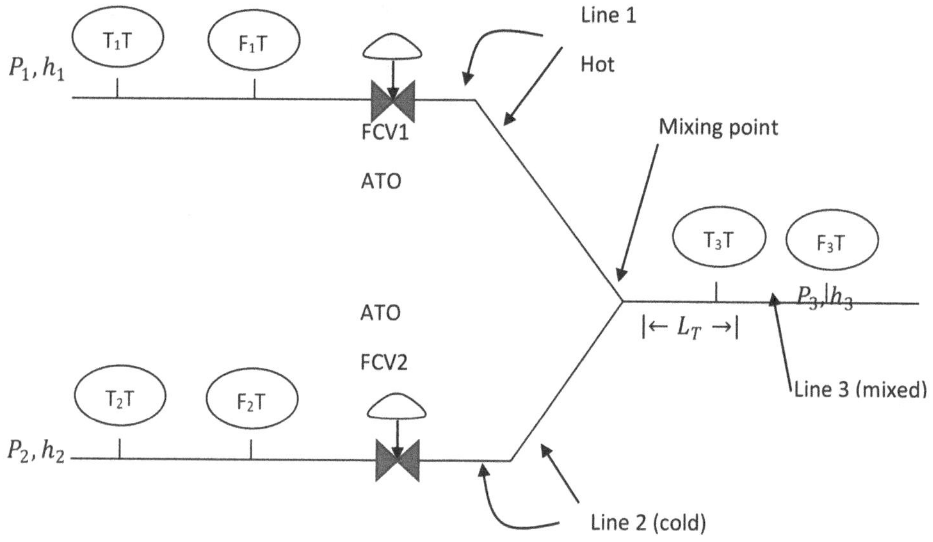

Figure B-1. Process illustration.

In-line mixing hot and cold water can be related to controlling the water temperature and flow rate in a shower, making it easy for nearly everyone to understand. However, the principles are the same if the process were gasoline blending, composition control, reagent addition, or many other common in-line mixing process applications.

Although simple, this process offers several challenges characteristic of process industry applications:

- It is nonlinear: (1) Equal-percentage valves have a nonlinear impact on flow rate, but the installed characteristic is more linear than it would be using valves with a linear characteristic. (2) The pressure friction loss of the turbulent flow is a nonlinear (ideally quadratic) response to flow rate. (3) The gain of how one valve affects temperature depends on the flow rate through the other valve.

- It is interacting: (1) Changing either valve affects the temperature and flow rate. (2) Further, changing the flow rate of one fluid changes the mixed flow rate, which changes the discharge line pressure drop, affecting the inflow rate of the other fluid.
- The process has drifting disturbances: (1) The inlet fluid pressures and temperatures change in time, causing disturbances. (2) The friction factors in the pipe segments, the flow rate, and the temperature sensor calibration all drift in time. (3) The valves have stiction.
- It is noisy: There is noise on the flow rate and temperature measurements.
- It is nonstationary: (1) The transport delay (dead time) on the temperature response changes with the flow rate. (2) The drifts to the fluid flow pressure-loss factors change the process response in time.
- The measurements all have calibration faults: At a steady state, the material and energy balances reveal an error.
- It has constraints: At a high flow rate set point, either high- or low-temperature set points cannot be simultaneously achieved; here, a controller must compromise on the two set points without windup.

Although this simulator includes many complications, the simulated process is relatively simple to understand (just material and energy balances, elementary fluid dynamics, and simple noise models) and simple to code.

Hot water enters through Line 1 from a header at pressure P_1 and elevation, h_1. Cold water enters through Line 2 from a header at elevation, h_2 and pressure P_2. Measured temperatures and flow rates of the inlet fluid could be transmitted to the control room as signals from transmitters T_1T and T_2T, and F_1T and F_2T; however, characteristic of chemical processes, only the minimal number of variables would be measured. Here, they would be mixed temperature and total flow rate. Flow control valves FCV1 and FCV2 are both modified equal percentage types with air-to-open (ATO) (fail-closed) actuators. The hot and cold fluids meet at the mixing point, where they enter Line 3 and exit at a pressure of P_3 and elevation h_3. Elevation differences are included in the simulator but ignored in the model because they have the same impact as the inlet and exit pressures, and the objective is to keep the model as simple as possible while retaining challenging features.

Mixed fluid temperature is measured by the sensor-transmitter (T_3T) at a distance L_3 downstream from the mixing point. Mixed flow rate and temperature are transmitted as F_3T and T_3T.

Measurements have errors. The true process value is biased by calibration drift and noise. Those are normal conditions and are included in the simulator. Sensor faults, such as catastrophic failure, are not included in the simulator. Inlet and exit pressures, P_1, P_2, and P_3, vary in time, which makes the flow rates vary even with no change in the signals to the control valves. And inlet temperatures also vary in time. These are all normal environmental influences.

B.2 Model Description

The fundamental mass and energy balances used to calculate mixed flow rate and temperature are

$$F_3 = F_1 + F_2 \tag{B-1}$$

$$T_3 = \frac{T_1F_1 + T_2F_2}{F_1 + F_2} \tag{B-2}$$

This, of course, assumes constant density and specific heat of the fluid (water). Adding temperature dependence on fluid properties increases complication; however, balancing sufficiency with perfection, this relatively inconsequential contribution to the control or data analysis aspects is ignored.

Because fluid acceleration has a much lower time constant than the valve or temperature sensor responses, in-line fluid acceleration ($F = ma$) is not considered. The fluid flow models represent a steady state because the fluid acceleration is relatively instantaneous. This simplifies the simulation, yet it retains the control and data analysis difficulties.

The steady-state fluid flow equations represent mechanical energy balances on the fluid. Here, ΣF is the sum of all forces trying to accelerate or decelerate the fluid. Each force is a pressure, ΔP, (or pressure drop) times the associated area, A. Because the fluid acceleration is considered negligible, the $\Sigma F = \Sigma \Delta P\, A = 0$. For flow Paths 1 and 2, which both merge with Line 3:

$$0 = P_1A_1 - \Delta P_{friction,1}A_1 - \Delta P_{valve,1}A_1 - \Delta P_{Bernoulli,mix\ point}A_1 - \Delta P_{friction,3}A_3 - P_3A_3 \tag{B-3}$$

$$0 = P_2A_2 - \Delta P_{friction,2}A_2 - \Delta P_{valve,2}A_2 - \Delta P_{Bernoulli,mix\ point}A_2 - \Delta P_{friction,3}A_3 - P_3A_3 \tag{B-4}$$

In Equation B-3, P_1 is the Line 1 entrance pressure driving the fluid and P_3 is the exit pressure. Each line has a cross-sectional area of A_i. The terminology is similar for Line 2, Equation B-4.

Here, the line losses, $\Delta P_{friction,i}$, are calculated by a conventional Moody-Darcy friction factor relation, with equivalent length accounting for the loss caused by devices and fittings in the pipeline.

$$\Delta P_{friction} = f\frac{L_{equ}}{D}\frac{1}{2g_c}\rho v^2 = f\frac{L_{equ}}{D}\frac{1}{2g_c}\rho\frac{F^2}{A^2} \tag{B-5}$$

where f is the friction factor, L_{equ} is the equivalent length of the piping system, D is the pipe diameter, ρ is the fluid density, and g_c is the dimensional unifier.

The simulator uses a Blasius model for the friction factor when the flow is turbulent, when the Reynolds Number is greater than 2,000, $Re = \frac{Dv\rho}{\mu} > 2{,}000$.

$$f = 0.32Re^{-1/4} \tag{B-6}$$

And simulation smoothly approaches the standard laminar flow model, $f = 64/Re$, for transition and laminar conditions using the following model.

$$f = e^{-Re/2000}\frac{64}{Re} + (1 - e^{-Re/2000})0.32Re^{-1/4} \tag{B-7}$$

The standard valve relation between flow rate and pressure drop is presented as follows:

$$\dot{Q} = Cv\, f(s)\sqrt{\Delta P_v / G\xi} \tag{B-8}$$

where $\xi = 1\text{Pa}$ and $G = \rho_{process\ fluid}/\rho_{water@STP}$. (Note: Often, the dimensional unifier, ξ is omitted from the equation, as g_c is usually omitted from $F = ma/g_c$ and from the conventional fluid mechanics equations.) The process industry uses a variety of temperatures as the standard temperature to assess specific gravity, G, within the range of 60° to 70°F.) The fluid is water, and the temperature impact on density has a negligible effect on the simulator, so $G = 1$.

Valve Cv values are a measure of valve capacity—the flow rate of water (here m^3/s) through the fully opened valve when there is a unit pressure drop (here, 1 Pa) across the valve. Following conventional practice, the valve Cv is chosen so the pressure drop across a fully open valve is about 50% of the line losses. Values of $f_i(s_i)$ in Equation B-8 are the valve characteristics, which indicate how the flow rate, F, changes with the valve stem position, s_i. Valve characteristics and stem positions have the same range:

$0 \le f_i \le 1$, and $0 \le s_i \le 1$. If the valve has a linear characteristic, then $f_i = s_i$. However, a valve with a linear characteristic, when installed in a pipeline, has a nonlinear gain because the pressure drop across the valve changes with pressure losses in the line, which changes with the flow rate. In practice, an equal-percentage characteristic is commonly chosen to help linearize the flow rate response to the valve stem position of an installed valve where the line losses change with the flow rate.

Rearranging the valve equation can determine the valve pressure drop required in Equation B-3 or B-4, given flow and valve properties:

$$\Delta P_v = \left(\frac{F}{Cvf(s)} \right)^2 G\xi \tag{B-9}$$

Equation B-10 presents a standard model of an ideal equal-percentage valve. In it, R is the valve rangeability.

$$f(s) = R^{s-1} \tag{B-10}$$

Note: In the nominal equal-percentage mathematical model, $f(s)$ does not have a zero value when $s = 0$, which means the valve would not close. Therefore, this simulator uses a modified equal percentage valve characteristic that represents industrial practice. If $s_i < 0.2$, the valve characteristic is modeled to be linear with s_i.

$$f(s) = \begin{bmatrix} R^{s-1}, & s \ge 0.2 \\ \frac{s}{0.2} R^{0.2-1}, & s < 0.2 \end{bmatrix} \tag{B-11}$$

The symbol R in the equal-percentage model of Equation B-10 is termed *rangeability*, and it indicates the ideal flow rate ratio between a fully opened and fully closed valve, which is a bit different from the alternate use of the same term representing the ratio of maximum flow rate to the minimum controllable flow rate. So, take caution when using the term.

It would be a rare confluence of conditions to have identical velocities in each of the three lines. At the mixing point, fluid flow rates likely change abruptly because of the flow combination and/or pipe area. Therefore, there is an associated pressure jump with the velocity change. Using a plug-flow model, the ideal Bernoulli effect is

$$\Delta P_{Bernoulli} = \frac{1}{2g_c} \rho \Delta v^2 = \frac{1}{2g_c} \rho \left[\left(\frac{F_{out}}{A_{out}} \right)^2 - \left(\frac{F_{in}}{A_{in}} \right)^2 \right] \tag{B-12}$$

Valves are considered to have air-operated actuators, and their stem positions are modeled as first-order responses to controller output changes. Valve time constants are $\tau_{V,i}$. The valve stem original differential equation is:

$$\tau \frac{ds}{dt} + s = O/100 \tag{B-13}$$

where O represents the 0% to 100% output (manipulated variable [MV]) of the controller.

Note: This valve model represents ideal, perfect, calibrations in the sequence of final element devices. Ideally, 0% to 100% from the controller is translated by 4 to 20 mA signals (electric current), converted into 3 to 15 psi (pneumatic) signals to the valve actuator, and then converted to 0 to 1 stem positions. For instance, ideally, when the controller output is 0%, the signal transmitted to the i/p device is 4 mA, the i/p output is 3 psig, and the valve actuator spring and diaphragm pressure perfectly balance and hold the stem at exactly closed. With experience, however, one commonly determines that the calibration is not ideal. For instance, the valve may not begin to open until the controller output is at 7%, or the valve is fully open when the controller output is 90%. Some controllers have limits of −6% and 106% to overcome calibration inaccuracy in the final element sequence. You could adjust the valve model to represent a nonideal calibration situation. Also, you could use alternate time constants for opening or closing the valve, representing pneumatic filling, and venting mechanisms. Again, such a step toward greater rigor in realism provides a complexity that exceeds the control relevance benefit. At the same time, including such transducer device calibration errors in the simulator would provide user understanding of the sequence of events that makes the final element change, which is usually missing in control instruction.

Nominally, the controller would send signals to the valves. Equation B-13 would determine the valve stem position, Equation B-11 would determine the valve characteristic, and Equation B-9 would determine the valve pressure drop. If the Line 1 and 2 flow rates were known, Equation B-1 would reveal the Line 3 flow rate. Then, Equations B-5 through B-12 would determine the pressure drops for each item. If the flow rates are correct, Equations B-3 and B-4 will balance. There is no convenient arrangement of the nonlinear equations to solve explicitly for F_1 and F_2, so this simulator uses Newton's secant root-finding technique initialized with extreme F values. Once F_1 and F_2 values are obtained, Equation B-2 calculates the mixed fluid temperature at the mixing point.

However, the mixed fluid temperature is measured downstream of the mix point. To simulate the measurement delay, the mixed fluid T is stored in a vector, and the fluid T at the downstream sensor is based on the plug-flow transport-delayed value. A "clock" technique is used to manage put and read locations in the array to minimize computational work. The transport delay is flow rate dependent, and if the flow profile is ideal plug flow, then $\theta = LA/F$. Accordingly, the temperature array item that represents the delayed T will change with the flow rate. The temperature sensors are modeled as third-order, which provides an approximation to both thermowell sensor dynamics and the not-really-plug-flow in-line fluid mixing.

In a basic 2 × 2 controller simulation with two single-input, single-output (SISO) controllers, the measured F_3 and T_3 values would be given to the controllers where they are compared to set points. The controllers would calculate desired MV values, O_1 and O_2, and send them to the valves.

Undesirably, the two SISO controllers are interactive. If the flow controller adjusts its valve to change the flow rate, this upsets the mixed temperature. Then, the temperature controller will adjust its valve, which upsets the mixed flow rate. Although relative gain array (RGA) techniques could identify the best loop pairing, the RGA values change with the temperature set point. When the T set point is high, the best pairing is T_3 to the cold valve. However, when the T set point is low, the best pairing switches to T_3 to the hot valve. A classic advanced regulatory solution to eliminate interaction is to use ratio control. Alternatively, double ratio control, feedforward control, or decouplers could be used to fix the interaction. Another option is to use a model-based controller to compensate for the interaction. In any case, if a controller strategy needs either additional devices or measurements of the F_1, F_2, T_1, and/or T_2 values, it adds installed equipment cost and maintenance expenses; the proposer of an advanced strategy should not expect it all to be free.

B.3 Deterministic Simulator

This section is termed *deterministic* because it represents the basic process features with no noise, disturbances, or changes in coefficient values. In the following figures, the upper graph indicates the signals from the controller to the valves, and the lower graph shows the three flow rates and mixed temperature responses.

Figure B-2 reveals the effects of changing the signal to the cold water valve while keeping the signal to the hot water valve at a fixed value. Note the following aspects: (1) The changes in the cold and total flow rates are fairly identical because the equal-percentage valves substantially linearize the response. (2) The hot water flow rate increases with each reduction in the cold flow because the back pressure from the mixed flow in Line 3 is reduced. This shows one form of interaction. (3) The mixed fluid temperature shows a nonlinear response. Although the signal makes identical 25% changes, the temperature rise in the 60 s period is about 5°C, but the temperature rise in the 190 s period is about 35°C, about seven times greater. (4) Finally, when the total flow rate is high, in the 15 s period, the delay for the temperature change is about 2 s, but when the total flow rate is low, in the 190 s period, the delay is about 10 s, a five-to-one change. This delay can be much larger when the hot valve is also at a lower position. From the maximum to a very low flow rate, the temperature delay can range from nearly zero to over 20 s.

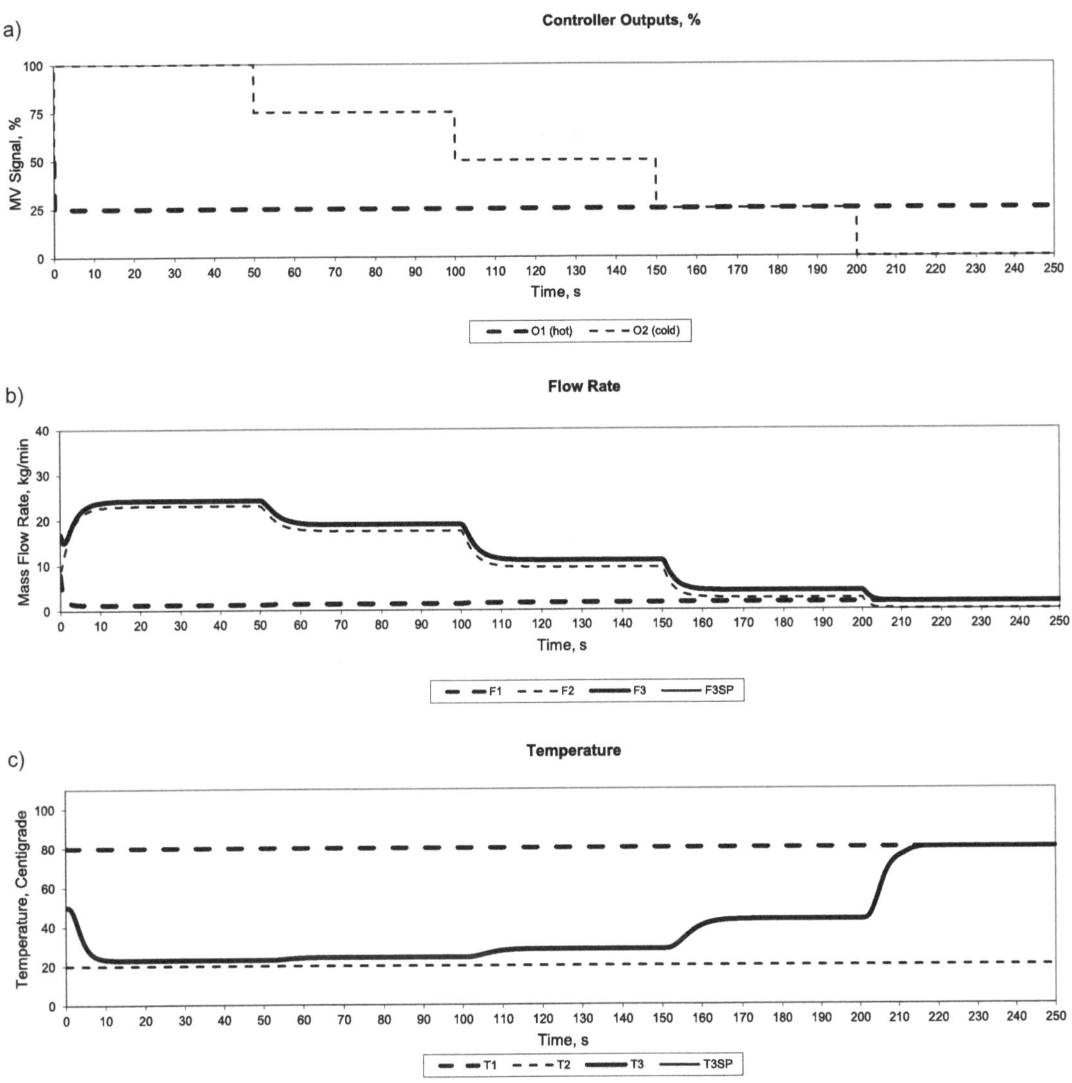

Figure B-2. Deterministic response to steps in one MV.

Alternate phenomena are revealed in Figure B-3. Here, the MV values start at the extremes of 0% and 100% and then simultaneously make 33% increments to the other extreme. If the attributes of Lines 1 and 2 were identical, the total flow rate would remain constant, but it drops with the change at 100 s and then rises when it reaches a steady state at about 120 s. Further, after each steady state, the total flow rate initially drops and then rises. These behaviors are because the valve dynamics, the valve *Cv* values, and the Line 1 and 2 properties are not identical.

Overall, the input-output sequence for the simulator is: Send a signal to a valve. This is the controller output (CO), or MV, and is often labeled with the symbol *m* or *u*. The simulator uses the symbol *O* for output. This causes the stem position to change in

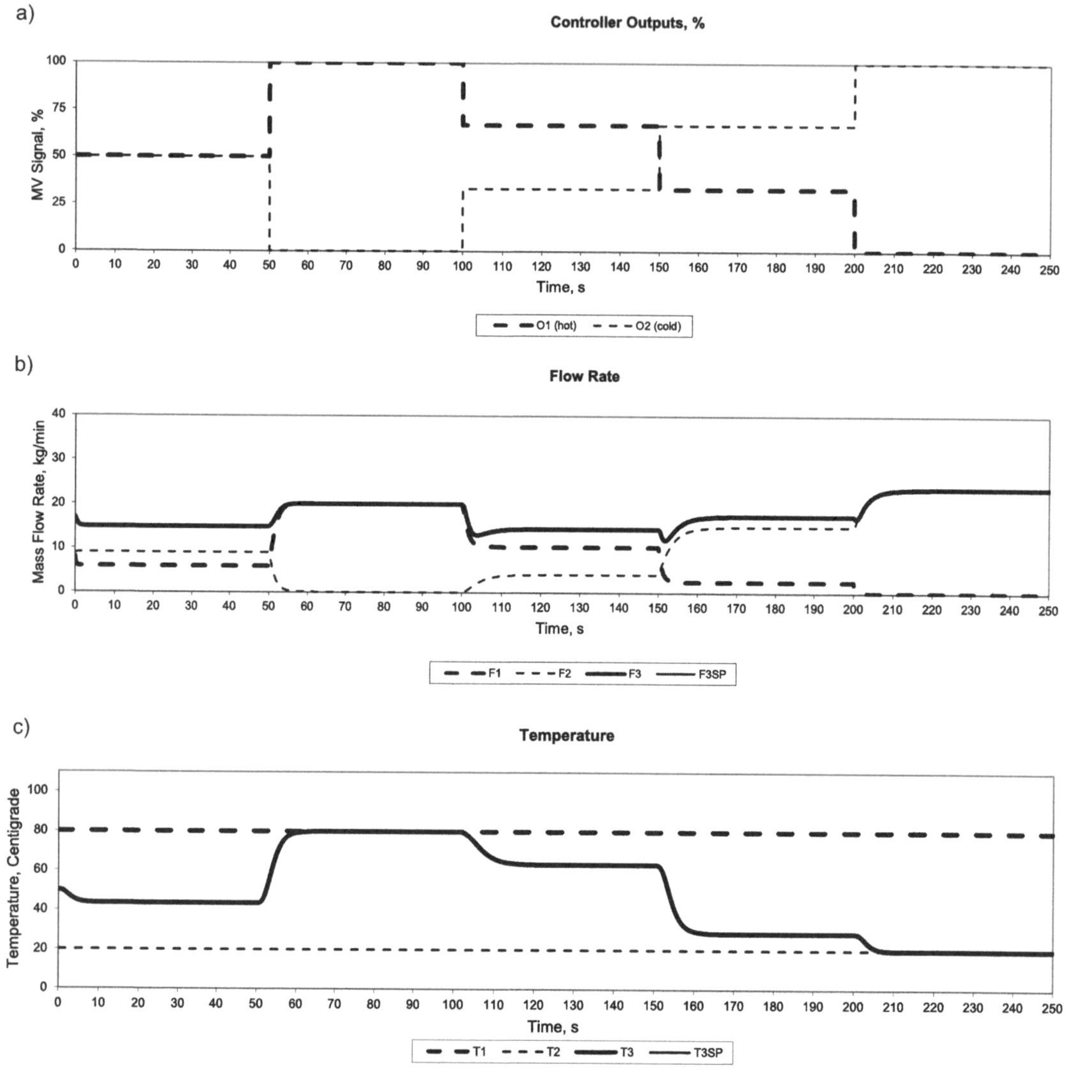

Figure B-3. Deterministic response to coordinated steps in both MVs.

a first-order manner, conflicted by stiction. The fluid flow rate is determined by the mechanical energy balance of the two fluid paths, but at each sampling, new drift values for model coefficients lead to different flow rate solutions. The mixed T is calculated from the thermal energy balance and is subject to the drifts in the inlet temperatures. The inlet temperatures are sensed at upstream sensors. The actual values must be transported to the mix point, and the measured values lagged. The simulator stores the inflow temperatures in an array, which considers the in-line transport to be plug flow with no axial mixing. The mixed T is also stored in a vector so the fluid T at the downstream sensor is based on the plug-flow, transport-delayed value. The temperature sensors are modeled as third-order, providing an approximation to both sensor dynamics and in-line fluid mixing. Finally, any measurement represents the true simulator value corrupted by measurement drift. Flow rate measurements are filtered to temper noise.

B.4 Coefficient Values and Units

For the valves:

O		Controller output, %
τ_{v1}	3	s
τ_{v2}	1	s
s		Fraction of full stroke, dimensionless
F		m^3/s
$f_i(s_i)$		Characteristic, nominally =%
R_{V1}	25	Rangeability, dimensionless
R_{V2}	22	Rangeability, dimensionless
Deadband	0.05	Fraction of full stroke, 5%
C_{v1}	2×10^{-5}	m^3/s @ 1 P_a
C_{v2}	2.5×10^{-5}	m^3/s @ 1 P_a

For the pipe system:

F		m^3/s
D_1	0.06	m
D_2	0.06	m
D_3	0.06	m
L_1	50	m
L_2	25	m
L_3	100	m
L_{T1}	2	m
L_{T2}	1	m
L_{T3}	1	m
$T_{h,in}$	70	C
$T_{c,in}$	15	C
P_1	500	kPa
P_2	550	kPa
P_3	150	kPa
$L_{equ\,1}$	150	m
$L_{equ\,2}$	75	m
$L_{equ\,3}$	400	m

ρ	1,000	kg/m^3
μ	0.001	Pa-s

For the sensors:

τ_i	1	s, each of the three sensors is third-order

For the measurement noise:

σ_T	0.5	K
σ_F	0.1	Fraction of F

For stochastic inputs:

τ_{P1}	100	s
R_{P1}	100	kPa
τ_{P2}	100	s
R_{P2}	100	kPa
τ_{Tc}	100	s
R_{Tc}	20	K
τ_{Th}	100	s
R_{Th}	25	K

B.5 Discussion

For convenience, fluid properties (density, viscosity, specific heat) are assumed independent of temperature and pressure at the values indicated. Including temperature dependence has an insignificant impact relative to other influences, and the complexity adds no benefit.

The control and measurements are on a 10 Hz (0.1 s) interval, a characteristic (but relatively fast) scan rate for industrial flow and temperature control.

The default output includes F_1, F_2, T_1, and T_2. However, it is not normal to measure such extraneous variables. Usually, the process owners only install devices to measure the essential T_3, and F_3 values. So, if you plan on using F_1, F_2, T_1, and T_2 in your controller, include a penalty for the cost (installation and maintenance) of the extra measurements.

A VBA version and user guide are accessible to the public at *www.r3eda.com*.

Many investigators are exploring model-based techniques where the models are phenomenological, as is this simulator. A rule of playing fair in the game of claiming credibility is that you cannot use the same model for the supervisory application as you use in the process simulator. Nature is always one step more complicated than human-devised models. Therefore, to credibly demonstrate your technique, be sure that your model for the application is not identical to the model in the process simulator. Change the coefficient values and functionalities (those in the valve, overall heat transfer coefficient model, etc.). Change dynamic relations such as time constants. Although you can observe variable values in the simulator, the only variables fair play permits your application to observe are the three noise-corrupted measured fluid flow rates and the three noise-corrupted measured temperatures. Reveal that your application model is similar to but reasonably different from the simulation model.

Although running the simulator without stochastic inputs and noise-corrupted measurements may be useful, be sure to test your procedure with those effects.

Exercises

Access the simulator from *www.r3eda.com*. There are nine trial conditions in the simulator. Enter a trial number in Cells(10, 18) and a controller type (model-based control or independent proportional-integral-derivative controllers) in Cells(9, 18). Or, you might enjoy writing your own.

B.1 Quantify the process dynamics by making output and disturbance steps in manual mode. Quantify how nonlinearity (process gain), nonstationarity (time constant and delay), and interaction (gains) change with conditions.

B.2 Seek to tune the proportional-integral-derivative (PID) controllers to minimize the integral of the squared error and travel in response to both set-point and disturbance changes.

B.3 Quantify the degradation of control at alternate set-point or inflow temperature conditions.

B.4 Quantify control during constrained conditions and the response when the constraint is removed.

B.5 Add noise and evaluate the impact on control.

B.6 Repeat with model-based control (MBC). Note the coordinated control action. Classic advanced regulatory control (ARC) would use decouplers, ratio, or double ratio with PID to effect the instant decoupling. Which approach seems preferable, ARC or MBC?

C

pH Neutralization Simulator

C.1 Introduction

This appendix offers a simulation model of a pH process designed to be used to test and demonstrate automation algorithms, such as control techniques, controller tuning, data reconciliation, steady-state detection, model adaptation, noise filtering, constraint prediction, fault detection, and neural network (NN) and fuzzy modeling. Based on a full-scale, nonideal continuous stirred tank reactor (CSTR), the simulator is nonlinear, with third-order lags, and subject to unmeasurable and measurable disturbances. Although the simulator is rich in process features, it is relatively simple to code. A VBA version and user guide are accessible to the public at *www.r3eda.com*.

Researchers commonly use simulators to guide developing novel automation and data processing techniques, to demonstrate their techniques, and to benchmark their techniques to other approaches. Instructors use simulators to aid learning. The simulators are often conveniently simple, which is important for ease of understanding, ease of coding, and immunity from execution errors. However, most simulators I have encountered in research publications are too idealized to establish the credibility of the automation technique for the practitioner; when phenomenological models are replaced with transfer functions, undesirably, mechanistic understanding is masked, not enhanced. Real processes are nonlinear and nonstationary, have measurement noise, and are subject to continually varying inputs and disturbances. Real processes use a concentrated reagent to neutralize the process fluid, not a 0.1N reagent to simulate a laboratory-convenient 1 L tank. This appendix offers a pH process simulator, which is convenient for use (coding, understanding, etc.) yet also reasonably grounded in process reality to legitimately guide technique development and establish credibility.

Many commercially important chemical reactions depend on acidity, as do biological, food, and waste treatment processes. Because water is about neutral, we do not want to discharge acids or bases and ruin the environment for the natural life-forms in rivers, lakes, or underground geology. Therefore, pH control is important and ubiquitous. Because pH control is also difficult, it has become a classic test case for control algorithms.

Many researchers use clever approximations to simplify the pH calculation, often using very simple pH process representations that are too trivialized to represent a legitimate and credible demonstration basis. Accordingly, the approach to pH simulation in this appendix is recommended as a benchmark challenge application for those who wish to explore pH control and related data analysis.

C.2 Model Description – Determining pH

C.2.1 About pH

pH is a measure of acid (or base) activity. An ideal definition is that pH is the negative of the base-10 logarithm of the hydrogen ion concentration.

$$pH = -\log_{10}([H^+]) \tag{C-1}$$

The brackets around a chemical species denote the concentration of that species in moles per liter.

More rigorously, pH is defined as

$$pH = -\log_{10}(\gamma[H^+]/[H^+]_0) \tag{C-2}$$

Here, γ is the activity coefficient, and $[H^+]_0$ is the reference state commonly taken to be 1 mol/L. If the activity coefficient is the ideal $\gamma = 1$ value, the two equations are identical. The ideal Equation C-1 is commonly used even though it appears dimensionally inconsistent with the nominal value of 1 mole/liter substituted for the symbol $[H^+]_0$.

The units of the hydrogen ion concentration are gram-moles per liter. One mole equals 6.02214… $\times$ 10^{23} number (Avogadro's number) of molecules. Atomic weight, from the periodic table of elements, is the mass in grams of one mole of that atom.

In pH applications, the hydrogen ion concentration can vary by many orders of magnitude (about 10 in practice); therefore, the logarithm keeps the pH within a convenient range of values.

For several reasons, pH is also nonlinear, and the sensitivity of the pH response to the control action can change by several orders of magnitude. Often, this is termed *highly nonlinear,* contrasting the magnitude of the nonlinear aspects of many process control applications (flow rate, distillation, etc.) with only a 5:1, or so, gain change. In addition, pH control is nonstationary, meaning that the sensitivity changes over time as the chemical composition of the inflowing material changes.

C.2.2 Dissociation Stoichiometry

A conventional acid molecule releases a hydrogen ion in water or an aqueous solution. For hydrochloric acid, for example, the chemical reaction is simply indicated as

$$HCl \leftrightarrow H^{+} + Cl^{-} \tag{C-3}$$

In water, the electrically neutral HCl molecule can dissociate into a positively charged hydrogen ion, H^+ (missing its electron), and a negatively charged chlorine ion, Cl^- (with an extra electron). The solution is electrically neutral. The double-headed arrow in the reaction equation indicates that the ions can recombine to make a neutral molecule.

The hydrogen atom only has one proton and one electron. So, the H^+ ion is just the extraordinarily tiny, one-proton nucleus of the H atom. Chemical equilibrium is nearly instantaneous relative to the speed of mixing or creating a solution.

> Note: This is an elementary one-step view of the reaction. A rigorous kinetic model might include intermediate species (e.g., the water-solvated H^+, H_3O^+) or catalytic pathways. However, the impact of many aspects of a commercial process (embodied in this model) does not justify greater rigor in a kinetic model.

Pure water also partially dissociates.

$$H_2O \leftrightarrow H^{+} + OH^{-} \tag{C-4}$$

Nominally, the H^+ of the concentration of pure water is 0.0000001 mol/L ($[H^+] = 10^{-7}$). This makes the pH of pure water $7 = -\log_{10}(10^{-7})$. The nominal water concentration is 55.5 mol/L (an easy calculation from density and molecular weight). This means that only 1 of about 555,000,000 water molecules are dissociated. Most of the water is undissociated H_2O.

Usually, acids are not fully dissociated in water. A generic representation of an acid is

$$HA \leftrightarrow H^+ + A^- \tag{C-5}$$

Here, the symbol A could represent the chloride part of hydrochloric acid or the acetic, formic part of those respective acids. If an acid totally dissociated, then $[HA]$, the concentration of the molecule in water, would be zero ($[HA] = 0$). If 0.001 mol of the acid were topped off with water to the 1 L mark and it totally dissociated, then the $[H^+]$ and $[A^-]$ values from the acid would be $[H^+] = [A^-] = 0.001$ mol/L. This is four orders of magnitude greater than the $[H^+]$ concentration in neutral water. If the H^+ contribution of the water is ignored, the solution is $pH = 3 = -\log_{10}(0.001)$.

However, not all acids fully dissociate. If 1/20 of the acid molecules dissociate in this 0.001 molar example, the remaining $[HA] = 0.00095$ mol/L, and $[H^+] = [A^-] = 0.00005$ mol/L. In this case, still ignoring the relatively small effect of water, the same nominal acid concentration of 0.001 mol/L results in a pH $= 4.3 = -\log_{10}(0.00005)$.

However, it is not that some of the HA molecules decide to dissociate and others decide to remain bonded. The water can solvate each of them. However, when a solvated H^+ and A^- "see" each other in the water, they are ionically attracted and may recombine. Because every H^+ is identical to any other, and every A^- is identical to any other, the two original parts of the molecule may not be the ones that recombine.

At equilibrium, the rate of HA molecules dissociating is equal to the rate of H^+ and A^- ions that are recombining. At equilibrium, the concentrations, $[HA]$, $[H^+]$, and $[A^-]$, remain constant. However, the condition (ion or molecule) of each atom is continually shifting. This could be termed *steady state*, but it is a dynamic, continually exchanging situation. Again, these reaction rates are so fast relative to the mixing dynamics that the chemical compositions can be considered to be in equilibrium.

To complicate the equilibrium, the H^+ from the water is identical to the H^+ from the acid. So, the recombining of the water might be from an H^+ from the acid, and the recombining of the acid might be from an H^+ from the water.

Note: This simulator uses simple acids and bases that dissociate with either the H^+ or OH^- ions. There are Bronsted and Lewis acids and bases that capture the water-produced ion, leaving the free OH^- or H^+ to upset the neutral pH balance.

One final reaction concept before the equations: Some acids have more than one H to release. A diprotic acid has two hydrogens to release. Sulfuric acid, H_2SO_4, is an example. A triprotic acid can release three protons (hydrogen ions). Phosphoric acid, H_3PO_4, is an example. Generically, a triprotic acid can be represented by H_3A, and the dissociation is considered to be sequential:

$$H_3A \leftrightarrow H^+ + H_2A^- \tag{C-6}$$

$$H_2A^- \leftrightarrow H^+ + HA^{-2} \tag{C-7}$$

$$HA^{-2} \leftrightarrow H^+ + A^{-3} \tag{C-8}$$

C.2.3 Reaction Equilibrium Models

A mathematical relation between the concentration of the H^+ ions and the amount of acids and bases in the solution depends on the reaction equilibrium relations. At equilibrium, the forward rate, r_f, of a reaction is the same as the backward rate, r_b. Ideally, concentration is used to represent chemical activity. The generic reaction is

$$XY \leftrightarrow X^+ + Y^- \tag{C-9}$$

Simplistically, the forward part, $XY \rightarrow X^+ + Y^-$, has a rate that depends on the probability that an XY molecule exists, which, ideally, is proportional to the concentration of XY, $r_f = k_f[XY]$. And, it has a backward part, $XY \leftarrow X^+ + Y^-$, with a rate that depends on the chance that a X^+ and a Y^- find each other in the solution, which, ideally, is dependent on the product of concentrations, $r_b = k_b[X^+]\,[Y^-]$. At equilibrium, the two rates are equal, $r_f = r_b$, and the ideal relation between all three concentrations is

$$\frac{k_f}{k_b} = K = \frac{[X^+][Y^-]}{[XY]} \tag{C-10}$$

Note: The rates of reaction, and therefore the equilibrium constant K, are temperature dependent. Balancing sufficiency with perfection, the temperature functionality (often adequately modeled by the Arrhenius relation, $K(T) = k_0 e^{-E/RT}$, is not included here.

Further, ideally, the probability of a molecule participating in a reaction is proportional to its concentration, but its propensity to react, its activity, also depends on its surrounding interactions. Rigorously, the reaction rate and equilibrium relations should use the activity of a molecule, which is usually modeled as the concentration times a factor (activity coefficient) related to its propensity to react. However, for this pH simulation study, including activity coefficients makes a small contribution to either the realism or the control difficulty, and the magnitude of the adjustment to pH is small relative to the uncertainty of compositions within industrial practice. So, ideal models are fully sufficient.

For a monoprotic acid, $HA \leftrightarrow H^+ + A^-$, the equilibrium relation is

$$K_a = \frac{[H^+][A^-]}{[HA]} \tag{C-11}$$

For water, $H_2O \leftrightarrow H^+ + OH^-$, the relation is similar

$$K = \frac{[H^+][OH^-]}{[H_2O]} \tag{C-12}$$

However, $[H_2O]$ is about 555,000,000 times greater than the concentration of H^+ or OH^- ions. If one out of 555,000,000 water atoms dissociate, "only" 554,999,999 remain. Accordingly, the concentration of H_2O can be considered independent of the reactions, constant at the nominal value. Then, the water equilibrium is simply written as

$$K_w = [H^+][OH^-] \tag{C-13}$$

Notes:

1. This representation of the equilibrium relation is termed the *ion product*.
2. The nominal value for the ion product of water is $K_w = 10^{-14}$. However, it ranges from about 10^{-15} (at nearly ice temperature) to about 10^{-8} at high temperatures (and associated high pressures). It also depends on the other ions in the solution and the isotope of the molecules. Therefore, in a simulation of normal temperatures with normal ion concentrations, pK_w values between 12.5 and 15.5 would also be reasonable.

For a diprotic acid, there is a sequence of dissociations, $H_2A \leftrightarrow H^+ + HA^-$ and $HA^- \leftrightarrow H^+ + A^{-2}$, and an equilibrium relation for each.

$$K_{a1} = \frac{[H^+][HA^-]}{[H_2A]} \tag{C-14}$$

$$K_{a2} = \frac{[H^+][A^{-2}]}{[HA^-]} \tag{C-15}$$

A base has similar relations. For a simple, single hydroxyl (monovalent) base, $BOH \leftrightarrow B^+ + OH^-$, the equilibrium model is

$$K_b = \frac{[B^+][OH^-]}{[BOH]} \tag{C-16}$$

Values of the equilibrium constants are properties of the particular acids, bases, and water (and their temperature). And temperature-dependent K values are easily accessible in many references. Some references give them as pK values, meaning $pK = -\log_{10}(K)$. The advantage is that pK values are in the convenient range of 0 to 14. This simulator has users inputting the pK version.

In contrast to the chemical properties of the molecules, the concentrations and the components in a mixture depend on the user's choice of the mixture.

C.2.4 Solving the Equilibrium Models

First, choose the nominal concentrations of your undissociated materials. If you are mixing in a continuous flow or fed-batch process (e.g., a titration), this will likely require material balances. See Section C.2.5.

Then, consider a unit volume of 1 L, a basis of 1 L. This might be a small portion of a 10,000 gal tank or 100 times more than a test tube sample. It is not necessary to have 1 L, but the 1 L concept simplifies the calculation because the stoichiometric coefficients are the same as concentrations.

This example will describe water, a diprotic acid, a monovalent titrant, and common ion salt. The reaction extent for each will be represented by w, x, y, z, and s. These are the reaction moles in the forward direction. If the reaction is in the backward direction, the reaction extent will have a negative value. When acids and bases are added to water, the recombination of H^+ and OH^- ions form water, the backward reaction dominates the water equation, and the value of z will be negative. The reaction set is

$$H_2A \leftrightarrow H^+ + HA^- \quad w \tag{C-17}$$

$$HA^- \leftrightarrow H^+ + A^{-2} \quad x \tag{C-18}$$

$$BOH \leftrightarrow B^+ + OH^- \quad y \tag{C-19}$$

$$H_2O \rightarrow H^+ + OH^- \quad z \tag{C-20}$$

$$B_2A \rightarrow 2B^+ + A^{-2} \quad s \tag{C-21}$$

In this example, a is the number of moles of acid in the 1 L sample, and b is that of the base. Here, the acid, water, and base provide w, x, y, and z reaction moles, leaving $a - w$ and $b - y$ moles of the undissociated acid and base. The s moles of the common ion salt are considered totally dissociated. The reaction set, with after-equilibrium moles of each species indicated, is shown in Table C-1.

Note: There is not an infinite number of moles of water, but there is so much more than the z moles that dissociate that the water concentration can be considered unchanged in the reactions.

Table C-1. Moles of species in the hypothetical 1 L sample.

H_2A	$(a - w)$
HA^-	$(w - x)$
BOH	$(b - y)$
H_2O	$(\infty - z)$
H^+	$(w + x + z)$
A^{-2}	$(x + s)$
B^+	$(y + 2s)$
OH^-	$(y + z)$

The four equilibrium relations are then obtained by substituting the mol/L from Table C-1 into the equilibrium model equations.

$$K_{a1} = \frac{(w+x+z)(w-x)}{(a-w)} \tag{C-22}$$

$$K_{a2} = \frac{(w+x+z)(x+s)}{(w-x)} \tag{C-23}$$

$$K_b = \frac{(y+2s)(y+z)}{(b-y)} \tag{C-24}$$

$$K_w = (w+x+z)(y+z) \tag{C-25}$$

These represent four equations with four unknowns. Solving the equations will give values for w, x, y, and z. Then, $(w + x + z)$ is the number of moles of H^+ ions in the 1 L basis, which is the $[H^+]$; so the $pH = -\log_{10}(w + x + z)$. It would be easy to solve for w, x, y, and z if the equation set was linear, but it is nonlinear. Further, the robustness of the solution method can be very sensitive to trial initial values, and many classic approaches to root-finding (like Newton's method) diverge when solving these relations.

A recommended solution is to use interval halving with a trial pH value.[1] With a guess of the pH value, pH_g, the definition of pH can be used to determine the hydrogen ion concentration: $(w+x+z)_g = [H^+]_g = 10^{-pH_g}$. The water equilibrium relation then provides the consequential hydroxyl ion concentration: $(y+z)_c = [OH^-]_c = K_w/[H^+]_g = K_w 10^{+pH_g}$.

1. J. Y. Choi, H. G. Pandit, R. R. Rhinehart, and R. J. Farrell, "A Process Simulator for pH Control Studies," *Computers & Chemical Engineering* 19, no. 5 (1995): 527–540.

This leaves three linear relations for the three unknowns w, x and y, the consequence of the pH guess.

$$(a - w_c)K_{a1} = [H^+]_g(w_c - x_c) \tag{C-26}$$

$$(w_c - x_c)K_{a2} = [H^+]_g(x_c + s) \tag{C-27}$$

$$(b - y_c)K_b = (y_c + 2s)[OH^-]_c \tag{C-28}$$

The subscript g means "guess," and the subscript c means "consequence." The equations are independent and linear in the unknowns w_c, x_c, and y_c.

With y_c so calculated, the z_c response can be obtained from the $[OH^-]_c$ consequence of the pH guess, $z_c = [OH^-]_c - y_c$. Finally, the response pH_c can be calculated as $pH_c = -\log_{10}(x_c + z_c)$. If the response pH is the same as the guess, you guessed at the right pH value. The initial guess was probably not the right value, but the right value is likely between the unnatural (but mathematically possible) pH extremes of –2 and 16. So, start with $pH_g = 16$ and $pH_g = -2$ and calculate the associated pH_c responses. Then, use interval halving to search for the pH_g that is close enough to the root value that makes the $pH_c = pH_g$. Because pH measurements might have a 0.1 unit certainty, convergence on a range of high and low pH_g of 0.01 is fully adequate for automation-testing simulations (or perhaps excessively precise). If the initial search range is 18 pH units, then interval halving will be within 0.01 pH units of the true root after 16 iterations (including the initial extremes).

With more complicated mixtures, the procedure is the same, and it has been validated against both data and rigorous electrolyte simulators (Choi et al. 1995).

Salt precipitates (formed by common metal or acid ion) create a nonlinear complication. This approach considers that all salts are totally dissociated and in solution. They do not precipitate and leave the reaction system. Similarly, products of the neutralization reactions may produce a gas that exceeds its solubility in water and escapes, like baking soda and vinegar generate CO_2. This also is not considered.

Equilibrium constants are temperature dependent and often depend on other electrolytes that confound solvation by water. The approach above uses nominal pK values. However, one can perturb all givens (concentrations, pK values). Balancing sufficiency and perfection, this simulator will perturb (1) the influent compositions, (2) the water pkW, and (3) pH sensor calibration error to generate alternate realizations of the resulting pH.

C.2.5 Determining Concentrations of Component Molecules

Material balances are fairly simple. What flows in, minus what flows out, is what accumulates.

$$In - Out = Accumulation \tag{C-29}$$

In a reaction system, one could include the internal generation (and depletion) in the balance. However, even though the dissociation and recombination reactions generate ions and perhaps salts, because the reaction species are invariant (a sulfate or sodium component is the same whether it is in a compound or as an ion), only the nominal inflowing neutral components (water, acids, bases, and salts) must be considered.

Note: Some people perform the material balances on the reaction invariants, the ionic molecules. The result is the same.

First, consider a well-mixed, continuous flow-through, constant-level tank, often termed a CSTR or CST (continuous stirred tank). Contents within a CST are considered to be ideally mixed. As soon as a drop of something enters, it is instantly, completely, and uniformly dispersed throughout the entire liquid volume. The tank contents change in time but are considered to be spatially uniform.

We will also consider that there are no density changes or volume changes from the mixing. The "improvement" by such scientific rigor is inconsequential within the basic model effects.

Typically, volumetric flow rates and concentrations are used to quantify the rate at which material flows in or out, but Equation C-29 represents the mass quantity, not the mass rate. If the mass is measured in grams, then the water balance on a vessel is

$$\rho F_{in,t} \Delta t - \rho F_{out,t} \Delta t = m_{t+\Delta t} - m_t \tag{C-30}$$

Here, m_t is the mass of water in the vessel at the current time, and Δt represents a small time interval. F_{in} and F_{out} represent the inflow and outflow during Δt, which is small enough to consider the flow rates are constant during the small interval. $m_{t+\Delta t}$ is the mass after a time interval of Δt. Dividing by Δt, and recognizing that in the limit of very small Δt, $(m_{t+\Delta t} - m_t)/\Delta t$ is the calculus differential $\frac{dm}{dt}$; and assuming that density is constant and dividing by ρ, the mass balance becomes

$$\frac{dV}{dt} = F_{in} - F_{out} \tag{C-31}$$

If the level in the tank is unchanged, then $\frac{dV}{dt}=0$, and the equation becomes $F_{out}=F_{in}$, which is the common basis for an ideal CST. However, if the process is a batch (laboratory titration with titrant added to a sample or an industrial tank batch being neutralized), then $F_{out}=0$, and the equation becomes $\frac{dV}{dt}=F_{in}$. Although this simulator is nominally a CST, the outlet flow from a neutralization process might be zero, purposely, to prevent discharging out-of-specification material by accumulating volume. Then, when *pH* is within specifications, a high outflow rate will return the tank level to a nominal value.

The material balance for any component, such as an acid, H_2A, is similarly derived from the overall mass balance. Using the symbol n to represent moles of the component as a measure of mass in the vessel

$$c_{in,t}F_{in,t}\Delta t - c_{out,t}F_{out,t}\Delta t = n_{t+\Delta t} - n_t = V_{t+\Delta t}c_{t+\Delta t} - V_t c_t \tag{C-32}$$

Again, the numerical difference becomes the calculus differential.

$$\frac{d(Vc)}{dt} = c_{in}F_{in} - c_{out}F_{out} \tag{C-33}$$

If the volume changes, then V is not a constant over time and the product rule yields

$$V\frac{dc}{dt} + c\frac{dV}{dt} = c_{in}F_{in} - c_{out}F_{out} \tag{C-34}$$

Notes:

1. c could be the concentration of an acid, base, salt, or water molecule.
2. In an ideal CSTR, the concentration in the tank is the same as the exiting concentration, $c = c_{out}$.
3. The equations describing how component concentration changes in time are presented as differential equations. However, a time-dependent simulator that permits continual changes must use a numerical method to approximate a solution. This simulator uses Euler's explicit method, which is adequate if the time interval is small relative to process time constants.

The material balances describe how reaction invariant specie concentrations change over time. Once material balances provide concentrations, the thermodynamic equilibrium calculations of Section C.2.4 provide pH.

In contrast with a CST concept that seems acceptable in laboratory-scale mixers, a full-scale mixer cannot instantaneously blend the entire fluid contents with the fluid that flows into it. There are zones of concentration, which might be layers or rings. Further, there are baffles that break the mixing vortex and dead zones behind the baffles. There are dead zones in tank ridges and sample and drain ports. One simple approach to modeling a nonideal CST is to consider a single dead zone volume that has a continual material exchange with the ideally mixed contents. The dead zone might be in several places or stratification levels that progressively exchange material. Regardless, the simple nonideal concept is to have one dead zone of a volume portion, v, of the overall volume, V. Then, the active volume of the CST is $v_a = V - v_{dz}$. The dead zone and the main mixing zones are considered to be CSTs with a flow exchange between them. If the volumes in both zones are constant, the material balances for each component become

$$v_a \frac{dc_a}{dt} = c_{in}F_{in} - c_aF_{out} + (c_{dz} - c_a)F_{exchange} \tag{C-35}$$

$$v_{dz} \frac{dc_{dz}}{dt} = (c_a - c_{dz})F_{exchange} \tag{C-36}$$

The simulator considers that the dead zone volume is constant but that the active volume will change with liquid inventory.

$$\frac{dv_a}{dt} = F_{in} + F_{caustic} - F_{out} \tag{C-37}$$

$$v_a \frac{dc_a}{dt} = c_{in}F_{in} + (c_{dz} - c_a)F_{exchange} - c_a\left(F_{in} + F_{caustic}\right) \tag{C-38}$$

$$v_{dz} \frac{dc_{dz}}{dt} = (c_a - c_{dz})F_{exchange} \tag{C-39}$$

C.3 Simulator Results

Figure C-1 illustrates a range of titration curves for diverse inflow compositions. The dashed and solid curves are drawn for visual distinction.

Note: Key features are (1) the extreme changes in gain (titration curve slope) often at neutrality, (2) the amount of reagent needed to neutralize the influent is not proportional to influent pH, and (3) the diverse locations and magnitude of inflections. All combine to confound control or empirical modeling.

Figure C-2 illustrates one 1,000-minute controlled run of the simulated neutralization CST. The controller is a set of heuristic IF-THEN rules. The effluent pH is a fairly horizontal trend but noisy and is maintained at a value just below 7 (to infrequently exceed 7). The trend also reveals discretization error (resolution limitations) on the measured value. The caustic reagent ranges from about 2 to 10 during the run, compensating for the influent flow rate and composition variations.

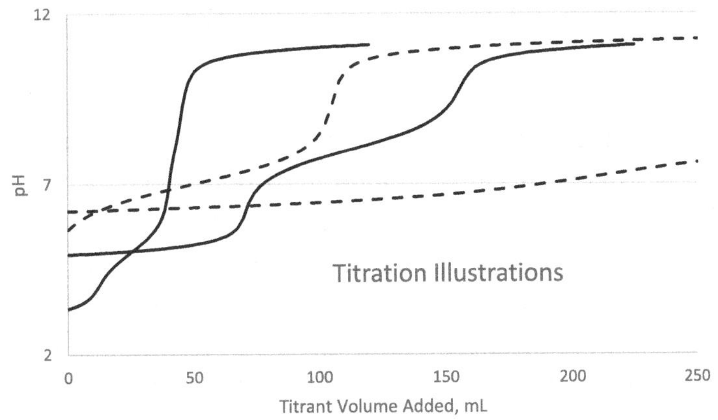

Figure C-1. Some representative titration curves.

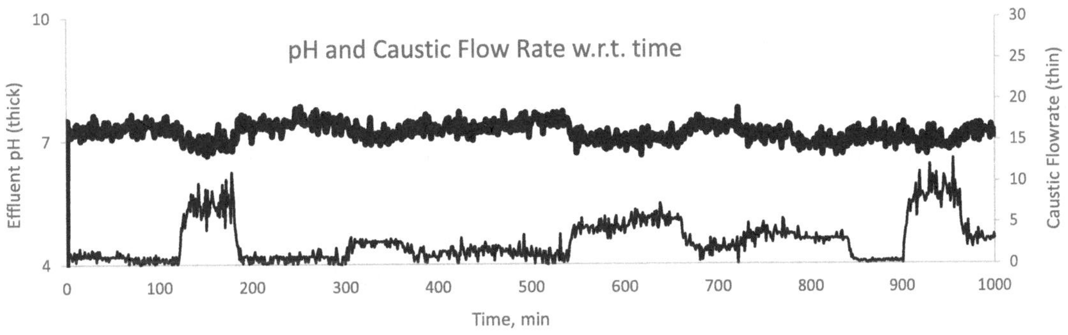

Figure C-2. A controlled response.

C.4 Discussion

This simulator represents a nonlinear interactive and nonstationary process subject to constraints. There are two controlled variables (pH and level) with corresponding manipulated variables (reagent inflow and discharge rate). pH gain and dynamics change with the operating conditions. Unlike many studies presented in the literature, the characteristics of this test case are not contrived to illustrate any difficulties; they are naturally expressed in this realistic and relatively simple simulator.

The simulator has an acidic inflow composed of water, two diprotic acids, one monoprotic acid, two dihydroxy bases, and pairs of common ion salts. The titrant is a concentrated solution of a strong monohydroxy base. The user can choose values for nominal inflow concentrations and *pK* values that represent species of interest to a particular application.

The inflow rate is measurable, but the value contains noise and calibration errors. The inflow concentrations are not measurable. The pH measurement represents that in the active mixing zone but is confounded by noise, calibration drift, and digital discretization (resolution).

This pH model does not consider solubility effects (molecule precipitates or gas effervescence) or activity effects (because of temperature or ion concentrations). Further, the simulator does not include triprotic acids or tri-hydroxyl bases. The simulator contains several stochastic changes and measurement noise, and has the tank level change in response to outflow control to prevent out-of-specification discharges. I believe this is rich in challenge features for generic tests on pH control systems. If you wish to include activity coefficient models or trivalent materials, you can easily edit the simulator to include such. If you wish to include solubility effects, adding the new species and solving the nonlinear relations takes a bit more effort.

A VBA version is accessible to the public at *www.r3eda.com.*

Although running the simulator without stochastic inputs and noise-corrupted measurements may be useful, be sure to test your procedure with those effects.

From the *CRC Handbook of Chemistry and Physics*[2] and some internet searches, here are a few nominal pKa values of dilute acids in water at 20°C (Table C-2).

Table C-2. Some characteristic pKa values.

Molecule	pKa1	pKa2
Acetic Acid	4.75	–
Ascorbic Acid	4.10	8.79
Aspartic Acid	3.86	9.82
Arsenic Acid (pKa3 = 11.60)	2.25	6.77
Carbonic Acid	6.37	10.25
Chloroacetic Acid	2.85	–
Citric Acid (pKa3 = 6.39)	3.14	4.77
H_2S	7.04	11.96
Hydrochloric Acid	(very strong; I use –1)	–
Lactic Acid	3.08	–
Maleic Acid	1.83	6.07
o-Phosphoric (pKa3 = 12.67)	2.12	7.21
o-Phthalic Acid	2.89	5.51
Oxalic Acid	1.23	4.19
Sulfuric Acid	(very strong; I use –1)	1.92

2. David R. Lide, ed., *CRC Handbook of Chemistry and Physics*, 72nd ed. (Boca Raton, FL: CRC Press, 1991).

Table C-3 includes a few nominal pKb values of dilute bases in water $pKb = 14 - pKa$.

Table C-3. Some characteristic pKb values.

Molecule	pKb1	pKb2
Ammonia	4.75	–
Analine	9.4	–
Calcium Hydroxide (very limited solubility)	1.4	2.43
Ethylamine	3.25	–
Hexamethlyenediamine	2.14	3.24
Lithium Hydroxide	−0.36	–
Nicotine	5.98	10.88
Potassium Hydroxide	0.5	–
Quinine	5.46	9.87
Sodium Hydroxide	0.2	–

Note: Various sources report significantly differing values for the hundredths digit. Also, recognize that the pK values depend on temperature, concentration, and the presence of other ion sources that could interact with the primary components. The information presented provides the reader with a range of characteristic values that are useful for general ideal simulation.

Exercises

You can access the simulator from my website, www.r3eda.com.

There are several trial conditions in the simulator. Enter a trial number in Cells(2,18) and the number of replicates in Cells(1,18). Enter nominal influent attributes in the green highlighted cells in Rows 3 through 11 and Columns 3 through 5. Or, you might enjoy writing your own. The controller in the simulator is simply a set of heuristic rules, such as an operator might manually perform. The objective is to neutralize the material so the tank effluent pH is between 6 and 9. If the tank pH is out of range, the control action is to stop effluent and accumulate influent in the tank until it is within discharge specification. When back in specification, the effluent must flow faster to return the tank to a nominal volume.

Too great a nominal volume means there is little capacity to retain out-of-specification material, leading to discharge violations. Too little a nominal volume means that process gain (pH response to events) is very high and control is difficult. You can enter values for nominal controller gain and target volume in Cells(1,21) and Cells(2,21).

For each replicate run, the tank fills to the target level, then effluent starts. As time progresses, the influent flow rate and compositions change. After a 1,000-minute simulation (approximately two 8-hour shifts), the controller is evaluated by several undesirables (time and volume of effluent in violation, a pH-weighted penalty, and the number of solenoid clicks to open and close the effluent). Because the influent changes are random, one 1,000-minute run will not provide a comprehensive evaluation. So, run multiple trials. Enter the number in Cells(1,18). I think 50 is a reasonable number, balancing execution time with getting relatively definitive results.

The larger the tank, the larger the mixing dilution of upsets and the easier the pH control. However, large tanks cost more and are more expensive to maintain.

C.1 Choose a desirable influent composition.

C.2 Quantify the process dynamics and nonlinearity.

C.3 Seek to tune the controllers to minimize violations and control undesirables.

C.4 Add noise and evaluate the impact on control.

C.5 Add your control algorithm to the simulator and run trials to quantify the goodness of the control metrics.

C.6 Seek a controller that minimizes the tank size, violations, and other control undesirables.

D

Heat Exchanger Simulator

D.1 Introduction

This appendix offers a simulation model of a process heat exchanger, which can be used to test and demonstrate automation algorithms, such as control techniques, controller tuning, data reconciliation, steady-state detection, model adaptation, noise filtering, constraint prediction, fault detection, and neural network or fuzzy modeling. Based on a full-scale exchanger, the simulator is nonlinear, with variable transport delay and third-order lags, and subject to unmeasurable and measurable disturbances. Although the simulator is rich in process features, it is relatively simple to code. A VBA version of the simulator and user guide are accessible to the public at *www.r3eda.com*.

The process heat exchanger simulator is convenient to use (coding, understanding, etc.) but is also reasonably grounded in an actual exchanger (see Figure D-1) to credibly guide technique development and establish credibility.

Figure D-1. Heat exchanger unit formerly in the unit operation lab of Oklahoma State University.

Researchers commonly use simulators to guide developing novel automation techniques, demonstrate their techniques, and benchmark their techniques to other techniques. Instructors use simulators to aid learning. The simulator is often conveniently simple, which is important for ease of understanding, ease of coding, and immunity from execution errors. However, most simulations are too idealized to be able to establish the technique's credibility for the practitioner. Further, using transfer function models in a process simulation misdirects the user from phenomenological process understanding. Real processes are nonlinear and nonstationary, have measurement noise, and are subject to continually varying inputs and disturbances.

D.2 Model Description

The heat exchanger is a liquid-liquid, countercurrent, single-pass type. The hot fluid is in the shell side and modeled as an oil. The process fluid being heated is modeled as water. The control objective is to control the exit water temperature by sending a signal to the flow control valve on the hot fluid entrance line. The control valve has a linear inherent characteristic, but it has a moderately quick-opening characteristic when installed in the line. The valve is air actuated, with a stem position response that is roughly first-order to the controller output, the manipulated variable (MV). Each of the entrance and exit lines has a thermowell to measure temperature. The measurement has a first-order lag and a delay because of fluid transport between the sensor and the heat exchanger.

The flow rate of the hot fluid is controlled. The uncontrolled inputs to the heat exchanger are the cold fluid flow rate, the hot and cold fluid temperatures, and ambient heat losses. These four variables continually fluctuate using an autoregressive moving average (ARMA [1,1]) model driven by normal (Gaussian) independent deviations with a zero mean and a standard deviation, NID(0,σ). The user can also make intentional changes to these variables.

A key and undesirable characteristic of a heat exchanger is heat transfer surface fouling. It might result from conditions such as scale buildup or a component that precipitates or chars when hot. As fouling progresses, the hot fluid flow rate must be increased to sustain the outlet fluid at its set point. Fouling typically progresses over time, or it may happen fairly rapidly if the inlet fluid contains a batch of the undesirable material. The simulator also changes the fouling factor.

The model can be used for more than just control; operators are interested in using it to identify fouling, schedule maintenance, and forecast constrained operation limits.

There are several stages to the heat exchanger model, which will be developed independently.

D.2.1 The Control Valve

The controller output, the MV, is u, in the 0% to 100% range. The valve stem position is x, the fraction of full travel, in the 0 to 1 range. Because of the air actuator, x has a nearly first-order response to u with a time constant of τ_v.

$$\tau_v \frac{dx}{dt} + x = .01u \tag{D-1}$$

Numerically,

$$x := \lambda_v 0.01u + (1 - \lambda_v)x \tag{D-2}$$

$$\lambda_v = 1 - e^{-\Delta t/\tau_v} \tag{D-3}$$

In Equation D-2, the := represents an assignment statement. On the right-hand side, x represents the valve stem position at the prior time step, Δt, which is used to predict the new value at the end of the time interval. Differentiated from an algebraic ARMA(1,1) model, subscripts are not needed in computer code. Use Equation D-2 to determine how x responds to u.

The valve equation describes how a steady-state flow rate responds to the stem position.

$$F = C_v f(x)\sqrt{\frac{\Delta P_v}{G\xi}} \tag{D-4}$$

Here, C_v is the valve capacity coefficient with the same units as F, the volumetric flow rate. ΔP_v is the pressure drop across the valve, G is the fluid specific gravity, and $\xi = 1$ is a dimensional unifier and has the same units as ΔP_v. $f(x)$ is the inherent valve characteristic, which reveals how the steady-state flow rate depends on the stem position when ΔP_v is constant, $0 \leq f(x) \leq 1$. For a linear valve, $f(x) = x$.

In a pipeline, the liquid mass needs time to accelerate from a prior velocity value to a new steady-state value. However, typical process flow lines are short enough so this acceleration time constant is much smaller than the valve actuator, τ_v, and the flow can be considered at the steady state associated with x (termed a *pseudo-steady-state model*).

However, as the flow rate changes, the friction loss pressure drop, ΔP_f, in the line changes. Assuming that the system pressure drop, ΔP_s, does not change

$$\Delta P_s = \Delta P_v + \Delta P_f \tag{D-5}$$

Using a classic turbulent flow model for losses in piping and fittings, with k representing the sum of all pipe and fitting influences,

$$\Delta P_f = k\frac{1}{2}\rho v^2 = k\frac{1}{2}\rho\frac{1}{A_{xc}^2}F^2 \tag{D-6}$$

where ρ is the fluid density, v is the fluid velocity, and A_{xc} is the pipe cross-sectional area.

Solving Equation D-4 for ΔP_v, inserting Equations D-4 and D-6 into D-5, combining constants into coefficients $b = G\xi/\Delta P_s, c = k\frac{1}{2}\rho\frac{C_v^2}{A_{xc}^2}/\Delta P_s$, using the linear inherent characteristic, and solving for F, converts the linear inherent characteristic to the installed characteristic.

$$F = C_v x\sqrt{\frac{1}{b + cx^2}} \tag{D-7}$$

Use Equation D-2 to determine how x responds to u, then Equation D-7 to determine how the controlled flow rate responds to x.

D.2.2 The Heat Exchanger

The design equation for a single-pass, countercurrent heat exchanger is

$$\dot{Q}_t = UA\Delta T_{lm} = UA\frac{(T_{h,i} - T_{c,o}) - (T_{h,o} - T_{c,i})}{ln\left(\frac{T_{h,i} - T_{c,o}}{T_{h,o} - T_{c,i}}\right)} \tag{D-8}$$

Here, $\dot{Q}_t$ is the rate at which heat is transferred, U is the overall heat transfer coefficient, A is the heat transfer area, and T represents fluid temperature. The subscripts h and c represent the hot and cold fluids, and the subscripts i and o represent inflow or outflow locations. ΔT_{lm} is called *delta-T-log-mean* and represents the effective average temperature difference along the tube length for heat transfer. Given values for UA and for $T_{h,i}$ and $T_{c,i}$, there are three unknowns in Equation D-8.

These are solved by using equations for the heat gained by the cold fluid and lost by the hot fluid, ideally modeled as

$$\dot{Q}_c = F_c \rho_c Cp_c \Delta T_c = F_c \rho_c Cp_c (T_{c,o} - T_{c,i}) \tag{D-9}$$

$$\dot{Q}_h = F_h \rho_h Cp_h (T_{h,i} - T_{h,o}) - \dot{Q}_{amb} \tag{D-10}$$

where Cp is the fluid specific heat, and $\dot{Q}_{amb}$ represents the losses from the heat exchanger to the environment, here credited only to the hot shell-side fluid.

Because $\dot{Q}_t = \dot{Q}_h = \dot{Q}_c$, Equations D-8 to D-10 can be solved to determine $\dot{Q}_t$, $T_{h,o}$, and $T_{c,o}$, given values for the inlet flow rates; fluid properties, $\dot{Q}_{amb}$, U, A; and the inlet temperatures.

The overall heat transfer coefficient depends on the hot and cold fluid heat transfer coefficients, h, the conduction resistance through the tube walls, r_{cond}, and fouling resistance, r_{foul}.

$$\frac{1}{U} = \frac{1}{h_h} + r_{cond} + \frac{1}{h_c} + r_{foul} \tag{D-11}$$

Experimentally, dimensionless correlations, such as the Dittus-Boelter relation for turbulent flow, show that $h \propto F^{0.8}$, so

$$U = \frac{1}{\frac{1}{h_{h,ref}}\left(\frac{F_{h,ref}}{F_h}\right)^{0.8} + r_{cond} + \frac{1}{h_{c,ref}}\left(\frac{F_{c,ref}}{F_c}\right)^{0.8} + r_{foul}} \tag{D-12}$$

The subscript "ref" means the value at a reference condition.

The equations do not seem to provide an explicit solution, so a root-finding approach is needed. Guess at $\dot{Q}_t$ as $\dot{Q}_g$, solve Equations D-9 and D-10 for the associated outlet temperatures, and then use Equation D-8 to determine if the associated $\dot{Q}_t$ is equal to $\dot{Q}_g$. The author finds that interval halving (bisection search) from high and low extreme possible values for $\dot{Q}_g$ is stable and fast. In the simulator, the low guess is that the cold fluid only rises by a small amount (1 degree). The high guess would be the minimum of either the hot fluid exiting very close to the cold inlet T or the cold fluid exiting very close to the hot inlet T.

This procedure determines the steady-state exit temperatures of the heat exchanger.

Experimentally, the dynamics of the heat exchanger are found to be nearly first-order and small relative to the transport delay of the fluid. This will be modeled as a Hammerstein model, and the lags in the temperature sensors can be modeled as first-order.

Assuming plug flow (nearly valid for turbulent flow) and adiabatic flow (no heat transfer to or from the thermally insulated piping system), the transport delay from the HX to the sensors is simply $\theta = V/F$, where V is the volume from the middle of the HX to the T sensor, and digitized to the number of simulation time steps,

$$N_{delay} = INT\left(\frac{V}{F\Delta t}\right) \tag{D-13}$$

The temperature of the fluid at the ith simulation interval would be the temperature exiting the HX N_{delay} samples prior. This is for both cold and hot fluids. Further, the inlet temperatures are sampled prior to the fluid entering the heat exchanger, so the fluid entering the HX at any particular sample is that at the sensor N_{delay} samples after.

The clock method is used to store and retrieve delayed values using an array of 100 elements. Instead of viewing it as a linear column array, consider that the array is circular, like the numbers on a clock, with the array element 100 adjacent to the 1, like the clock hour 12 is adjacent to the 1. At each sampling, increment the store index, i_{put}, by 1, and if $i_{put} = 101$, then reset it to 1. Store the value in the $x(i_{put})$ array element. Determine the i_{read} index value as $i_{put} - N_{delay}$. If i_{read} is negative, reset it using $i_{read} := i_{read} + 100$.

Finally, all the temperature sensors have a lag to respond to the process fluid. This lag is modeled as a first-order lag.

$$\tau_{tw}\frac{dT_{sensor}}{dt} + T_{sensor} = T_{fluid} \tag{D-14}$$

Numerically,

$$T_{sensor} := \lambda_{tw} T_{fluid} + (1 - \lambda_{tw}) T_{sensor} \tag{D-15}$$

$$\lambda_{tw} = 1 - e^{-\Delta t/\tau_{tw}} \tag{D-16}$$

To determine the outlet temperatures, use Equation D-12 to determine the value of U, then use the N_{delay} past inlet temperatures in the interval halving procedure to determine the HX outlet temperatures. Then, use Equation D-15 on the delayed outlet temperatures to determine the sensor T.

D.3 Coefficient Units and Values

For the valve:

u		Controller output, %
τ_v	1	s
x		Fraction of full stroke, dimensionless
F		m³/h
C_v	500	m³/h
b	1	Dimensionless
c	5	Dimensionless

For the heat exchanger:

$\dot{Q}$		kJ/h
U		kJ/(h m² K)
A	0.25	m²
$T_{h,in}$	500	K
$T_{c,in}$	285	K
$h_{h,ref}$	700,000	kJ/(h m² K)
$h_{c,ref}$	700,000	kJ/(h m² K)
$F_{h,ref}$	100	m³/h
$F_{c,ref}$	20	m³/h
r_{cond}	0.000,001	(h m² K)/kJ
r_{foul}	0	(h m² K)/kJ
Cp_h	3	kJ/(kg K)
ρ_h	900	kg/m³
Cp_c	4	kJ/(kg K)
ρ_c	1,000	kg/m³
$\dot{Q}_{amb}$	50,000	kg/h
V	0.01	m³
τ_{tw}	3	s

For measurement noise:

σ_T	0.5	K
σ_F	0.1	Fraction of F

For stochastic inputs:

τ_{Fc}	100	s
R_{Fc}	10	m³/h
τ_{Tc}	100	s
R_{Tc}	50	K
τ_{Th}	100	s
R_{Th}	100	K
$\tau_{\dot{Q}amb}$	100	s
$R_{\dot{Q}amb}$	10,000	kJ/h

D.4 Simulator Results

The installed valve characteristic is illustrated in Figure D-2. Although the valve's inherent characteristic is linear, when installed, the valve has a mildly quick-opening functionality.

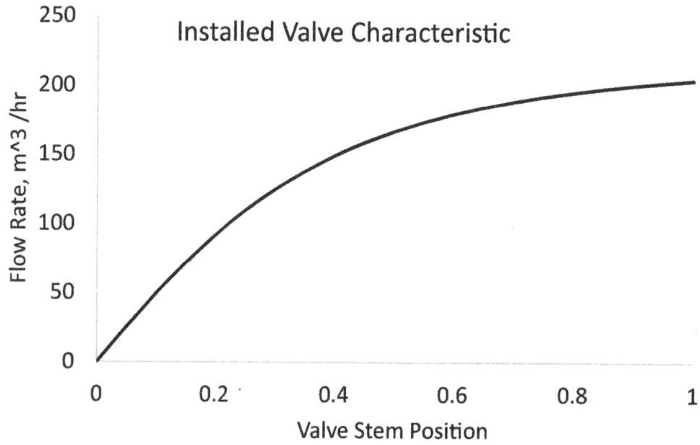

Figure D-2. Installed valve characteristic.

Source: From "A Heat Exchanger Simulator for Testing and Demonstrating Automation Algorithms"[1] by R. R. Rhinehart. Used with permission of the American Automatic Control Council (AACC).

Figure D-3a presents a sequence of steps in the MV, and Figure D-3b indicates the measured temperature responses when there is no measurement noise or input ARMA disturbances. The upper traces in Figures D-3b, D-3c, and D-3d are the exit hot

1. R. R. Rhinehart, "A Heat Exchanger Simulator for Testing and Demonstrating Automation Algorithms," *2022 American Control Conference (ACC)* proceedings, paper FrA12.6, Atlanta, GA (2022): 3789–3794. doi: 10.23919/ACC53348.2022.9867196.

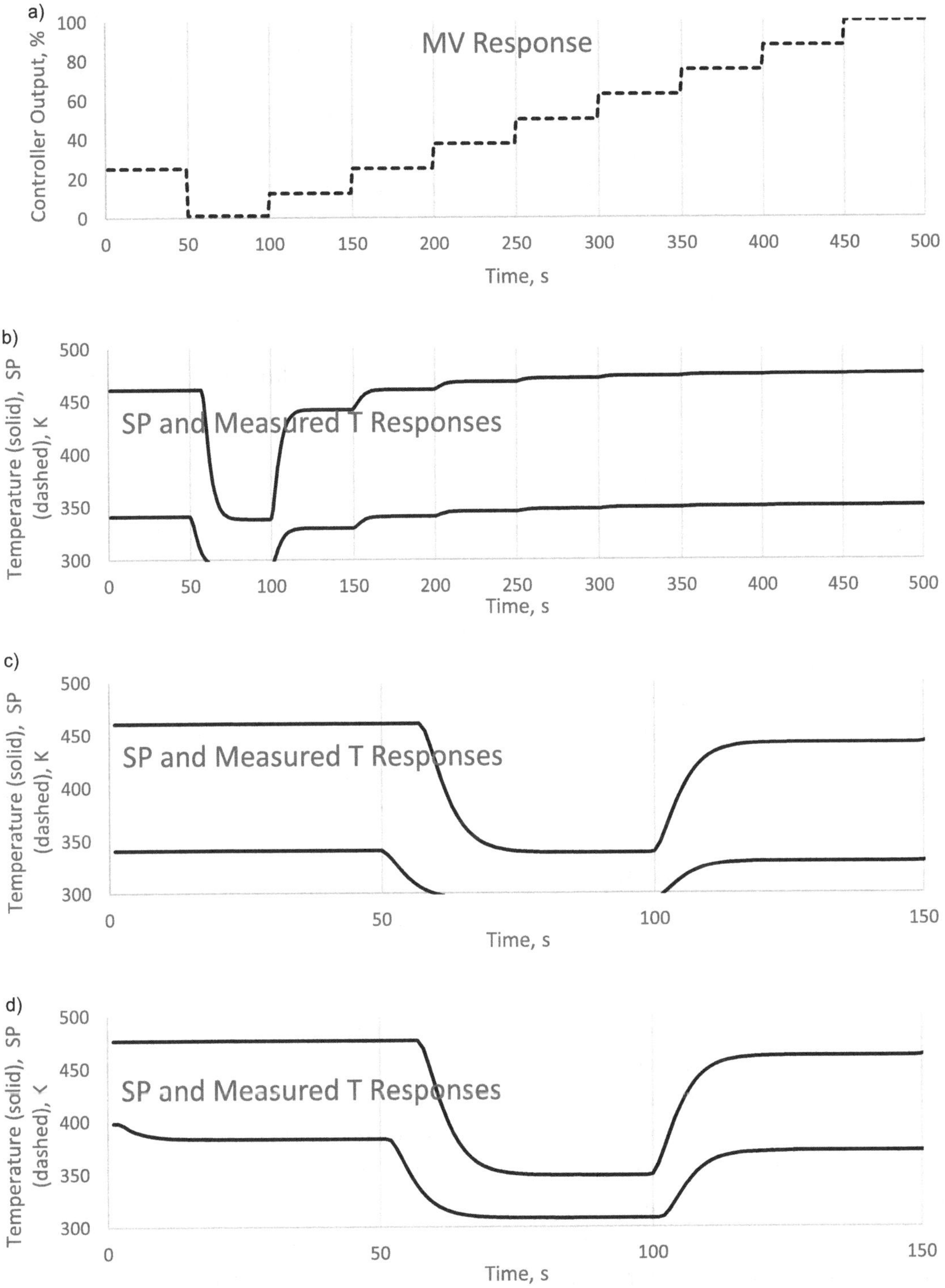

Figure D-3. (a) Steps in the MV, (b) temperature responses, (C) revealing delay and lags and (d) repeated trial with lower cold flow rate.

Source: From "A Heat Exchanger Simulator for Testing and Demonstrating Automation Algorithms" by R. R. Rhinehart. Used with permission of the American Automatic Control Council (AACC).

temperatures, and the lower traces are the exit cold temperatures. Because the controller is in MAN mode, the set point follows the exit cold fluid temperature; therefore, the dashed set point line is masked by the cold exit temperature line. In the low flow rate range, the gain is about 4.5 K/% and the high range gain is about 0.5 K/%, about a 9:1 range. Reasons for the nonlinear gain include the installed valve characteristic, the flow rate dependence on the overall heat transfer coefficient, and the heat exchanger itself.

Figure D-3c reveals the detail at the step at 100 s, indicating a several-second delay on the slower-flowing hot exit temperature.

For the same input sequence, but with one-third of the cold fluid flow rate, Figure D-3d shows the same detail as Figure D-3c.

> Note: At the lower cold fluid flow rate, the delay in cold outlet T is about three times longer, and the hot fluid T response dynamics are substantially the same. However, both outlet temperatures are substantially higher, and the gains and delays depend on the two flow rates.

Figure D-4a illustrates a sequence of steps in the influences. The upper solid trace is the measured hot inlet fluid temperature, and the lower solid trace is the measured cold inlet fluid. Each makes step changes and recoveries, but the lags are because of the sensor dynamics. The dashed lines are the inlet cold (lower) and hot fluid (upper) flow rates. Then, Figure D-4b reveals the process responses. Because the controller is in MAN mode, the set point follows the exit cold fluid temperature; therefore, the dashed set point line is masked by the cold exit temperature line.

Although each temperature input step has the same magnitude, the changes in the response variables are not the same. The responses depend on flow rates and temperature differences. The gains of the process also depend on influence conditions.

Figure D-5 presents two nonstationary effects. The signal to the flow control valve is fixed. At a time of 80 s, the fouling factor begins a ramp increase, holds steady at 200 s, and then drops back to zero at 350 s. The drop in the lower trace (cold outlet T) and the slight increase in the hot temperature result from the fouling factor change. However, throughout the simulation, stochastic influences on the inputs on inlet pressures and temperatures are also continually drifting to create irregular outlet T responses. Because the controller is in MAN mode, the set point follows the exit cold fluid temperature; therefore, the dashed set point line is masked by the cold exit temperature line.

Not illustrated here, the gains also change with the fouling.

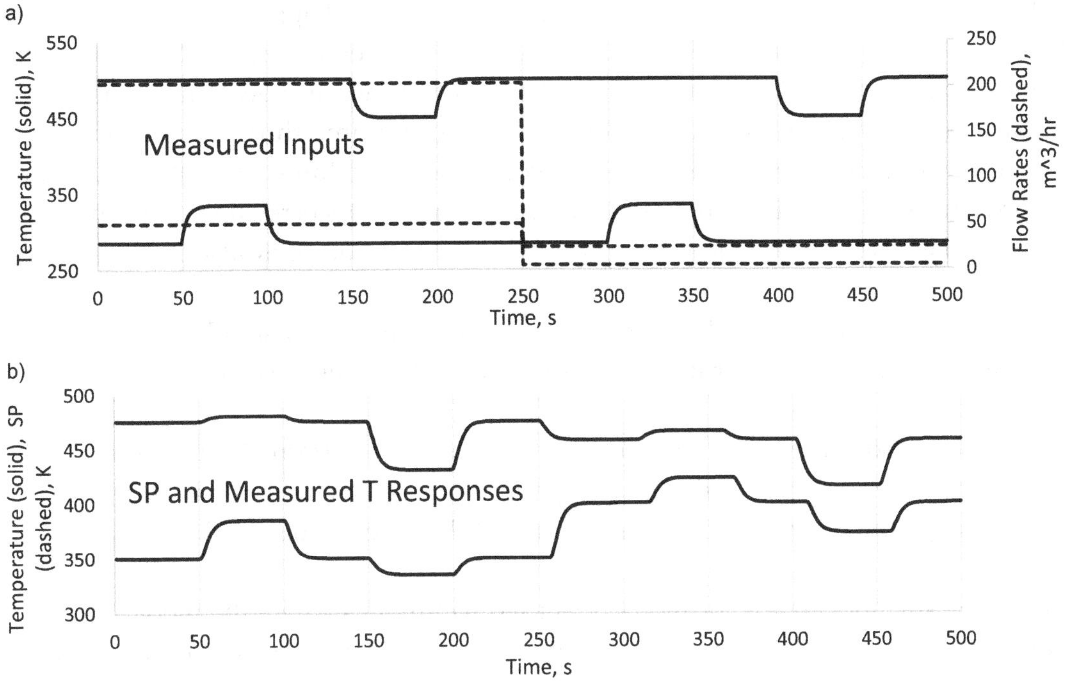

Figure D-4. (a) Step inputs in influence variables and (b) response variables to Figure D-4a.

Source: From "A Heat Exchanger Simulator for Testing and Demonstrating Automation Algorithms" by R. R. Rhinehart. Used with permission of the American Automatic Control Council (AACC).

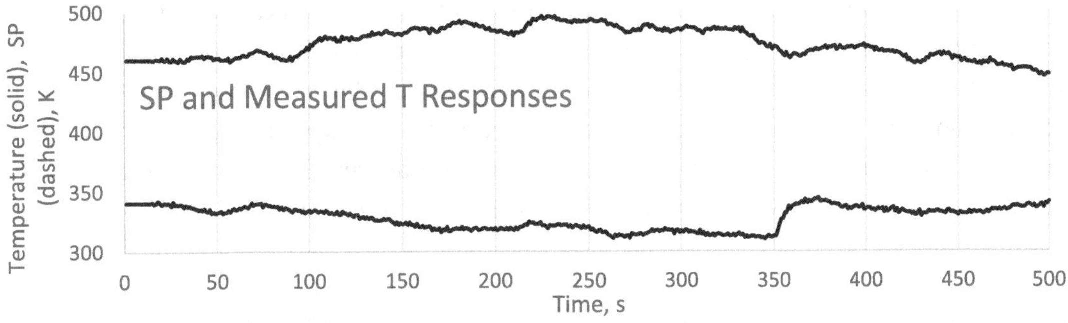

Figure D-5. Measured response to fouling and stochastic inputs.

Source: From "A Heat Exchanger Simulator for Testing and Demonstrating Automation Algorithms" by R. R. Rhinehart. Used with permission of the American Automatic Control Council (AACC).

D.5 Discussion

This is a nonlinear interactive and nonstationary process. Gains and dynamics change with the operating conditions. However, the characteristics of this test case are not contrived to illustrate any difficulties that procedures for control strategy, tuning, fault detection, steady-state identification, and so on might encounter. They are naturally expressed in this realistic, yet relatively simple, simulator.

Although this is a single-input, single-output (SISO) control application, control is difficult because of the changes in gains and delays. Further, backing out fouling factor values from data is difficult because (1) the noise on measurements has a nonlinear impact and (2) transient conditions continually require time shifts in matching input to output variable values. The same issues confound procedures for fault identification and constrained operation forecasting. Because there is autocorrelation in the measurements, not just ideal noise, identifying steady-state conditions may also be challenging.

Although the simulator is rich in process features, it is relatively simple to code. A VBA simulator and user guide for it are accessible to the public at *www.r3eda.com*.

A user could add disruptive features such as flow rate patterns (steps, ramps, oscillations, or ARMA drifts), fouling factor patterns, or sensor calibration faults, or they could change the severity and time constants of the several noise and drift variables.

Many investigators are exploring model-based techniques where the models are phenomenological, as is this simulator. To have credible conclusions from a technique simulation, you cannot use the same model for the supervisory application as you use in the process simulator. Nature is always one step more complicated than human-devised models. Therefore, to credibly demonstrate your technique, be sure that your model for the application is not identical to the model in the process simulator. Change the coefficient values and functionalities (those in the valve, overall heat transfer coefficient model, etc.). Change the dynamic relations, such as time constants. Note that although you can observe all variable values in the simulator, the only variables your application is permitted to observe are the two noise-corrupted measured fluid flow rates and the four noise-corrupted measured temperatures. Reveal that your application model is similar to but reasonably different from the simulation model.

Although it may be useful to run the simulator without stochastic inputs and noise-corrupted measurements, as illustrated in Figures D-3 and D-4, be sure to test your procedure with those confounding effects.

Exercises

Visit *www.r3eda.com* to access my Excel VBA simulator. Or, write your own. The controller in the simulator is proportional-integral-derivative (PID).

D.1 Choose trial conditions to quantify the process's nonlinearity and nonstationarity (delay and impact of changes in fouling).

D.2 Tune the controller at nominal conditions and evaluate the impact of control on a range of other conditions.

D.3 Simulate and evaluate control at constrained conditions and recovery.

D.4 Write your own model-based control (you might even decide to explore gain-scheduled PI) and design trials, and compare them to the PID simulator.

D.5 This file also uses incremental model adjustment to make the model better match the process, which is useful for process condition monitoring and constraint forecasting. Explore this feature.

D.6 Explore Exercises D.2 through D.5 with various noise and disturbance conditions.

E Nomenclature

E.1 Abbreviations

ABCD	A, B, C, and D Matrix of Modern (or State-Space) Control
APC	advanced process control
ARC	advanced regulatory control
ARIMA	autoregressive integrating moving average
ARL	average run length
ARMA	autoregressive moving average
AUTO	automatic mode
AuxV	auxiliary variable
Btu	British thermal unit
CDF	cumulative distribution function
CHD	Cyclic Heuristic Direct
CO	controller output, the same as MV
CPI	chemical process industry (or industries)
CPU	central processing unit
CST	continuous flow stirred tank (ideal)
CSTR	continuous flow stirred tank reactor (ideal)
CUSUM	cumulative sum
CV	controlled variable
DCS	distributed control system
DDC	direct digital control
DMC	dynamic matrix control
DOF	degree of freedom

DP	dynamic programming
DV	decision variable
EC	equal concern (factor)
EHS&LP	Environmental Health, Safety, and Loss Prevention
erf	external reset feedback
FIR	finite impulse response
FL	fuzzy logic
FLC	fuzzy logic control
FO	first-order
FoF	first-order filter
FOPDT	first-order plus dead time
GIGO	garbage in, garbage out
GMC	generic model control
GMC-SS	generic model control with a steady-state model
H:	hypothesis
HJ	Hooke-Jeeves
HMI	human-machine interface
H_o:	null hypothesis
HPC	horizon predictive control
HX	heat exchanger
IAE	integral of the absolute value of the error
IdCom	identify and command
IE	integral of the error
IMC	internal model control
IMPOL	incremental model parameter adjustment online
irf	internal reset feedback
ISE	integral of the squared error
ITAE	integral of the error times time
ITSE	integral of the square of the error times time
LF	Leapfrogging
LHS	left-hand side
LM	Levenberg-Marquardt
L-MPC	linear model predictive control
LP	linear programming
MA	moving average
MAN	manual mode
MBC	model-based control
MIMO	multiple-input, multiple-output
MISO	multiple-input, single-output

MPC	model predictive control
MV	manipulated variable, usually the same as controller output (CO), but it might represent the final element influence on the process in modeling
N2F	now-to-future (model)
nAPC	nonlinear advanced process control
nIAE	normalized integral of the absolute value of the error
NID(μ, σ)	normally and independently distributed with mean μ and standard deviation σ
nIE	normalized integral of the error
nISE	normalized integral of the squared error
nITAE	normalized integral of the error times time
nITSE	normalized integral of the square of the error times time
NLP	Natural Language Processing (fuzzy logic and expert systems)
NMBC	nonlinear model-based control
N-MPC	nonlinear model predictive control
NN	neural network
NOFE	number of function evaluations
NR	Newton-Raphson
ODE	ordinary differential equation
OF	objective function
P2N	past-to-now (model)
PFC	predictive functional control
PI	proportional-integral
PID	proportional-integral-derivative
PLC	programmable logic controller
PMBC	process-model-based control
pmm	process-model mismatch
PS	possible solution
PV	process variable
QAD	quarter-amplitude damped
R	range, ratio statistic, or valve rangeability
RHS	right-hand side
RoC	rate of change
RTO	real-time optimization
SIS	safety instrumented system
SISO	single-input single-output
SOPDT	second-order plus dead time
SPC	statistical process control
SQ	successive quadratic

SS	steady state
SSID	steady-state identification
TS	transient state
TSID	transient state identification
TSK	Takagi-Sugeno-Kang
w.r.t.	with respect to
yN2F	now-to-future modeled value
yP2N	past-to-now modeled value

E.2 Symbols

:= The "equal" sign in an assignment statement. It is not an algebraic equal; it means to calculate the value from the right-hand side instructions and assign that value to the variable name on the left-hand side.

E.3 Variables

$\hat{e}$ = the Laplace transform of the variable e
e = the actuating error, $e = y_{SP} - y_{process}$
e = 2.7128281... the base of the natural or Naperian logarithm
y_{SP} = the set point
$y_{process}$ = the process measured value
y_p = the process measured value
$y_m = \tilde{y}$ the model output representing the process value
m = controller output (CO or MV)
u = controller output (CO or MV)
b = controller adjustable bias, typically the integral value
μ = mean
σ = standard deviation
Δ = difference, threshold
ε = small difference
ρ = density
τ = time constant

Bibliography

Alam, S. Samir, R. R. Rhinehart, K. A. High, and P. J. Gemperline. "On-Line Optimization of a Batch Reaction—Development and Experimental Demonstration." In *Proceedings of the 2004 American Control Conference*, vol. 6, paper FrM11.1. Boston, MA (2004): 4861–4866. doi: 10.23919/ACC.2004.1384619.

Alford, J. A., B. N. Hrankowsky, and R. R. Rhinehart. "Data Filtering in Process Automation Systems." *InTech* 65, no. 4 (July–August 2018): 14–19.

Box, G. E. P., and Mervin E. Muller. "A Note on the Generation of Random Normal Deviates." *Annals of Mathematical Statistics* 29, no. 2 (1958): 610–611.

Cao, S., and R. R. Rhinehart. "An Efficient Method for On-Line Identification of Steady-State." *Journal of Process Control* 5, no. 6 (1995): 363–374.

Cao, S., and R. R. Rhinehart. "Critical Values for a Steady-State Identifier." *Journal of Process Control* 7, no. 2 (1997): 149–152.

Choi, J. Y., H. G. Pandit, R. R. Rhinehart, and R. J. Farrell. "A Process Simulator for pH Control Studies." *Computers & Chemical Engineering* 19, no. 5 (1995): 527–540.

Choi, J. Y., and R. R. Rhinehart. "Nonlinear Process-Model-Based-Control of a Batch Polymerization Reactor." In *Proceedings of the 1997 American Control Conference*, vol. 5. Paper FM15-1, Cat. No. 97CH36041. Albuquerque, NM (1997): 3455–3459. doi: 10.1109/ACC.1997.612109.

Crowe, E. L., F. A. Davis, and M. W. Maxfield. *Statistics Manual.* New York: Dover Publications, 1955.

Cutler, C. R., and D. L. Ramaker. "Dynamic Matrix Control—A Computer Control Algorithm." *Proceeding of JACC,* San Francisco, CA (1980), Paper WP5-b, American Automatic Control Council.

Dhir, S., K. J. Morrow, Jr., R. R. Rhinehart, and T. F. Wiesner. "Dynamic Optimization of Hybridoma Growth in a Batch Reactor." *Biotechnology and Bioengineering* 67, no. 2 (2000): 197–205.

Dutta, P., and R. R. Rhinehart. "Experimental Comparison of a Novel, Simple, Neural Network Controller and Linear Model Based Controllers." In *Proceedings of the 1995 American Control Conference—ACC'95,* vol. 3. Paper TA10-3. Seattle, WA (1995): 1787–1789. doi: 10.1109/ACC.1995.531192.

Garcia, C. E., and M. Morari. "Internal Model Control I. A Unifying Review and Some New Results." *Industrial and Engineering Chemistry Process Design and Development* 21 (1986): 308–323.

Govindarajan, A., S. K. Jayaraman, V. Sethuraman, P. R. Raul, and R. R. Rhinehart. "Cascaded Process Model Based Control: Packed Absorption Column Application." *ISA Transactions* 53, no. 2 (2014): 391–401.

Guo, B., A. Jiang, X. Hua, and A. Jutan. "Nonlinear Adaptive Control for Multivariable Chemical Processes." *Chemical Engineering Science* 56 (2001): 6781–6791.

Gupta, A., and R. R. Rhinehart. "Experimental Comparison of Advanced Control Techniques on a Lab-Scale Distillation Column." In *Proceedings of the 1995 American Control Conference—ACC'95,* vol. 5. Paper FA10-4. Seattle, WA (1995): 3188–3191. doi: 10.1109/ACC.1995.532104.

Incropera, F. P., and D. P. DeWitt. *Introduction to Heat Transfer.* New York: John Wiley & Sons, 1985.

Iyer, M. S., T. F. Wiesner, and R. R. Rhinehart. "Dynamic Re-optimization of a Fed-Batch Fermenter." *Biotechnology and Bioengineering* 63, no. 1 (April 5, 1999): 10–21.

Joshi, N. V., P. Murugan, and R. R. Rhinehart. "Experimental Comparison of Control Strategies." *Control Engineering Practice* 5, no. 7 (1997): 885–896.

Kramida, Alexander, Karen Olsen, and Yuri Ralchenko. *Periodic Table: Atomic Properties of the Elements.* Version 14. Gaithersburg, MD: National Institute of Standards and Technology (NIST), 2019. Accessed July 19, 2023. https://www.nist.gov/pml/periodic-table-elements.

Lee, P. L., and G. R. Sullivan. "Generic Model Control." *Computers and Chemical Engineering* 12, no. 6 (1998): 573–580.

Lee, P. L., R. B. Newell, and G. R. Sullivan. "Generic Model Control—A Case Study." *Canadian Journal of Chemical Engineering* 67, no. 3 (June 1989): 478–484.

Lide, David R., ed. *CRC Handbook of Chemistry and Physics.* 72nd ed. Boca Raton, FL: CRC Press, 1991. (There are more recent publications and many other sources of data.)

Mah, R. S., G. M. Stanley, and D. M. Downing. "Reconciliation and Rectification of Process Flow and Inventory Data." *Industrial & Engineering Chemistry Process Design and Development* 15, no. 1 (1976): 175–183.

Mahuli, S. K., R. R. Rhinehart, and J. B. Riggs. "Experimental Demonstration of Nonlinear Model-Based In-Line Control of pH." *Journal of Process Control* 2, no. 3 (1992): 145–153.

Mahuli, S. K., R. R. Rhinehart, and J. B. Riggs. "Nonlinear Model-Based In-Line Control of Wastewater pH: A Laboratory Study." *ISA Transactions* 32 (1993): 241–245.

Mahuli, S. K., R. R. Rhinehart, and J. B. Riggs. "pH Control Using a Statistical Technique for Continuous On-Line Model Adaptation." *Computers & Chemical Engineering* 17, no. 4, (1993): 309–317.

Manimegalai-Sridhar, U., A. Govindarajan, and R. R. Rhinehart. "Demonstration of Leapfrogging for Implementing Nonlinear Horizon Predictive Control on a Heat Exchanger." *ISA Transactions* 60 (2016): 218–227.

Manimegalai-Sridhar, U., A. Govindarajan, and R. R. Rhinehart. "Improved Initialization of Players in Leapfrogging Optimization." *Computers & Chemical Engineering* 60 (2014): 426–429.

Montandon, A. G., R. M. Borges, and H. M. Henrique. "Experimental Application of a Neural Constrained Model Predictive Controller Based on Reference System." *Latin America Applied Research* 38, no. 1 (January 2008).

Muthiah, N., and R. Russell Rhinehart. "Evaluation of a Statistically-Based Controller Override on a Pilot-Scale Flow Loop." *ISA Transactions* 49, no. 2 (2010): 154–166.

Ou, J., and R. R. Rhinehart. "CV Damping vs. MV Suppression for MPC Tuning." In *Proceedings of the 2002 American Control Conference,* vol. 4. Paper TM14-1, IEEE Cat. No. CH37301. Anchorage, AK (2002): 2695–2700. doi: 10.1109/ACC.2002.1025194.

Ou, J., and R. R. Rhinehart. "Grouped Neural Network Model Predictive Control." *Control Engineering Practice* 11, no. 7 (2003): 723–732.

Ou, J., and R. R. Rhinehart. "Grouped Neural Network Modeling for Model Predictive Control." *ISA Transactions* 41, no. 2 (April 2002): 195–202.

Ou, J., G. Narayanswamy, and R. R. Rhinehart. "External Reset Feedback for Generic Model Control." *ISA Transactions* 37, no. 3 (1998): 189–199.

Pandit, H. G., and R. R. Rhinehart. "Experimental Demonstration of Constrained Process-Model-Based Control of a Non-Ideal Distillation Column." In *1992 American Control Conference* proceedings. Chicago, IL (1992): 630–631. doi: 10.23919/ACC.1992.4792144.

Pandit, H. G., R. R. Rhinehart, and J. B. Riggs. "Experimental Demonstration of Nonlinear Model-Based Control of a Nonideal Binary Distillation Column." In *1992 American Control Conference* proceedings. Chicago, IL (1992): 625–629. doi: 10.23919/ACC.1992.4792143.

Parekh, M., M. Desai, H. Li, and R. R. Rhinehart. "In-Line Control of Nonlinear pH Neutralization Based on Fuzzy Logic." *IEEE Transactions on Components, Packaging, and Manufacturing Technology,* Part A, 17, no. 2 (June 1994): 192–201.

Paruchuri, V. P., and R. R. Rhinehart. "Experimental Demonstration of a Process Model-Based Flow Controller." *Advances in Instrumentation and Control* 48, Part 2 (1993): 529–540. (Proceedings of the ISA/93 Technical Conference, Chicago, IL, September 1993.)

Paruchurri, V. P., and R. R. Rhinehart. "Experimental Demonstration of Nonlinear Model-Based Control of a Heat Exchanger." In *Proceedings of 1994 American Control Conference—ACC '94,* vol. 3. Paper FP14-5:20. Baltimore, MD (1994): 3533–3537. doi: 10.1109/ACC.1994.735239.

Paruchurri, V. P., and R. R. Rhinehart. "Model-Based Flow Control Boosts Accuracy, Eases Tuning." *InTech* 42, no. 4 (April 1995): 52–56.

Puxty, G., M. Maeder, R. R. Rhinehart, S. Alam, S. Moore, and P. J. Gemperline. "Modeling of Batch Reactions with In-Situ Spectroscopic Measurements and Calorimetry." *Journal of Chemometrics* 19 (2005): 329–340.

Ramchandran, S. B. "Consider Steady-State Models for Process Control." *Chemical Engineering Progress* 93, no. 2 (February 1998): 75–81.

Raul, P. R., H. Srinivasan, S. Kulkarni, M. Shokrian, G. Shrivastava, and R. R. Rhinehart. "Comparison of Model-Based and Conventional Controllers on a Pilot-Scale Heat Exchanger." *ISA Transactions* 52, no. 3 (2013): 391–405.

Rhinehart, R. R. "A CUSUM-Type On-Line Filter." *Process Control and Quality* 2, no. 2 (February 1992): 169–176.

Rhinehart, R. R. "A Heat Exchanger Simulator for Testing and Demonstrating Automation Algorithms." In *2022 American Control Conference (ACC)* proceedings. Paper FrA12.6. Atlanta, GA (2022): 3789–3794. doi: 10.23919/ACC53348.2022.9867196.

Rhinehart, R. R. "A Statistically Based Filter." *ISA Transactions* 41, no. 2 (April 2002): 167–175.

Rhinehart, R. R. *Engineering Optimization: Applications, Methods, and Analysis*. Hoboken, NJ: John Wiley & Sons, 2018.

Rhinehart, R. R. *Nonlinear Regression Modeling for Engineering Applications: Modeling, Model Validation, and Enabling Design of Experiments*. Hoboken, NJ: John Wiley & Sons, 2016.

Rhinehart, R. R. "Optimization Algorithm Selection for Process Applications." *InTech* 66, no. 6 (November/December 2019): 18–23.

Rhinehart, R. R. and R. M. Bethea. *Applied Engineering Statistics*. 2nd ed. Boca Raton, FL: Taylor & Francis CRC Press, 2022.

Rhinehart, R. R., and J. B. Riggs. "Process Control Through Nonlinear Modeling." *Control* 3, no. 7 (July 1990): 86–90.

Rhinehart, R. R., and J. B. Riggs. "Two Simple Methods for On-Line Incremental Model Parameterization." *Computers & Chemical Engineering* 15, no. 3 (1991): 181–189.

Rhinehart, R. R., M. L. Darby, and H. L. Wade. "Editorial—Choosing Advanced Control." *ISA Transactions* 50, no. 1 (2011): 2–10.

Rhinehart, R. R., M. Su, and U. M. Sridhar. "Leapfrogging and Synoptic Leapfrogging: A New Optimization Approach." *Computers & Chemical Engineering* 40 (May 11, 2012): 67–81.

Rhinehart, R. R., R. M. Felder, and J. K. Ferrell. "Process-Model Based Control of a Pilot-Scale Fluidized Bed Coal Gasification Reactor." *Advances in Instrumentation and Control* 43, Part 1 (October 1988): 341–349.

Richalet, J., and D. O'Donovan. *Predictive Functional Control: Principles and Industrial Applications.* London: Springer, 2009. doi: 10.1007/978-1-84882-493-5.

Richalet, J., A. Rault, J. L. Testud, and J. Papon. "Model Predictive Heuristic Control: Applications to Industrial Processes." *Automatica* 14, no. 5 (1978): 413–428.

Riggs, J. B., and R. R. Rhinehart. "A Comparison Between Two Nonlinear Process Model Based Controllers." *Computers & Chemical Engineering* 14, no. 10 (October 1990): 1075–1081.

Riggs, J. B., and R. R. Rhinehart. "Method for pH Control." US Patent 4,940,551, issued July 10, 1990.

Shrowti, N., K. Vilankar, and R. R. Rhinehart. "Type-II Critical Values for a Steady-State Identifier." *Journal of Process Control* 20, no. 7 (2010): 885–890.

Skach, A., P. Raul, and R. R. Rhinehart. "Process-Model Based Differential Pressure Control on a Packed Tower Absorber." In *Proceedings of the ISA Automation Week Conference.* Submission 0040-000221. Mobile, AL (2011).

Subawalla, H., and R. R. Rhinehart. "Experimental Comparison of Model-Based and Conventional Pressure Control for a Plasma Reactor." In *Proceedings of 1994 American Control Conference—ACC '94*, vol. 3. Paper FM14-1:30. Baltimore, MD (1994): 3122–3126. doi: 10.1109/ACC.1994.735146.

Subawalla, H., V. P. Paruchurri, A. Gupta, H. G. Pandit, and R. R. Rhinehart. "Comparison of Model-Based and Conventional Control: A Summary of Experimental Results." *Industrial and Engineering Chemistry Research* 35, no. 10 (October 1996): 3547–3559.

von Neumann, J. "Distribution of the Ratio of the Mean Square Successive Difference to the Variance." *Annals of Mathematical Statistics* 12 (1941): 367–395.

Index

Printed in the USA/Agawam, MA
September 17, 2026
910661.228